T0211711

Communications in Computer and Information Science 509

More information about this series at http://www.springer.com/series/7899

Eric Pinson · Fernando Valente
Begoña Vitoriano (Eds.)

Operations Research and Enterprise Systems

Third International Conference, ICORES 2014
Angers, France, March 6–8, 2014
Revised Selected Papers

Editors
Eric Pinson
IMA, LISA
Angers
France

Fernando Valente
Polytechnic Institute of Setúbal
Setúbal
Portugal

Begoña Vitoriano
Complutense University
Madrid
Spain

ISSN 1865-0929 ISSN 1865-0937 (electronic)
Communications in Computer and Information Science
ISBN 978-3-319-17508-9 ISBN 978-3-319-17509-6 (eBook)
DOI 10.1007/978-3-319-17509-6

Library of Congress Control Number: 2015937937

Springer Cham Heidelberg New York Dordrecht London

Printed on acid-free paper

Springer International Publishing AG Switzerland is part of Springer Science+Business Media (www.springer.com)

Preface

The present book includes extended and revised versions of a set of selected papers from the Third International Conference on Operations Research and Enterprise Systems (ICORES 2014), held in ESEO, Angers, Loire Valley, France, during March 6–8, 2014, which was organized by the Institute for Systems and Technologies of Information, Control, and Communication (INSTICC) and held in cooperation with the ACM Special Interest Group on Applied Computing (ACM SIGAPP). ICORES is also technically cosponsored by the Portuguese Association of Operational Research (Apdio).

The purpose of the International Conference on Operations Research and Enterprise Systems (ICORES) is to bring together researchers, engineers, and practitioners interested in the theory and applications in the advances and applications in the field of operations research. Two simultaneous tracks were held, covering on one side domain-independent methodologies and technologies and on the other side practical work developed in specific application areas.

ICORE 2014 received 96 paper submissions from 31 countries, in all continents. To evaluate each submission, a double-blind paper review was performed by the Program Committee, whose members are highly qualified researchers in ICORES topic areas. Based on the classifications provided, only 38 papers were selected for oral presentation (20 full papers and 18 short papers) and 21 papers were selected for poster presentation. The full paper acceptance ratio was about 21% and the total oral acceptance ratio (including full papers and short papers) was 40%. These strict acceptance ratios show the intention to preserve a high-quality forum which we expect to develop further next year.

We would like to highlight that ICORES 2014 included also three plenary keynote lectures given by internationally distinguished researchers, namely: Nicolas Zufferey (University of Geneva, Switzerland), Marie-Christine Costa (ENSTA - ParisTech, France), and Paul Williams (London School of Economics, UK).

We would like to express our appreciation to all of them and in particular to those who took the time to contribute with a paper to this book.

We must thank the authors, whose research and development efforts are recorded here. We also thank the keynote speakers for their invaluable contribution and for taking the time to synthesize and prepare their talks. Finally, special thanks to all the members of the INSTICC team, whose collaboration was fundamental for the success of this conference.

October 2014

Eric Pinson
Fernando Valente
Begoña Vitoriano

Organization

Conference Co-chairs

Eric Pinson	IMA, LISA, Angers, France
Fernando Valente	Polytechnic Institute of Setúbal, Portugal

Program Chair

Begoña Vitoriano	Complutense University, Spain

Organizing Committee

Helder Coelhas	INSTICC, Portugal
Vera Coelho	INSTICC, Portugal
Ana Guerreiro	INSTICC, Portugal
André Lista	INSTICC, Portugal
Andreia Moita	INSTICC, Portugal
Raquel Pedrosa	INSTICC, Portugal
Vitor Pedrosa	INSTICC, Portugal
Cláudia Pinto	INSTICC, Portugal
Susana Ribeiro	INSTICC, Portugal
Sara Santiago	INSTICC, Portugal
Mara Silva	INSTICC, Portugal
José Varela	INSTICC, Portugal
Pedro Varela	INSTICC, Portugal

Program Committee

El-Houssaine Aghezzaf	Ghent University, Faculty of Engineering and Architecture, Belgium
Javier Alcaraz	Universidad Miguel Hernández de Elche, Spain
Maria Teresa Almeida	ISEG, UTL, Portugal
Lionel Amodeo	University of Technology of Troyes, France
Ronald Askin	Arizona State University, USA
Lyes Benyoucef	Aix-Marseille University, France
Jean-Charles Billaut	École Polytechnique de l'Université François-Rabelais de Tours, France
Christian Blum	IKERBASQUE and University of the Basque Country, Spain
Ralf Borndörfer	Zuse Institute Berlin, Germany
Endre Boros	Rutgers University, USA

Ahmed Bufardi	École Polytechnique Fédérale de Lausanne, Switzerland
Alfonso Mateos Caballero	Universidad Politécnica de Madrid, Spain
Jacques Carlier	Université de Technologie de Compiègne, France
José Manuel Vasconcelos Valério de Carvalho	Universidade do Minho, Portugal
Bo Chen	University of Warwick, UK
John Chinneck	Carleton University, Canada
Mikael Collan	Lappeenranta University of Technology, Finland
Xavier Delorme	Ecole Nationale Supérieure des mines de Saint-Etienne, France
Marc Demange	ESSEC Business School, France
Clarisse Dhaenens	French National Institute for Research in Computer Science and Control, France
Tadashi Dohi	Hiroshima University, Japan
Nikolai Dokuchaev	Curtin University, Australia
Christophe Duhamel	Université Blaise Pascal, Clermont-Ferrand II, France
Pankaj Dutta	Indian Institute of Technology Bombay, India
Gintautas Dzemyda	Vilnius University, Lithuania
Andrew Eberhard	RMIT University, Australia
Ali Emrouznejad	Aston University, UK
Nesim Erkip	Bilkent University, Turkey
Gerd Finke	Grenoble Institute of Technology G-SCOP Laboratory, France
Jörg Fliege	University of Southampton, UK
Bernard Fortz	Université libre de Bruxelles, Belgium
Robert Fullér	Óbuda University, Hungary
Heng-Soon Gan	The University of Melbourne, Australia
Michel Gendreau	École Polytechnique de Montréal, Canada
Giorgio Gnecco	IMT - Institute for Advanced Studies - Lucca, Italy
Juan José Salazar Gonzalez	Universidad de La Laguna, Spain
Armando Guarnaschelli	INGAR CONICET-UTN, Argentina
Christelle Guéret	University of Angers, France
Nalan Gulpinar	The University of Warwick, UK
Jin-Kao Hao	University of Angers, France
Mhand Hifi	EPROAD EA 4669, University of Picardie Jules Verne, France
Han Hoogeveen	Universiteit Utrecht, The Netherlands
Johann Hurink	University of Twente, The Netherlands
Josef Jablonsky	University of Economics, Prague, Czech Republic
Joanna Józefowska	Poznań University of Technology, Poland
Joaquim Júdice	Instituto Telecomunicações, Portugal
Hans Kellerer	University of Graz, Austria
Tamás Kis	Hungarian Academy of Sciences, Hungary
Jesuk Ko	Gwangju University, Korea

Pradeep Kumar	Indian Institute of Technology Roorkee, India
Philippe Lacomme	Université Blaise Pascal, Clermont-Ferrand II, France
Sotiria Lampoudi	Liquid Robotics Inc., USA
Dario Landa-Silva	University of Nottingham, UK
Pierre L'Ecuyer	Universite de Montreal, Canada
Janny Leung	The Chinese University of Hong Kong, Hong Kong
Benjamin Lev	Drexel University, USA
Shan Li	Zicklin School of Business, Baruch College (City University of New York), USA
Abdel Lisser	The University of Paris-Sud 11, France
Pierre Lopez	LAAS-CNRS, Université de Toulouse, France
Helena Ramalhinho Lourenço	Universitat Pompeu Fabra, Spain
Prabhat Mahanti	University of New Brunswick, Canada
Viliam Makis	University of Toronto, Canada
Arnaud Malapert	Université Nice Sophia Antipolis CNRS, France
Patrice Marcotte	Université de Montréal, Canada
Concepción Maroto	Universidad Politécnica de Valencia, Spain
Pedro Coimbra Martins	Polytechnic Institute of Coimbra, Portugal
Ana Meca	Universidad Miguel Hernández de Elche, Spain
Nimrod Megiddo	IBM Almaden Research Center, USA
Marta Mesquita	Universidade de Lisboa, Portugal
Michele Monaci	Università degli Studi di Padova, Italy
Jairo R. Montoya-Torres	Universidad de La Sabana, Colombia
Young Moon	Syracuse University, USA
Dolores Romero Morales	University of Oxford, UK
José Oliveira	Universidade do Minho, Portugal
Linet Özdamar	Yeditepe University, Turkey
Selin Özpeynirci	Izmir University of Economics, Turkey
Sophie Parragh	University of Vienna, Austria
Vangelis Paschos	University Paris-Dauphine, France
Ulrich Pferschy	University of Graz, Austria
Cynthia A. Phillips	Sandia National Laboratories, USA
Diogo Pinheiro	Brooklyn College of the City University of New York, USA
Eric Pinson	IMA, LISA, Angers, France
Selwyn Piramuthu	University of Florida, USA
Jan Platos	VSB - Technical University of Ostrava, Czech Republic
Caroline Prodhon	Charles Delaunay Institute, France
Günther Raidl	Vienna University of Technology, Austria
Andres Ramos	Universidad Pontificia Comillas, Spain
Celso Ribeiro	Universidade Federal Fluminense, Brazil
Andre Rossi	Université de Bretagne-Sud, France
Stefan Ruzika	University of Koblenz and Landau, Germany

Ahti Salo	Aalto University, Finland
Marcello Sanguineti	University of Genoa, Italy
Cem Saydam	University of North Carolina at Charlotte, USA
Marc Sevaux	Université de Bretagne-Sud, France
Patrick Siarry	Paris-Est University, France
Francis Sourd	SNCF, France
Kathryn Stecke	University of Texas at Dallas, USA
Vadim Strijov	Computer Center of the Russian Academy of Sciences, Russian Federation
Thomas Stützle	Université libre de Bruxelles, Belgium
David Sundaram	University of Auckland, New Zealand
Jacques Teghem	Faculté Polytechnique de Mons, Belgium
Alexis Tsoukiàs	CNRS, France
Alexander Vasin	Lomonosov Moscow State University, Russian Federation
Begoña Vitoriano	Complutense University, Spain
Maria Vlasiou	Eindhoven University of Technology, The Netherlands
Hsiao-Fan Wang	National Tsing Hua University, Taiwan
Ling Wang	Tsinghua University, China
Gerhard-Wilhelm Weber	Middle East Technical University, Turkey
Dominique de Werra	École Polytechnique Fédérale de Lausanne, Switzerland
Marino Widmer	University of Fribourg, Switzerland
Margaret Wiecek	Clemson University, USA
Gerhard Woeginger	Eindhoven University of Technology, The Netherlands
Yiqiang Zhao	Carleton University, Canada
Sanming Zhou	University of Melbourne, Australia
Konstantinos Zografos	Lancaster University Management School, UK

Auxiliary Reviewer

Pasi Luukka	Lappeenranta University of Technology, Finland

Invited Speakers

Nicolas Zufferey	University of Geneva, Switzerland
Marie-Christine Costa	ENSTA - ParisTech, France
Paul Williams	London School of Economics, UK
Jean Philippe Vial	University of Geneva, Switzerland

Contents

Applications

Invited Paper

Learning Tabu Search for Combinatorial Optimization

Nicolas Zufferey[1] (✉) and David Schindl[2]

[1] Geneva School of Economics and Management, GSEM, University of Geneva, Blvd du Pont-d'Arve 40, 1211 Geneva, Switzerland
n.zufferey@unige.ch

[2] Geneva School of Business Administration, Rte de Drize 7, 1227 Carouge, Switzerland
david.schindl@hesge.ch

Abstract. In this paper, a new type of local search algorithm is proposed, called *Learning Tabu Search* and denoted *LTS*. It is assumed that any solution of the considered problem can be represented with a list of *characteristics*. *LTS* involves a learning process relying on a *trail* system. The trail system is based on the idea that if some *combinations* of characteristics *often* belong to *good* solutions during the search process, such combinations of characteristics should be favored when generating new solutions. It will be showed that *LTS* obtained promising results on a refueling problem in a railway network.

Keywords: Tabu search · Combinatorial optimization · Learning process

1 Introduction

As exposed in [9], modern methods for solving complex optimization problems are often divided into exact methods and metaheuristic methods. An exact method guarantees that an optimal solution will be obtained in a finite amount of time. Among the exact methods are branch-and-bound, dynamic programming, Lagrangian relaxation based methods, and linear and integer programming based methods [6]. However, for a large number of applications and most real-life optimization problems, which are typically NP-hard [3], such methods need a prohibitive amount of time to find an optimal solution. For these difficult problems, it is preferable to quickly find a satisfying solution. If solution quality is not a dominant concern, then a simple heuristic can be employed, while if quality occupies a more critical role, then a more advanced metaheuristic procedure is warranted. There are mainly two classes of metaheuristics: local search and population based methods. The former type of algorithm works on a single solution (e.g., descent local search, simulated annealing, tabu search, and variable neighborhood search), while the latter makes a population of solutions evolve (e.g., genetic algorithms, scatter search, ant colonies, adaptive memory algorithms). At each iteration of a local search, a *neighbor* solution is generated

E. Pinson et al. (Eds.): ICORES 2014, CCIS 509, pp. 3–11, 2015.
DOI: 10.1007/978-3-319-17509-6_1

from the *current* solution by performing a modification on the current solution, called a *move*. The reader interested in a recent book on metaheuristics is referred to [4].

In this work is proposed a generalized version of the learning tabu search (*LTS*) proposed in [8]. *LTS* is a new type of local search algorithm, easy to adapt within the tabu search framework. It is assumed that any solution of the considered problem can be represented with a list of *characteristics*. *LTS* involves a learning process relying on a *trail* system. The trail system is based on the idea that if some *combinations* of characteristics *often* belong to *good* solutions during the search process, such combinations of characteristics should be favored when generating new solutions.

This paper is organized as follows. In Sect. 2 is proposed and discussed a generic version of *LTS*. In Sect. 3, a successful adaptation of *LTS* is described for a refueling problem in a railway network. The paper ends up with a conclusion in Sect. 4, where *LTS* is positioned according to the main existing metaheuristics.

2 Learning Tabu Search (*LTS*)

In this section are formally presented the general framework of local search techniques, the descent local search, tabu search, and *LTS*, which is a tabu search involving a learning process relying on a trail system.

Let f be an objective function which has to be minimized. At each step of a *local search*, a *neighbor* solution s' is generated from the *current* solution s by performing a specific modification on s, called a *move*. All solutions obtained from s by performing a move are called *neighbor solutions* of s. The set of all the neighbor solutions of s is denoted $N(s)$. First, a local search needs an initial solution s_0 as input. Then, the algorithm generates a sequence of solutions $s_1, s_2, \ldots$ in the search space such that $s_{r+1} \in N(s_r)$. The process is stopped for example when an optimal solution is found (if it is known), or when a time limit is reached. Famous local search algorithms are: the descent method (where at each step, the best move is performed and the process stops when a local optimum is reached), simulated annealing, variable neighborhood search, and tabu search. Note that to escape from a local optimum, in *tabu search*, when a move is performed from a current solution s_r to a neighbor solution s_{r+1}, it is forbidden to perform the reverse of that move during *tab* (parameter) iterations. Such forbidden moves are called *tabu* moves. Formally, the solution s_{r+1} is computed as $s_{r+1} = \arg \min_{s \in N'(s_r)} f(s)$, where $N'(s)$ is a subset of $N(s)$ containing solutions which can be obtained from s by performing a non tabu move. Many variants of this basic tabu search algorithm can be found in [5]. More generally, the reader is referred to [4] for a recent book on metaheuristics, and to [9] for guidelines to efficiently design a metaheuristic.

A tabu search with a learning process relying on a *trail* system is now proposed and denoted *LTS*. Let (P) be the problem under consideration. It is assumed that $C = \{c_1, c_2, \ldots, c_N\}$ is the set of all the possible *characteristics* (i.e. specific features or attributes) of solutions of (P). Further, it is assumed that any solution of (P) with n characteristics can be denoted $s = \{c_{(1)}, c_{(2)}, \ldots, c_{(n)}\}$, where $c_{(i)}$ is the ith characteristic of solution s. Therefore, with each solution s can be associated a set $IN(s)$ (resp. $OUT(s)$) of characteristics which belong (resp. do not belong) to s. If s can be represented with a vector of size n, then $c_{(i)}$ could simply be the ith component of the vector. At each iteration, in order to generate a neighbor solution s' of the current solution s, a basic move m consists in one of the following options: (1) *add* a characteristic c to s (i.e. move c from $OUT(s)$ to $IN(s)$, which is denoted $m(c^+)$), (2) *drop* a characteristic c from s (i.e. move c from $IN(s)$ to $OUT(s)$, which is denoted $m(c^-)$), (3) *switch* two characteristics c and c' between $IN(s)$ and $OUT(s)$ (i.e. move c from $OUT(s)$ to $IN(s)$ and move c' from $IN(s)$ to $OUT(s)$, which is denoted $m(c \leftrightarrow c')$). The straightforward notation $s' = s + m$ can thus be used. Then, it is forbidden (i.e. tabu) to perform the reverse move for *tab* (parameter) iterations. More precisely: (1) if $m(c^+)$ is performed, it is forbidden to perform $m(c^-)$; (2) if $m(c^-)$ is performed, it is forbidden to perform $m(c^+)$; (3) if $m(c \leftrightarrow c')$ is performed, it is forbidden to perform $m(c' \leftrightarrow c)$.

The trail system relies on the idea that if some *combinations* of characteristics *often* belong to *good* solutions during the search process, such combinations of characteristics should be favored when generating new solutions. If one considers combinations of k characteristics, the trail $tr(c_{(i_1)}, c_{(i_2)}, \ldots, c_{(i_k)})$ associated with characteristics $c_{(i_1)}, c_{(i_2)}, \ldots, c_{(i_k)}$ indicates if it is a good idea to have such k characteristics *together* in a solution, according to the observation of the history of the search.

Let $\Pi_k^n(s)$ be the set of all possible combinations of k characteristics of a solution s containing $n > k$ characteristics. Consider that $s' = s+m$. The trail $Tr(s')$ associated with solution s' can be defined as follows: (1) $\sum_{\pi \in \Pi_{k-1}^n(s)} tr(c, \pi)$ if $m(c^+)$ is considered; (2) $-\sum_{\pi \in \Pi_{k-1}^n(s)} tr(c, \pi)$ if $m(c^-)$ is considered; (3) $\sum_{\pi \in \Pi_{k-1}^n(s)} tr(c, \pi) - \sum_{\pi \in \Pi_k^n(s)} tr(c', \pi)$ if $m(c \leftrightarrow c')$ is considered. For instance, one can remark that if $Tr(s')$ associated with a move $m(c^+)$ is large, it means that c is in average attracted by the characteristics of the current solution s (because it was often observed in the past that the combination of c with $k-1$ characteristics of s leads averagely to good solutions).

The updating of the trail values is now discussed. Let $\rho \in [0, 1]$ be a parameter representing an *evaporation* coefficient. A *cycle* of size I (parameter) is defined as a sequence of I iterations of tabu search. Every I iterations, the best solution $\hat{s}$ of the last cycle is used to update the trail system as follows: $tr(c_{(i_1)}, c_{(i_2)}, \ldots, c_{(i_k)}) = \rho \cdot tr(c_{(i_1)}, c_{(i_2)}, \ldots, c_{(i_k)}) + \Delta tr(c_{(i_1)}, c_{(i_2)}, \ldots, c_{(i_k)})$, where $\Delta tr(c_{(i_1)}, c_{(i_2)}, \ldots, c_{(i_k)})$ is a *reinforcement* term proportional to the quality of $\hat{s}$ if the k characteristics $c_{(i_1)}, c_{(i_2)}, \ldots, c_{(i_k)}$ appear in $\hat{s}$, and $\Delta tr(c_{(i_1)}, c_{(i_2)}, \ldots, c_{(i_k)}) = 0$ otherwise. An alternative would be to say that $\Delta tr(c_{(i_1)}, c_{(i_2)}, \ldots, c_{(i_k)})$ is proportional to: (1) the number of times that characteristics $c_{(i_1)}, c_{(i_2)}, \ldots, c_{(i_k)}$ jointly appear in the solutions visited within the cycle, (2) the average quality of the solutions of the cycle having characteristics $c_{(i_1)}, c_{(i_2)}, \ldots, c_{(i_k)}$ together.

At each iteration of *LTS*, it is proposed to first randomly choose a random set A (resp. D and S) of non tabu neighbor solutions obtained with add (resp. drop and switch) moves, such that $|A| = |D| = |S|$. Note that the size of A is an important and sensitive parameter to tune. Then, let A_q (resp. D_q and S_q) be the set containing the q (parameter) solutions of A (resp. D and S) with the largest trail values. The performed move among $A_q \cup D_q \cup S_q$ is then the best one according to the objective function f of the considered problem. This technique is particularly relevant if it is cumbersome to evaluate a solution with f, as f is only used to evaluate a sample of solutions which rank highly according to the trail function Tr (which does not require a lot of computation). In other words, at each iteration of *LTS*, the objective function f (or its associated incremental computation) is only used to evaluate $3 \cdot q$ *good* neighbor solutions according to Tr, which is likely to be more promising than to evaluate a *random* sample of $3 \cdot q$ neighbor solutions.

In order to help to visit new regions of the solution space, the following diversification mechanism (denoted *DIV*) is introduced, relying on a different

---

Algorithm 1. *LTS*: Learning Tabu Search.

Initialization

1. construct an initial solution s;
2. set $s^\star = s$ and $f^\star = f(s)$;
3. set $iter = 0$ (iteration counter);
4. set *DIV* as closed (i.e. not active);

While a time limit is not reached, **do**:

1. from s, generate a set A (resp. D and S) of non tabu neighbor solutions obtained with add (resp. drop and switch) moves, such that $|A| = |D| = |S|$;
2. if *DIV* is closed, identify the sets $A_q \subseteq A$ (resp. $D_q \subseteq D$ and $S_q \subseteq S$) containing the solutions with the q largest trail values;
3. if *DIV* is open, identify the sets $A_q \subseteq A$ (resp. $D_q \subseteq D$ and $S_q \subseteq S$) containing the solutions with the q smallest trail values;
4. select the neighbor solution: set $s' = arg \min_{s'' \in A_q \cup D_q \cup S_q} f(s'')$;
5. update the current solution: set $s = s'$;
6. update the tabu status: the reverse move is forbidden for *tab* iterations;
7. update the best encountered solution: if $f(s) < f^\star$, set $s^\star = s$ and $f^\star = f(s)$;
8. update the iteration counter: set $iter = iter + 1$;
9. update the status of *DIV*:
   (a) open *DIV* if t_1 iterations without improving $s^\star$ have been performed;
   (b) close *DIV* if it has been performed during t_2 consecutive iterations, or if $s^\star$ has been improved within the current application of *DIV*;
10. if $(iter \mod I) = 0$, update the trail system with the best solution $\hat{s}$ among the last I iterations;

Return $s^\star$

---

use of the trail system. Generally, the trail system is used to favor good moves which were often performed in the previous cycles. This is also the case for the trail systems used in the various types of ant algorithms [10]. In contrast, to diversify the search, it is proposed here to perform good moves which were not often performed in the previous cycles. More precisely, assuming that a neighbor solution s' can be generated from the current solution s with move m (i.e., $s' = s + m$), it is diversifying to perform m if the corresponding $Tr(s')$ is small. Each iteration of *DIV* is performed as above, but the set A_q (resp. D_q and S_q) contains the q solutions of A (resp. D and S) with the smallest trail values. *DIV* relies on two sensitive parameters t_1 and t_2: it is triggered if t_1 iterations without improving $s^\star$ (the best encountered solution during the search) have been performed, and it is performed until one of the following conditions is satisfied: (1) $s^\star$ has been improved; (2) a sequence of t_2 iterations of *DIV* have been performed.

LTS can now be formulated in Algorithm 1.

3 *LTS* for a Refueling Problem in a Railway Network

The problem consists in optimizing the refueling costs of a fleet of locomotives over a railway network [7]. It is assumed that there is only one source of fuel: fueling trucks, located at yards. A solution of the problem has two important components: choose the number of trucks contracted at each yard, and determine the refueling plan of each locomotive (i.e. the quantity of fuel that must be dispensed into each locomotive at every yard). Such components are respectively called the *truck assignment problem* (TAP) and the *fuel distribution problem* (FDP) in this paper. The constraints are the following: the capacity of the tank of each locomotive is limited, as well as the maximum amount of fuel a truck can provide the same day; a locomotive can not be refueled at its destination yard; there is a maximum number of times (which is two) a train can stop to be refueled (excluding the origin); it is forbidden to run out of fuel. The encountered costs are the weekly operating cost of each fueling truck, the fuel price per gallon associated with each yard (which can vary from yard to yard because of the differences in distribution, marketing costs and other factors), and the fixed cost associated with each refueling. A solution satisfying all the above constraints is called *feasible*. The problem consists in finding a feasible solution minimizing the sum of the costs.

If the number of trucks is known for every yard (i.e., if a solution of the TAP is provided), it is possible to exactly solve the associated FDP with the *flow algorithm* proposed in [7]. To tackle the TAP, various solution methods have been proposed in [8]: a descent local search *DLS*, a regular tabu search *TS*, and a learning tabu search *LTS*. The common features of *DLS* and *TS* are the following. From a current solution s, a move simply consists in adding a contracted truck to a yard (*add move*), or in removing a contracted truck from a yard (*drop move*). When a move is performed, it is evaluated by the use of the *flow algorithm*. During the evaluation, all the costs are considered: the refueling costs for the used gallons of fuel, the fixed refueling costs and the contracting costs of the trucks. At each iteration, $|A|$ add moves and $|D|$ drop moves are generated (with the tuning $|A| = |B| = 10$). The resulting descent local search *DLS* for the TAP is presented in Algorithm 2. In contrast with *DLS*, *TS* does not stop when it reaches a local optimum, but returns the best encountered solution within a predefined time limit. When an add (resp. drop) move is performed on yard y_j, it is forbidden to consider yard y_j for a drop (resp. add) move for *tab* iterations (parameter tuned to 10).

Algorithm 2. *DLS*: Descent Local Search for the TAP.

Construct an initial solution s;

While a local optimum is not reached, **do**:

1. in solution s, randomly choose a set D containing yards for which drop moves are allowed; for any yard $y_j \in D$ and from s, remove a truck from it, and apply the *flow algorithm* to evaluate such a drop candidate move;
2. in solution s, randomly choose a set A of yards; for any yard $y_j \in A$ and from s, add a truck to it, and apply the *flow algorithm* to evaluate such an add candidate move;
3. from s, perform the best move among the $\mid A \cup D \mid$ above candidate moves, and rename the resulting solution as s;

In order to adapt *LTS* to the considered problem, one mainly has to define what is a *characteristic* and to set a *learning* process based on a *trail* system. Let x and y be two yards. A characteristic is simply a pair (x, y) of yards such that there is at least a truck on x and at least a truck on y. The trail $tr(x, y)$ associated with x and y aims to indicate if it is a good idea to have trucks on both yards x and y in the same solution.

Every I (parameter tuned to 50) iterations of *LTS*, such trails are globally updated with $\hat{s}$ (the best solution of the cycle) as follows (with ρ tuned to 0.9): $tr(x, y) = \rho \cdot tr(x, y) + \Delta tr(x, y)$, where $\Delta tr(x, y)$ is the number of trucks on x and y, computed only if $\hat{s}$ has trucks on both x and y (it is 0 otherwise). A move can be denoted by (x, s), indicating that a truck is added to or removed from yard x, in the current solution s. Let $Tr(x, s)$ be its associated trail value. It is straightforward to set $Tr(x, s) = \sum_{y \in s} tr(x, y)$ if it is used as follows. If (x, s) is an add move (i.e. add a truck to yard x of solution s), among the possible add moves, it is interesting to select a move with a large $Tr(x, s)$ value (because the history of the search seems to indicate that having trucks on yard x, as well as on the yards which already contain trucks in the current solution s, is a good idea). On the contrary, if (x, s) is a drop move (i.e. remove a truck from yard x of solution s), among the drop moves, it is better to select a move with a small $Tr(x, s)$ value.

The way to select a move at each iteration is now described. Remember that in *DLS* and in *TS*, the performed move is the best among the ones in the set

$| A \cup D |$, with $|A| = |D| = 10$. In *LTS*, two sets A and D of size 20 are first randomly chosen. Then, let A_q (resp. D_q) be the subset of A (resp. D) containing the q (parameter tuned to 10) moves with the largest (resp. smallest) trail values. Note that computing the trail value of a move is much quicker to compute than the value of the resulting neighbor solution, as the flow algorithm is requested for it. The performed move among $A_q \cup D_q$ is the best one according to the objective function of the problem (i.e., the sum of the costs). Therefore, for *DLS*, *TS* and *LTS*, the performed move has the best objective function value among a sample of 20 evaluated solutions. This will allow to better measure the impact of the trail system on the search. Note that if *DIV* is open (i.e., the diversification mechanism is activated), A_q (resp. D_q) is the subset of A (resp. D) containing the q moves with the smallest (resp. largest) trail values.

Now are presented results on the 57 linear instances and the 15 non linear instances instances associated with [8]. The non linear instances are based on the following idea. If several trucks from the same company are contracted for the same yard, the company is likely to propose discounted prices for that yard. The proposed algorithms were tested on an Intel Quad-core i7 @ 3.4 GHz with 8 GB DDR3 of RAM memory. A common time limit $T = 60$ min is imposed to each proposed method (i.e., *DLS*, *TS* and *LTS*). Note that *DLS* is restarted from scratch each time a local optimum is found, in order to be able to apply it during T minutes, and thus to perform a fair comparison with the two other metaheuristics.

For *DLS*, *TS* and *LTS*, Fig. 1 compares the evolution of the average best encountered solution value during 60 min on the linear instances (57 instances, 10 runs per instance). On the one hand, one can observe that *LTS* clearly outperforms the other methods from the beginning to the end of the hour of computation. This obviously indicates that the learning process (i.e. the trail system) introduced to *TS* to derive *LTS* is relevant. On the other hand, one can remark that *TS* is better than *DLS* in the first 30 min, and then both methods have a very comparable behavior.

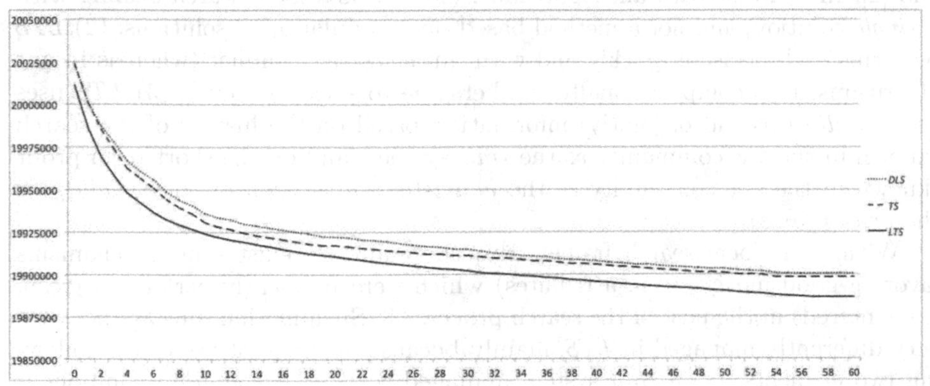

Fig. 1. Evolution of *DLS*, *TS* and *LTS* on the linear instances.

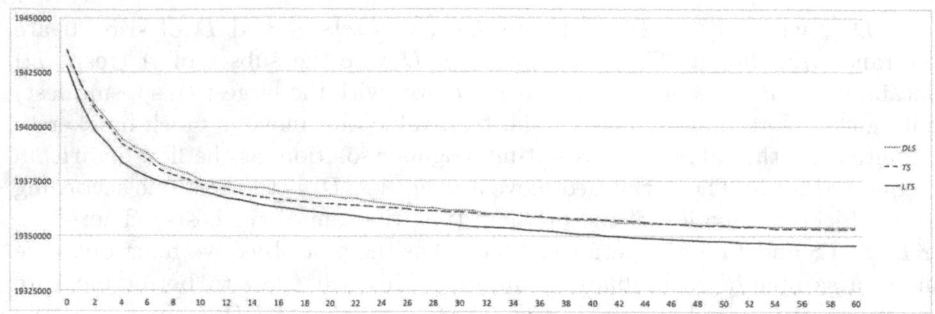

Fig. 2. Evolution of *DLS*, *TS* and *LTS* on the non linear instances.

Figure 2 is similar to Fig. 1 but is associated with the non linear instances. The same observations as before can be made: *LTS* outperforms *TS* and *DLS*, *TS* is better than *DLS* during half an hour, *TS* and *DLS* are comparable during the second half hour.

4 Conclusion

In this paper, a new type of local search is presented, called *Learning Tabu Search* and denoted *LTS*. It was showed that *LTS* was successfully adapted to a refueling problem in a railway network, which was initially motivated by the *Railway Applications Section of INFORMS*.

LTS involves a learning process relying on a *trail* system. The trail system is based on the idea that if some *combinations* of characteristics *often* belong to *good* solutions during the search process, such combinations of characteristics should be favored when generating new solutions.

Even if the concept of trail system also exists in ant algorithms, it is managed very differently in *LTS*. Recent overviews of ant algorithms can be found in [1] and [2]. In contrast with ant algorithms: (1) *LTS* is a local search dealing with a *single* solution, and not a method based on a population of solutions; (2) *LTS* performs each decision quickly and with an *aggressive* manner (whereas in ant algorithms, it is computationally cumbersome to select a move); (3) *LTS* uses *sequentially* (instead of jointly) information based on the history of the search (known in the ant community as the *trail* system) and on the short term profit (known in the ant community as the *heuristic information* or the *visibility* or the *greedy force*).

Within the local search frameworks, there already exist some mechanisms favoring good moves (resp. attributes) which were frequently performed (resp. encountered) in the past of the search process [4]. Such mechanisms are however very differently managed in *LTS*, mainly because of the joint use of the following two elements: (1) a *trail* system managed with an *evaporation* component and a *reinforcement* component; (2) *combinations* of characteristics are handled (instead of *individually* considering each characteristic).

Therefore, *LTS* can be clearly positioned according to the existing metaheuristics. Among the possible future works, it would be interesting to apply *LTS* to other kinds of problems, for example to find out for each of them which value of k is the most relevant.

References

1. Blum, C.: Ant colony optimization: introduction and recent trends. Phys. Life Rev. **2**(4), 353–373 (2005)
2. Dorigo, M., Birattari, M., Stuetzle, T.: Ant colony optimization—artificial ants as a computational intelligence technique. IEEE Comput. Intell. Mag. **1**(4), 28–39 (2006)
3. Garey, M., Johnson, D.S.: Computer and Intractability: A Guide to the Theory of NP-Completeness. Freeman, San Francisco (1979)
4. Gendreau M., Potvin, J.-Y.: Handbook of metaheuristics. In: International Series in Operations Research & Management Science, vol. 146, pp. 573–597. Springer, New York (2010)
5. Glover, F., Laguna, M.: Tabu Search. Kluwer Academic Publishers, Boston (1997)
6. Nemhauser, G., Wolsey, L.: Integer and Combinatorial Optimization. Wiley, New York (1988)
7. Schindl, D., Zufferey, N.: Solution methods for fuel supply of trains. Inf. Syst. Oper. Res. **51**(1), 22–29 (2013)
8. Schindl, D., Zufferey, N.: A learning tabu search for a truck allocation problem with linear and nonlinear cost components. Nav. Res. Logistics **61**(1), 42–45 (2015)
9. Zufferey, N.: Metaheuristics: some principles for an efficient design. Comput. Technol. Appl. **3**(6), 446–462 (2012)
10. Zufferey, N.: Optimization by ant algorithms: possible roles for an individual ant. Optim. Lett. **6**(5), 963–973 (2012)

Methodologies and Technologies

Four Serious Problems and New Facts of the Discriminant Analysis

Shuichi Shinmura(✉)

Faculty of Economics, Seikei University, Tokyo, Japan
shinmura@econ.seikei.ac.jp

Abstract. The discriminant analysis is essential knowledge in science, technology and industry. But, there are four serious problems. These are resolved by Revised IP–OLDF and k–fold cross validation.

Keywords: Minimum number of misclassifications (MNM) · Revised IP-OLDF · SVM · LDF · Logistic regression · k-fold cross validation

1 Introduction

Fisher [1] described the linear discriminant function (**LDF**), and founded the discriminant theory. Following this, the quadratic discriminant function (**QDF**) and multi-class discrimination using Mahalanobis distance were proposed. These functions are based on the variance-covariance matrices, and are easily implemented in the statistical software packages. They can be used in many applications. However, real data rarely satisfy Fisher's assumptions. Therefore, it is well known that logistic regression is better than LDF and QDF, because it does not assume a specific theoretical distribution, such as a normal distribution. In addition to this, the discriminant rule is very simple: If $y_i{*}f(\mathbf{x_i}) > 0$, $\mathbf{x_i}$ is classified to class1/class2 correctly. If $y_i{*}f(\mathbf{x_i}) < 0$, $\mathbf{x_i}$ is misclassified. There are four serious problems hidden in this simplistic scenario [22].

(1) **Problem 1**

We cannot properly discriminate between cases where $\mathbf{x_i}$ lies on the discriminant hypeplane ($f(\mathbf{x_i}) = 0$). This **unresolved problem** has been ignored until now. The proposed **Revised IP-OLDF** is able to treat this problem appropriately. Indeed, except for Revised IP-OLDF, no functions can correctly count the number of misclassifications (NM). These functions should count the number of cases where $f(\mathbf{x_i}) = 0$, and display this alongside the NM in the output.

(2) **Problem 2**

Fisher's LDF and QDF cannot recognize linear separable data (where the Minimum NM (MNM) = 0). This fact was first found when **IP-OLDF** was applied to Swiss bank note data [3]. In this paper, the determination of pass/fail in exams is used because it is trivially linear-separable and we can obtain it easily. We show that, in many cases, the NMs of LDF and QDF are not zero. Next, 100 re-samples of these data are generated, and the mean error rates are obtained by 100-fold cross validation. The mean error rates

E. Pinson et al. (Eds.): ICORES 2014, CCIS 509, pp. 15–30, 2015.
DOI: 10.1007/978-3-319-17509-6_2

of LDF are 6.23 % higher than that of Revised IP-OLDF in the validation samples of Table 7.

(3) **Problem 3**

If the variance-covariance matrix is singular, Fisher's LDF and QDF cannot be calculated because the inverse matrices do not exist. The LDF and QDF of JMP [9] are solved by the generalized inverse matrix technique. In addition to this, **RDA** [4] is used if QDF causes serious trouble with dirty data. However, RDA and QDF do not work properly for the special case in which the values of features belonging to one class are constant. If users can choose proper options for a **modified RDF** developed for this special case, it works better than QDF and LDF in Table 5.

(4) **Problem 4**

Some statisticians misunderstand that the discriminant analysis is the inferential statistical method as same as the regression analysis, because it is derived from Fisher's assumption. But there are no standard error (SE) of the discriminant coefficients or error rate, and variable selection methods such as stepwise methods and statistics such as Cp and AIC. In this paper, we propose "k-fold cross validation for small samples" and new variable selection method, the minimum mean error rates of which is chosen as the best model. In future works (**Future1**), generalization ability and 95 % confidence intervals of all LDFs are proposed.

In this research, two Optimal LDFs (**OLDFs**) based on the MNM criterion are proposed. The above three problems are solved by **IP-OLDF** and **Revised IP-OLDF** completely. IP-OLDF [13–15] reveals the following properties.

Fact (1) Relation between LDFs and NMs. IP-OLDF is defined on the data and discriminant coefficient spaces. Cases of $\mathbf{x_i}$ correspond to linear hyper-planes ($H_i(\mathbf{b}) = y_i^* ({}^t\mathbf{x_i b} + 1) = 0$) in the p-dimensional discriminant coefficient space that divide the space into two half-planes: the plus half-plane ($H_i(\mathbf{b}) > 0$) and minus half-plane ($H_i(\mathbf{b}) < 0$). Therefore, the coefficient space is divided into a finite convex polyhedron by $H_i(\mathbf{b})$. Interior point $\mathbf{b}_j$ of the convex polyhedron corresponds to the discriminant function $f_j(\mathbf{x}) = {}^t\mathbf{b}_j\mathbf{x} + 1$ on the data space that discriminates some cases properly and misclassifies others. This means that each interior point $\mathbf{b}_j$ has a unique NM. The "**Optimal Convex Polyhedron (OCP)**" is defined as that with the MNM. Revised IP-OLDF [16] can find the interior point of OCP directly, and solves the unresolved problem (**Problem 1**) because there are no cases on the discriminant hyperplane ($f(\mathbf{x_i}) = 0$). If $\mathbf{b}_j$ is on a vertex or edge of the convex polyhedron, however, the unresolved problem cannot be avoided because there are some cases on $f(\mathbf{x_i}) = 0$.

Fact (2) Monotonous decrease of MNM ($\mathbf{MNM_p \geq MNM_{(p+1)}}$). Let MNM_p be the MNM of p features (independent variables). Let $MNM_{(p+1)}$ be the MNM of the (p + 1) features formed by adding one feature to the original p features. MNM decreases monotonously ($MNM_p \geq MNM_{(p+1)}$), because OCP in the p-dimensional coefficient space is a subset of the (p + 1)-dimensional coefficient space [18]. If $MNM_p = 0$, all MNMs including p features are zero. Swiss bank note data consists of genuine and counterfeit bills with six features. IP-OLDF finds that this data is linear-separable according to two features (X4, X6). Therefore, 16 models including these two features have MNMs = 0. Nevertheless, Fisher's LDF and QDF cannot recognize that this data is linear-separable, presenting a serious problem. In this paper, we show that Revised IP-OLDF can resolve the above three problems, and is superior to Fisher's LDF, logistic regression, and Soft-margin SVM (S-SVM) [28] under 100-fold cross validation [20, 21] of the pass/fail determinations of exams [19] and their re-sampled data.

2 Discriminant Functions

2.1 Statistical Discriminant Functions

Fisher defined LDF to maximize the variance ratio (between/within classes) in Eq. (1). This can be solved by non-linear programming (NLP).

$$\mathrm{MIN} = {}^t\mathbf{b}(\mathbf{m}_1 - \mathbf{m}_2)^t(\mathbf{m}_1 - \mathbf{m}_2)\mathbf{b}/{}^t\mathbf{b}\,\Sigma\mathbf{b}; \tag{1}$$

If we accept Fisher's assumption, the same LDF is obtained in Eq. (2). This equation defines LDF explicitly, whereas Eq. (1) defines LDF implicitly. Therefore, statistical software packages adopt this equation. Some statisticians misunderstand that discriminant analysis is the same as regression analysis. Discriminant analysis is independent of inferential statistics, because there are no SEs of the discriminant coefficients and error rates (Problem 4). Therefore, the leave-one-out (LOO) method [6] was proposed to choose the proper discriminant model.

$$\text{Fisher's LDF}: \mathrm{f}(\mathrm{x}) = {}^t\{\mathbf{x} - (\mathbf{m}_1 + \mathbf{m}_2)/2\}\,\Sigma^{-1}(\mathbf{m}_1 - \mathbf{m}_2) \tag{2}$$

Most real data does not satisfy Fisher's assumption. When the variance-covariance matrices of two classes are not the same ($\Sigma_1 \neq \Sigma_2$), the QDF defined in Eq. (3) can be used. The Mahalanobis distance (Eq. (4)) is used for the discrimination of multi-classes, and the Mahalanobis-Taguchi [25] method is applied in quality control.

$$\mathrm{QDF}:\ \mathrm{f}(\mathbf{x}) = {}^t\mathbf{x}(\Sigma_2^{-1} - \Sigma_1^{-1})\mathbf{x}/2 + ({}^t\mathbf{m}_1\,\Sigma_1^{-1} - {}^t\mathbf{m}_2\,\Sigma_2^{-1})\mathbf{x} + \mathrm{c} \tag{3}$$

$$\mathrm{D} = \mathrm{SQRT}\left({}^t(\mathbf{x} - \mathbf{m})\Sigma^{-1}(\mathbf{x} - \mathbf{m})\right) \tag{4}$$

These functions are applied in many areas, but cannot be calculated if some features remain constant. There are three cases. First, some features that belong in both classes are the same constant. Second, some features that belong in both classes are different but constant. Third, some feature that belongs to one class is constant. Most statistical software packages exclude all features in these three cases. On the other hand, JMP enhances QDF using the generalized inverse matrix technique. This means that QDF can treat the first and second cases correctly, but cannot handle the third case properly.

Recently, the logistic regression in Eq. (5) has been used instead of LDF and QDF for two reasons. First, it is well known that the error rate of logistic regression is often less than those of LDF and QDF, because it is derived from real data instead of some normal distribution that is liberated from reality. Let 'P' be the probability of belonging to a group of diseases. If the value of some feature is increasing/decreasing, 'P' increases from zero (normal group) to one (group of diseases). This representation is very useful in medical diagnosis, as well as for ratings in real estate and bonds. On the contrary, Fisher's LDF assumes that cases near to the average of the diseases are representative cases of the diseases group. Medical doctors never permit this claim.

$$\mathrm{Log}(\mathrm{P}/(1-\mathrm{P})) = \mathrm{f}(\mathbf{x}) \tag{5}$$

2.2 Before and After SVM

Stam [24] summarized Lp-norm discriminant methods until 1997, and answers the question of "Why have statisticians rarely used Lp-norm methods?" He gives four reasons: Communication, promotion and terminology; Software availability; Relative accuracy of Lp-norm classification methods: Ad hoc studies; and the Accuracy of Lp-norm classification methods: decision theoretic justification. While each of these reasons is true, they are not important. The most important reason is that there is no comparison between these methods with statistical discriminant functions, because discriminant analysis was established by Fisher before mathematical programming (MP) approaches. There are two types of MP applications. The first is modeling by MP, such as for portfolio selection [26], and the second is catch-up modeling, such as for the regression and discriminant analysis. Therefore, the latter type should be compared with preceding results. No statisticians use Lp-norm methods, because there is no research indicating that Lp-norm methods are superior to statistical methods. Liitschwager and Wang [7] defined a model based on the MNM criterion. There are several mistakes, but the most important one is the restriction on the discriminant coefficients. Only one discriminant coefficient should be fixed to $-1/1$. There is no need to fix the other $(k-1)$ coefficients in the range $[-1, 1]$.

Vapnik proposed three different SVM models. The hard-margin SVM (H-SVM) indicates the discrimination of linear separable data. H-SVM is defined to maximize the distance of the "Support Vector (SV)" in order to obtain "good generalization" by NLP, which is similar to "not overestimating the validation data in statistics." H-SVM is redefined to minimize (1/"distance of SV") in Eq. (6). This is solved by quadratic programming (QP), which can only be used for linear separable data. This may be why investigation of linear separable data has been ignored. We statisticians misunderstand that discrimination of linear separable data is very easy. In statistics, there was no technical term for linear separable data. However, the condition "MNM = 0" is the same as being linear-separable. Note that "NM = 0" does not imply the data is linear-separable. It is unfortunate that there has been no research into linear separability.

$$\begin{aligned} &\mathrm{MIN} = \|\mathbf{b}\|^2/2; \quad y_i * ({}^t\mathbf{x_i}\mathbf{b} + b_0) \geq 1; \quad \mathbf{b} : \mathrm{p-discriminant\ coefficients.} \\ &y_i = 1/-1 \text{ for } \mathbf{x_i} \in \mathrm{class1/class2}.\ \mathbf{x_i} : \mathrm{p-features(independent\ variables).} \end{aligned} \tag{6}$$

Real data are rarely linear-separable. Therefore, S-SVM has been defined in Eq. (7). S-SVM permits certain cases that are not discriminated by SV ($y_i*({}^t\mathbf{x_i}\mathbf{b} + b_0) < 1$).

The second objective is to minimize the summation of distances of misclassified cases (Σe_i) from SV. These two objects are combined by defining some "penalty c." The Markowitz portfolio model to minimize risk and maximize return is the same as S-SVM. However, the return is incorporated as a constraint, and the objective function minimizes risk. The decision maker chooses a solution on the efficient frontier. On the contrary, S-SVM does not have a rule to determine c properly; nevertheless, it can be solved by an optimization solver. (Kernel-SVM is omitted from the research.)

$$\mathrm{MIN} = ||\mathbf{b}||^2/2 + c * \Sigma e_i; \quad y_i * ({}^t\mathbf{x}_i\mathbf{b} + b_0) \geq 1 - e_i; \tag{7}$$

c: penalty c to combine two objectives. e_i: non-negative value.

2.3 IP-OLDF and Revised IP-OLDF

Shinmura and Miyake [12] developed the heuristic algorithm of OLDF based on the MNM criterion. This solves the five features (5-features) model of Cephalo Pelvic Disproportion (CDP) data that consisted of two groups having 19 features. SAS was introduced into Japan in 1978, and three technical reports about the generalized inverse matrix, the sweep operator [5], and SAS regression applications [8] are related to this research. LINDO was introduced to Japan in 1983. Several regression models are formulated by MP [10], e.g., least-squares problems can be solved by QP, and LAV (Least Absolute Value) regression is solved by LP. Without a survey of previous research, the formulation of IP-OLDF can be defined as in Eq. (8). This notation is defined on p-dimensional coefficient space, because the constant is fixed to 1. In pattern recognition, the constant is a free variable. In this case, the model is defined on (p + 1)-coefficient space, and we cannot elicit the same deep knowledge as with IP-OLDF. This difference is very important. IP-OLDF is defined on both p-dimensional data and coefficient spaces. We can understand the relation between the NM and LDF clearly. The linear equation $H_i(\mathbf{b}) = y_i * ({}^t\mathbf{x}_i\mathbf{b} + 1) = 0$ divides p-dimensional space into plus and minus half-planes ($H_i(\mathbf{b}) > 0$, $H_i(\mathbf{b}) < 0$). If $\mathbf{b}_j$ is in the plus half-plane, $f_j(\mathbf{x}) = y_i*({}^t\mathbf{b}_j\mathbf{x} + 1)$ discriminates $\mathbf{x}_i$ correctly, because $f_j(\mathbf{x}_i) = y_i*({}^t\mathbf{b}_j\mathbf{x}_i + 1) = y_i *({}^t\mathbf{x}_i\mathbf{b}_j + 1) > 0$. On the contrary, if $\mathbf{b}_j$ is included in the minus half-plane, $f_j(\mathbf{x})$ cannot discriminate $\mathbf{x}_i$ correctly, because $f_j(\mathbf{x}_i) = y_i*({}^t\mathbf{b}_j\mathbf{x}_i + 1) = y_i *({}^t\mathbf{x}_i\mathbf{b}_j + 1) < 0$. The n linear equations $H_i(\mathbf{b})$ divide the coefficient space into a finite number of convex polyhedrons. Each interior point of a convex polyhedron has a unique NM that is equal to the number of minus half-planes. We define the OCP as that for which NM is equal to MNM. If $\mathbf{x}_i$ is classified correctly, $e_i = 0$ and $H_i(\mathbf{b}) \geq 0$ in Eq. (8). If there are p cases on $f(\mathbf{x}_i) = 0$, we can obtain the exact MNM. However, if there are over (p+1) cases on $f(\mathbf{x}_i)=0$, this causes the unresolved problem. If $\mathbf{x}_i$ is misclassified, $e_i = 1$ and $H_i(\mathbf{b}) \geq -10000$. This means that IP-OLDF chooses the discriminant hyper-plane $H_i(\mathbf{b}) = 0$ for correctly classified cases, and $H_i(\mathbf{b}) = -10000$ for misclassified cases according to a 0/1 decision variable. if IP–OLDF chooses a vertex having p cases, it chooses the OCP correctly. If it chooses a vertex having over (p+1) cases, it may not choose the OCP. In addition to this defect, IP-OLDF must be solved for the three cases where the constant is equal to 1, 0, −1, because we cannot determine the sign of y_i in advance. Combinations of $y_i = 1/{-1}$ for $\mathbf{x}_i \in$ class1/class2 are decided by the data, not the analyst.

$$\text{MIN} = \Sigma e_i; \quad H_i(\mathbf{b}) \geq -M * e_i; \quad M: 10{,}000 \text{ (Big M constant)}. \tag{8}$$

The Revised IP-OLDF in Eq. (9) can find the true MNM, because it can directly find the interior point of the OCP. This means there are no cases where $y_i*({}^t\mathbf{x}_i\mathbf{b} + b_0) = 0$. If $\mathbf{x}_i$ is discriminated correctly, $e_i = 0$ and $y_i*({}^t\mathbf{x}_i\mathbf{b} + b_0) \geq 1$. If $\mathbf{x}_i$ is misclassified, $e_i = 1$ and $y_i*({}^t\mathbf{x}_i\mathbf{b} + b_0) \geq -9999$. It is expected that all misclassified cases will be extracted to alternative SVs, such as $y_i*({}^t\mathbf{x}_i\mathbf{b} + b_0) = -9999$. Therefore, the discriminant scores of misclassified cases become large and negative, and there are no cases where $y_i*({}^t\mathbf{x}_i\mathbf{b} + b_0) = 0$. This means that $\mathbf{b}$ is interior point of OCP defined by IP-OLDF.

$$\text{MIN} = \Sigma e_i; \quad y_i * ({}^t\mathbf{x}_i\mathbf{b} + b_0) \geq 1 - M * e_i; \quad b_0: \text{ free decision variable}. \tag{9}$$

If e_i is a non-negative real variable, we utilize Revised LP-OLDF, which is an L1-norm LDF. Its elapsed runtime is faster than that of Revised IP-OLDF. If we choose a large positive number as the penalty c of S-SVM, the result is almost the same as that given by Revised LP-OLDF, because the role of the first term of the objective value in Eq. (7) is ignored. Revised IPLP-OLDF is a combined model of Revised LP-OLDF and Revised IP-OLDF. In the first step, Revised LP-OLDF is applied for all cases, and e_i is fixed to 0 for cases that are discriminated correctly by Revised LP-OLDF. In the second step, Revised IP-OLDF is applied for misclassified cases in the first step. Therefore, Revised IPLP-OLDF can obtain an estimate of MNM faster than Revised IP-OLDF [17, 23], but it is unknown to be free from the unresolved problem.

3 The Unresolved Problem (Problem 1)

3.1 Perception Gap of This Problem

About the unresolved problem, there are several understandings. Most researchers treat the cases $\mathbf{x}_i$ on $f(\mathbf{x}_i) = 0$ in class1. There is no explanation of why it makes sense. Some statisticians explain that it is decided stochastically, because the statistics is a sturdy of probability. This explanation seems theoretically at first glance, but it is nonsense by two reasons. Statistical software adopt the former decision rule because many papers and researchers adopt this rule. In the medical diagnosis, medical doctors strive to judge the patients near by the discriminant hyper-plane. If they know second explanation, they are deeply disappointed in the discriminant analysis. Until now, all LDFs such as Fisher's LDF, logistic regression, H-SVM and S-SVM cannot treat this problem properly. IP-OLDF reveals that only interior points of convex polyhedron can resolve this problem. It can find the vertex of true OCP if data is general position and it stop the optimization choosing p cases on the discriminant hyper-plane. But, it may not find the true MNM if data is not general position and it choose over (p + 1) cases on the discriminant hyper-plane. Revised IP-OLDF can find the interior point of the OCP directly. We cannot judge whether other LDFs choose the interior point, edge or vertex of the convex polyhedron. This is confirmed by checking the number of cases $\mathbf{x}_i$ that satisfy $|f(\mathbf{x}_i)| \leq 10^{-6}$ if we consider $|f(\mathbf{x}_i)| \leq 10^{-6}$ is zero. If this number is zero, this function chooses the interior point of the convex polyhedron. If this number 'm' isn't zero, this LDF chooses the vertex or edge of the convex polyhedron, and true NM has a possibility of increase up to 'm'

3.2 The Student Data

The student data[1] is proper for us to discuss about the unresolved problem. Fifteen students (y_i = 'F') fail the exam and twenty five students (y_i = 'P') pass the exam in Table 1. X1 is sturdy hours/day and X2 is expenditure (10,000 yen)/month. X3 is drinking days/week, X4 is sex and X5 is smoking/non-smoking. In the case that IP-OLDF discriminates two classes by (X1, X2), the discriminant hyper-plane of IP-OLDF is X2 = 5. Eight students (X2 > 5) are discriminated to the fail group correctly, four students are on X2 = 5 and three student (X2 < 5) are misclassified into the pass group. On the other hands, twenty one students (X2 < 5) are classified into the pass group correctly and four students are on X2 = 5. Nevertheless IP-OLDF cannot discriminate eight students on X2 = 5, it returns MNM = 3. Revised IP-OLDF can find three discriminant hyper-plane: X2 = 0.006*X1 + 4.984, X2 = 0.25*X1 + 3.65, X2 = 0.99*X1 + 212. And, true MNM = 5. S-SVM (SVM4, c = 10^4) is X2 = X1 + 1, and NM = 6. There is a student having the value of (4, 5) on the discriminant hyper-plane. Therefore, we had better estimated NM = 7. This data is tiny and toy data, but it is useful for the evaluation of the discriminant functions and it is easy for us to understand by scatter plots with two features.

Table 1. The student data.

y_i	F	F	F	F	F	F	F	F	F	F	F	F	F	F	F	P	P	P	P	P
X1	3	1	3	3	2	1	4	3	5	2	3	2	3	3	5	6	9	4	3	2
X2	10	8	7	7	6	6	6	6	5	5	5	5	3	2	2	5	5	5	5	4
y_i	P	P	P	P	P	P	P	P	P	P	P	P	P	P	P	P	P	P	P	P
X1	5	12	4	10	7	5	7	3	7	7	7	6	3	6	6	8	5	10	9	5
X2	4	4	4	4	4	4	3	3	3	3	3	3	3	3	3	3	3	2	2	2

4 The Discrimination of Linear Separable Data (Problem 2)

4.1 The Importance of This Problem

The purpose of discriminant analysis is to discriminate two classes or objects properly. For this purpose, the discrimination of linear separable data is very important, because we can evaluate the result very clearly. If some LDFs cannot discriminate linear separable data properly, these LDFs should not be used. It is very strange that there is

[1] This data was used for the description of three statistical books using SAS, SPSS and JMP. It is download from (http://sun.econ.seikei.ac.jp/~shinmura/). Click Tab of "Data Archive" and double click "aoyama.xls".

no research about the discrimination of the linear separable data. H-SVM implies us this discrimination very clearly. But it can be applied only for linear separable data. This may be the reasons why there is no good research about linear separable data until now. Some statistician believes that LDF based on MNM criterion is foolish method, because it over fits for the training samples and its generalization ability may be wrong for the validation samples without examination by real data.

IP-OLDF finds Swiss bank note data is linear separable by two features (X4, X6) and MNMs of 16 models including (X4, X6) are zero. Until now, nobody realize this fact. And, we think it is difficult for us to find linear separable data from real data. But, we can easily obtain two kinds of good research data. First, the pass/fail determination by scores. This is explained in 4.2. Second, every real data is changed to linear separable data by enlarging the distance between the mean of the two classes. The Swiss bank note data consisted of two kinds of bills: 100 genuine and 100 counterfeit bills. There were six features: X1 was the length of the bill (mm); X2 and X3 were the width of the left and right edges (mm), respectively; X4 and X5 were the bottom and top margin widths (mm), respectively; X6 was length of the image diagonal (mm). A total of 63 (= 2^6-1) models were investigated. According to Shinmura [18], of the 63 total models, 16 of them including two features (X4, X6) have MNMs of zero; thus, they are linearly separable. The 47 models that remain are not linearly separable. This data is adequate whether or not LDFs can discriminate linearly separable data correctly.

Table 2 shows four results. Upper right (B) is original bank data. Upper left (A) is data expanded to 1.25 times the average distance. Lower left (C) and right (D) are data that are reduced to 0.75 and 0.5 times the average distance. Fisher's LDF is independent of the inferential statistics. But, if we treat $y_i = 1/-1$ as object value and data is analyzed by the regression analysis, obtained regression coefficients are proportional to the discriminant coefficients of Fisher's LDF by the plug-in rule. The stepwise methods can be used formally. 'p' is the number of features by the forward stepwise method. 'Var.' is the selected features. From p = 1 to p = 6, X6, X4, X5, X3, X2 and X1 are selected in this order by the forward stepwise method. In the regression analysis, Mallow's Cp statistics and AIC are used as variable selection. Usually, the model with minimum of $|Cp - (p + 1)|$ and AIC are recommended. By this rule, Cp statistics choose the same full model. On the other hand, AIC chooses 4-features model (X3, X4, X5, X6) in data 'A'. AIC chooses 5-features model (X2, X3, X4, X5, X6) in other three data.

This table tells us two important facts. We can easily obtained the linear separable data from the real data. The same result as the bank data are observed by the student data, iris data and CPD data those are not linear-separable. Second fact is as follows: "Cp and AIC" choose almost same models, nevertheless 1-feature (X6) model is linear separable in 'A'. And, 2-features (X4, X6) model is linear separable in 'B'. The models selected by "Cp and AIC" are independent from the linear-separablility. Some statisticians don't permit this result by the plug-in rule. On the contrary, they consider Fisher's LDF is the inferential statistics, because it is derived by the Fisher's assumption. This confusion is new problem and is future work (**Future2**).

4.2 Pass/Fail Determination

The pass/fail determination of exam scores makes good research data, because it can be obtained easily, and we can find a trivial discriminant function. My theoretical research starts from 1997 and ends in 2009 [18]. My applied research began in 2010. I negotiated with the National Center for University Entrance Examinations (NCUEE), and borrowed research data consisting of 105 exams in 14 subjects over three years. I finished analyzing the data at the end of 2010, and obtained 630 error rates for Fisher's LDF, QDF, and Revised IP-OLDF. However, NCUEE had requested me not to present the results on March 2011. Therefore, I explain new research results using my statistical exam results. The reason for the special case of QDF and RDA (**Problem 3**) is resolved at the end of 2012. The course consists of one 90 min lecture per week for 15 weeks. In 2011, the course only ran for 11 weeks because of power shortages in Tokyo caused by the Fukushima nuclear accident. Approximately 130 students, mainly freshmen, attended the lectures. Midterm and final exams consisted of 100 questions with 10 choices. Two kinds of pass/fail determinations were discriminated by 100 item scores, and four testlet scores as features. If the pass mark is 50 points, we can easily obtain a trivial discriminant function (f = T1 + T2 + T3 + T4–50). If f ≥ 0 or f < 0, the student passes or fails the exam, respectively. In this case, students on the discriminant hyper-plane pass the exam, because their score is exactly 50. This indicates that there is no unresolved problem because the discriminant rule is decided by features.

Table 2. Swiss bank data [18].

		A: The distance *1.25				B: Original Bank Data			
Var.	p	Cp	AIC	MNM	LDF	Cp	AIC	MNM	LDF
1-6	6	**7.0**	-863	0	0	**7.0**	-779	0	0
2-6	5	5.3	-865	0	0	5.3	**-781**	0	0
3-6	4	10.5	**-896**	0	0	10.3	-776	0	0
4-6	3	10.9	-859	0	0	10.7	-775	0	0
4, 6	2	118.8	-779	0	0	107.0	-699	0	3
6	1	313.9	-679	0	1	292.0	-604	2	2

		C: The distance * 0.75				D: The distance * 0.5			
Var.	p	Cp	AIC	MNM	LDF	Cp	AIC	MNM	LDF
1-6	6	**7.0**	-676	1	2	**7.0**	-543	5	12
2-6	5	5.3	**-678**	1	2	5.3	**-545**	6	12
3-6	4	9.8	-673	1	1	8.9	-541	7	13
4-6	3	10.1	-673	1	2	8.8	-541	8	14
4, 6	2	97.9	-601	4	6	78.7	-482	16	19
6	1	253.8	-517	6	8	184.4	-417	53	56

4.3 Discrimination by Four Testlets

Table 3 shows the discrimination of four testlet scores as features for 10 % (from third column to seventh column) and 90 % (after eighth column) levels of the midterm exams. The results of 50 % level are omitted. 'p' denotes the number of features selected by the forward stepwise method. In 2010, T4, T2, T1, and T3 are entered in the model selected by the forward stepwise method. The MNM of Revised IP-OLDF and NM of logistic regression are zero in the full model, which means the data is linear-separable in four features. NMs of LDF and QDF are 9 and 2. This means LDF and QDF cannot recognize linear separability. In 2011, Revised IP-OLDF and logistic regression can recognize that the 3-features model (T2, T4, T1) is linear-separable. In 2012, the 2-features model (T4, T2) is linear-separable. T4 and T2 contain easy questions, and T1 and T3 consist of difficult questions for fail group students. This suggests the possibility that pass/fail determination using Revised IP-OLDF can elicit the quality of the test problems and understanding of students in the near future (**Future3**).

Table 3. NMs of four discriminant functions by forward stepwise in midterm exams at the 10 % (from 3rd column to 7th column) and 90 % levels (after 8th column).

	p	Var.	MNM	Logi.	LDF	QDF	Var.	MNM	Logi.	LDF	QDF
2010	1	T4	6	9	11	11	T3	10	37	24	24
	2	T2	2	6	11	9	T4	5	10	20	11
	3	T1	1	3	8	5	T1	**0**	**0**	20	10
	4	T3	**0**	**0**	9	2	T2	0	0	20	11
2011	1	T2	9	17	15	15	T3	6	7	14	14
	2	T4	4	9	11	9	T4	1	1	14	6
	3	T1	**0**	**0**	9	10	T1	**0**	**0**	13	5
	4	T3	0	0	9	11	T2	0	0	14	9
2012	1	T4	4	8	14	12	T3	8	30	12	12
	2	T2	**0**	**0**	11	9	T1	5	12	9	9
	3	T1	0	0	12	8	T4	3	3	10	3
	4	T3	0	0	12	1	T2	**0**	**0**	11	3

Table 4. Summary of error rates of Fisher's LDF and QDF.

		10 %		50 %		90 %	
		LDF	QDF	LDF	QDF	LDF	QDF
Midterm	10	7.5	1.7	2.5	5.0	**16.7**	9.2
	11	7.0	8.5	**2.2**	2.3	10.5	6.7
	12	9.9	**0.8**	4.9	4.8	13.6	7.1
Final	10	4.2	1.7	3.3	4.2	3.3	**10.8**
	11	11.9	2.9	2.9	3.6	3.6	8.6
	12	8.7	2.3	2.3	2.3	13.0	4.5

Table 4 shows a summary of the 18 error rates derived from the NMs of Fisher's LDF and QDF for the linear separable model. Ranges of the 18 error rates of LDF and QDF are [2.2 %, 16.7 %] and [0.8 %, 10.8 %], respectively. Error rates of QDF are lower than those of LDF. At the 10 % level, the six error rates of LDF and QDF lie in the ranges [4.2 %, 11.9 %] and [0.8 %, 8.5 %], respectively. Clearly, the range at the 50 %

level is less than for the 10 % and 90 % levels. Miyake and Shinmura [27] followed Fisher's assumption, and surveyed the relation between population and sample error rates. One of their results suggests that the sample error rates of balanced sample sizes such as 50 % level are close to the population error rates. The above results may confirm this. These results suggest a serious drawback of LDF and QDF based on the variance-covariance matrices. We can no longer trust the error rates of LDF and QDF. Until now, this fact has not been discussed, because there is little research using linear separable data. From this point on, we had best evaluate discriminant functions using linear separable data, because the results are very clear. In genome discrimination, researchers try to estimate the variance-covariance matrices using small sample sizes and large numbers of features. These efforts may be meaningless and lead to incorrect results.

5 Problem 3 (Discrimination of 44 Japanese Cars)

The special cases found in NCUEE exams are confirmed by my exams, also. It is resolved in Nov., 2012. It needs three years because I never doubt the algorithm of QDF and surveyed by the multivariate approach. I checked all features by t-test of two classes, before I abandon the survey. The special case above is more easily explained by the discrimination of 44 Japanese cars.[2] Let us consider the discrimination of 29 regular cars and 15 small cars. Small cars have a special Japanese specification. They are sold as second cars or to women, because they are cost efficient. The emission rate and capacity of small cars are restricted. The emission rate of small and regular cars ranges from [0.657, 0.658] and [0.996, 3.456], respectively. The capacity (number of seats) of small and regular cars are 4 and [5, 8], respectively.

Table 5. Discrimination of Japanese small and regular cars.

p	Var.	t	LDF	QDF[1]	MNM[2]	$\lambda = \gamma = 0.8$	0.5	0.2	0.1
1	Emission	11.37	2	**0**	**0**	2	1	1	**0**
2	Price	5.42	1	**0**	**0**	4	1	0	**0**
3	Capacity	8.93	1	29	**0**	3	1	0	**0**
4	CO_2	4.27	1	29	**0**	4	1	0	**0**
5	Fuel	−4.00	0	29	**0**	5	1	0	**0**
6	Sales	−0.82	0	29	**0**	5	1	0	**0**

[1]If we add small noise to the constant (capacity of small cars), "NMs = 29" are changed to zero.
[2]MNM and NMs of logistic regression are zero.

Table 5 shows the forward stepwise result. At first, "emission" enters the model because the t-value is high. The MNM and NMs of QDF are zero. LDF cannot recognize linear separability. Next, 'price' enters the 2-features model, although the t-value of 'price' is less than that of 'capacity'. In the third step, QDF misclassifies all

[2] This data is open to the paper about DEA (Table 1 in Page 4. http://repository.seikei.ac.jp/dspace/handle/10928/402).

29 regular cars as small cars after "capacity" is included in the 3-features model. This is because the capacity of small cars is fixed to four persons. It is very important that only QDF and RDA are adversely affected by this special case. LDF and the t-test are not affected, because these are computed from the pooled variance of two classes. Modified RDA offers two options such as λ and γ. Four trials show that $\lambda = \gamma = 0.1$ is better than others. JMP division is expected to show the guideline of two options.

6 K-fold Cross Validation (Problem 4)

Usually, the LOO method is used for model selection of the discriminant analysis. In this research, "k-fold cross validation for small sample sizes" is proposed, as it is more powerful than the LOO method. In near future, these results will reveal generalization abilities and the 95 % CIs of Revised IP-OLDF, Revised IPLP-OLDF, Revised LP-OLDF, H-SVM, S-SVM ($c = 10^4$, 1), logistic regression and Fisher's LDF.

6.1 Hundred-Fold Cross Validation

In the regression analysis, we benefit from inferential statistics, because the SE of regression coefficients, and model selection statistics such as Cp, AIC and BIC, are known a priori. On the other hand, there is no SE of discriminant coefficients and model selection statistics in the discriminant analysis. Therefore, users of the discriminant analysis and SVMs often use the LOO method. Let the sample size be n. One case is used for validation, and the other (n − 1) cases are used as training samples. We evaluate n sets of training and validation samples. If we have a large sample size, we can use k-fold cross validation. The sample is divided into k subsamples. We can evaluate k combinations of the training and validation samples. On the other hand, bootstrap or re-sampling methods can be used with small sample sizes. In this research, large sample sets are generated by re-sampling, and 100-fold cross validation is proposed using these re-sampled data. In this research, "100-fold cross validation for small sample sizes" is applied as follows: (1) We copy 100 times the data from midterm exams in 2012 using JMP. (2) We add a uniform random number as a new variable, sort the data in ascending order, and divide into 100 subsets. (3) We evaluate eight functions such as Revised IP-OLDF, Revised LP-OLDF, Revised IPLP-OLDF, H-SVM, S-SVM ($c = 10^4$ and 1), Fisher's LDF and logistic regression by 100-fold cross validation using these 100 subsets.

Revised IP-OLDF and S-SVM are analyzed by LINGO [11], developed with the support of LINDO Systems Inc. Logistic regression and LDF are analyzed by JMP, developed with the support of the JMP division of SAS Japan. There is merit in using 100-fold cross validation because we can easily calculate the 95 % CIs of the discriminant coefficients and NMs (or error rates). The LOO method can be used for model selection, but cannot obtain the 95 % CIs. These differences are quite important for analysis of small samples.

6.2 LOO and K–Fold Cross Validation

Table 6 shows the results of the LOO method and NMs in the original data. 'Var.' shows the suffix of four testlet scores named 'T'. Only 11 models were showed, because four 1-feature models were omitted from the table. The MNM of the 2-features model (T2, T4) in No. 6 is zero, as are those of the 4-features model (T1-T4) in No. 1, and the two 3-features models of (T1, T2, T4) in No. 2 and (T2, T3, T4) in No. 3. The NMs of logistic regression and SVM4 (c = 10^4) are zero in these four models, but NMs of SVM1 (c = 1) are 2 and 3 in No. 2 and No. 6, respectively. It is often observed that S-SVM cannot recognize linear separability when the penalty c has a small value. The LOO method recommends models in No. 3 and No. 6 because these NMs are minimum.

Table 6. LOO and NMs in original test data.

No	Var.	LOO	LDF	Logi	MNM	SVM4	SVM1
1	1–4	14	12	**0**	**0**	**0**	**0**
2	1,2,4	13	12	**0**	**0**	**0**	2
3	2,3,4	**11**	11	**0**	**0**	**0**	**0**
4	1,3,4	15	15	2	2	3	3
5	1,2,3	16	16	6	4	6	6
6	2,4	**11**	11	**0**	**0**	**0**	3
7	1,4	16	16	6	3	6	6
8	3,4	14	13	3	3	4	4
9	1,2	18	17	12	7	7	7
10	2,3	16	11	11	6	11	11
11	1,3	22	21	15	7	10	10

Table 7 shows the results given by Revised IP-OLDF (RIP), SVM4, LDF, and logistic regression (Logi.). The results of SVM1, Revised LP-OLDF and Revised IPLP-OLDF are omitted. First column shows the same No. in Table 6. After four linear separable models, the ranges of seven models are showed. 'MEAN1' column denotes the mean error rate in the training sample. Revised IP-OLDF and logistic regression can recognize linear separability for four models. For SVM4, only model No.1 has an NM of zero. The mean error rates of all Fisher's LDF are over 9.48 %. 'MEAN2' column denotes the mean error rate in the validation sample. Only two models (No.2 and No. 6) of Revised IP-OLDF have NMs of zero and are selected as the best models. The NMs of other functions are greater than zero, and those of LDF are over 9.91 %.

We can conclude that Fisher's LDF is the worst of these four LDFs. Some statisticians believe that NMs of Revised IP-OLDF is less suitable for validation samples, because it over fits for the training samples. On the other hand, Fisher's LDF does not lead to overestimation, because it assumes a normal distribution. These results show that the presumption of 'overestimation' is wrong. We may conclude that real data does not obey Fisher's assumption. To build a theory based on an incorrect assumption will lead to incorrect results [2]. 'Diff.' is the difference between MEAN2 and MEAN1. We think the small absolute value of 'Diff.' implies there is no overestimation. In this sense,

Fisher's LDF is better than the other functions, because all values are less than 0.9. However, only high values of the training samples lead to small values of 'Diff.'

'Diff1' denotes the value of (MEAN1 of seven LDFs - MEAN1 of Revised IP-OLDF) in the training samples, and 'Diff2' is the value of (MEAN2 of seven LDFs - MEAN2 of Revised IP-OLDF) in the validation samples. All values of 'Diff1 and Diff2' of SVM4, Fisher's LDF and logistic regression are greater than zero. The maximum values of 'Diff1' given by SVM4, LDF and logistic regression are 2.33, 11.34 and 3.13 %, respectively. And the maximum values of 'Diff2' given by these functions were 1.7, 10.55 and 1.62 %, respectively. It is concluded that Fisher's LDF was not as good as Revised IP-OLDF, S-SVM, and logistic regression by 100-fold cross validation. Therefore, we had better chosen the model of Revised IP-OLDF with minimum value of M2 as the best model. Two models such as (T1, T2, T4) and (T2, T4) are zero. In this case, we had better chosen 2-features model (T2, T4), because of the principle of parsimony or Occam's razor. The values of 'MEAN2' of Revised IP-OLDF, SVM4, Fisher's LDF and logistic regression are 0 %, 1.7 %, 9.91 % and 0.91 %, respectively. This implies that the mean error rates of Fisher's LDF is 9.91 % higher than the best model of Revised IP-OLDF in the validation sample.

Table 7. Comparison of four functions.

RIP	MEAN1	MEAN2	Diff.		
1	**0**	0.07	0.07		
2	**0**	**0**	0		
3	**0**	0.03	0.03		
6	**0**	**0**	0		
4,5,7–11	[0.79,4.94]	[0.03,7.21]	[0.03,2.39]		
SVM4	MEAN1	MEAN2	Diff.	Diff1	Diff2
1	**0**	0.81	0.81	0	0.74
2	0.73	1.62	0.90	0.73	1.62
3	0.13	0.96	0.83	0.13	0.93
6	0.77	**1.70**	0.93	0.77	**1.70**
4,5,7-11	[1.65,6.85]	[3.12,8.02]	[0.66,1.65]	[0.78,2.33]	[0.59,1.36]
LDF	MEAN1	MEAN2	Diff.	Diff1	Diff2
1	9.64	10.54	0.90	9.64	10.47
2	9.89	10.55	0.66	9.89	10.55
3	**9.48**	10.09	0.61	9.48	10.06
6	9.54	**9.91**	0.37	9.54	9.91
4,5,7-11	[10.81,16.28]	[11.03,16.48]	[0.16,0.6]	[7.97,11.34]	[**6.23**,9.61]
Logi	MEAN1	MEAN2	Diff.	Diff1	Diff2
1	**0**	0.77	0.77	0	0.70
2	**0**	1.09	1.09	0	1.09
3	**0**	0.85	0.85	0	0.82
6	**0**	**0.91**	0.91	0	0.91
4,5,7-11	[1.59,7.65]	[2.83,8.04]	[0.35,1.34]	[0.8,3.13]	[0.39,1.62]

In 2014, these results are recalculated using LINGO Ver.14. The elapsed runtimes of Revised IP-OLDF and SVM4 are 3 min 54 s and 2 min 22 s, respectively. The elapsed runtimes of LDF and logistic regression by JMP are 24 min and 21 min, respectively. Reversals of CPU time have occurred for this time.

7 Conclusions

In this research, we have discussed three problems of discriminant analysis. Problem 1 is solved by Revised IP-OLDF, which looks for the interior points of the OCP directly. Problem 2 is theoretically solved by Revised IP-OLDF and H-SVM, but H-SVM can only be applied to linear separable data. Error rates of Fisher's LDF and QDF are very high for linear separable data. This means that these functions should not be used for important discrimination tasks, such as medical diagnosis and genome discrimination. Problem 3 only concerns QDF. This problem was resolved by a t-test after three years of investigation, and can be solved by adding a small noise term to variables. Now, JMP offers a modified RDA, and if we can choose proper parameters, it may be better than LDF and QDF.

However, these conclusions are confirmed by the training samples. In many researches, statistical users have small sample sizes, and cannot evaluate the validation samples. Therefore, "k-fold cross validation for small samples" is proposed. This method confirms the same above conclusion by the validation samples. Many discriminant functions are developed using various criteria after Warmack and Gonzalez [29]. The mission of discrimination should be based on the MNM criterion. Statisticians have tried to develop functions based on the MNM criterion, but this can now be achieved by Revised IP-OLDF using MIP. It is widely believed that Revised IP-OLDF leads to overestimations, but Fisher's LDF is worse for validation samples. Comparison of eight LDFs are examined for future work (**Future4**) by 100-fold cross validation.

References

1. Fisher, R.A.: The use of multiple measurements in taxonomic problems. Ann. Eugenics **7**, 179–188 (1936)
2. Fisher, R.A.: Statistical Methods and Statistical Inference. Hafner Publishing Co., New York (1956)
3. Flury, B., Rieduyl, H.: Multivariate Statistics: A Practical Approach. Cambridge University Press, UK (1988)
4. Friedman, J.H.: Regularized discriminant analysis. J. Am. Stat. Assoc. **84**(405), 165–175 (1989)
5. Goodnight, J.H.: SAS Technical report—the sweep operator: its importance in statistical computing—(R100). SAS Institute Inc., Cary, US (1978)
6. Lachenbruch, P.A., Mickey, M.R.: Estimation of error rates in discriminant analysis. Technometrics **10**, 1–11 (1968)
7. Liitschwager, J.M., Wang, C.: Integer programming solution of a classification problem. Manage. Sci. **24**(14), 1515–1525 (1978)

8. Sall, J.P.: SAS Regression Applications. SAS Institute Inc, Cary (1981). Japanese version is translated by Shinmura, S
9. Sall, J.P., Creighton, L., Lehman, A.: JMP Start Statistics, 3rd edn. SAS Institute Inc, Cary (2004). Japanese version is edited by Shinmura, S
10. Schrage, L.: LINDO—An Optimization Modeling System. The Scientific Press, South San Francisco (1991). Japanese version is translated by Shinmura, S., Takamori, H
11. Schrage, L.: Optimization Modeling with LINGO. LINDO Systems Inc, Chicago (2006). Japanese version is translated by Shinmura, S
12. Shinmura, S., Miyake, A.: Optimal linear discriminant functions and their application. COMPSAC **79**, 167–172 (1979)
13. Shinmura, S.: Optimal linear discriminant functions using mathematical programming. J. Jpn. Soc. Comput. Stat. **11/2**, 89–101 (1998)
14. Shinmura, S.: A new algorithm of the linear discriminant function using integer programming. New Trends. Probab. Stat. **5**, 133–142 (2000)
15. Shinmura, S.: New algorithm of discriminant analysis using integer programming. In: IPSI 2004 Pescara VIP Conference CD-ROM, pp. 1–18 (2004)
16. Shinmura, S.: Overviews of discriminant function by mathematical programming. J. Jpn. Soc. Comput. Stat. **20**(1-2), 59–94 (2007)
17. Shinmura, S.: Improvement of CPU time of revised IPLP-OLDF using linear programming. J. Jpn. Soc. Comput. Stat. **22**(1), 37–57 (2009)
18. Shinmura, S.: The Optimal Linear Discriminant Function. Union of Japanese Scientist and Engineer Publishing, Japanese (2010)
19. Shinmura, S.: Problems of discriminant analysis by mark sense test data. Jpn. Soc. Appl. Stat. **40**(3), 157–172 (2011)
20. Shinmura, S.: Beyond Fisher's Linear Discriminant Analysis—New World of Discriminant Analysis. In: 2011 ISI CD-ROM, pp. 1–6 (2011)
21. Shinmura, S.: Evaluation of Optimal Linear Discriminant Function by 100-fold cross validation. In: 2013 ISI CD-ROM, pp. 1–6 (2013)
22. Shinmura, S.: End of Discriminant Functions Based on Variance Covariance Matrices. In: 2014 ICORE, pp. 5–16 (2014)
23. Shinmura, S.: Improvement of CPU time of linear discriminant functions based on MNM criterion by IP. Stat. Optim. Inf. Comput. **2**, 114–129 (2014)
24. Stam, A.: Non-traditinal approaches to statistical classification: some perspectives on Lp-norm methods. Ann. Oper. Res. **74**, 1–36 (1997)
25. Taguchi, G., Jugulu, R.: The Mahalanobis-Taguchi Strategy—A Pattern Technology System. Wiley, New York (2002)
26. Markowitz, H.M.: Portfolio Selection, Efficient Diversification of Investment. Wiley, New York (1959)
27. Miyake, A., Shinmura, S.: Error rate of linear discriminant function. In: de Dombal, F.T., Gremy, F. (eds.) Decision Making and Medical Care, North-Holland Publishing Company, pp. 435–445. North-Holland Publishing Company, New York (1976)
28. Vapnik, V.: The Nature of Statistical Learning Theory. Springer, Heidelberg (1995)
29. Warmack, R.E., Gonzalez, R.C.: An algorithm for the optimal solution of linear inequalities and its application to pattern recognition. IEEE Trans. Comput. **C-2212**, 1065–1075 (1973)

Simulated Annealing Algorithm for Job Shop Scheduling on Reliable Real-Time Systems

Daniil A. Zorin(✉) and Valery A. Kostenko

Lomonosov Moscow State University, Moscow, Russia
juan@lvk.cs.msu.su, kost@cs.msu.su

Abstract. This paper is devoted to the problem of designing a computational system utilizing the minimal number of processors to ensure that the program is executed before the deadline. The program is represented by a direct acyclic graph where vertices correspond to jobs. The system is supposed to tolerate both hardware and software faults. The schedule of the program execution does not include the exact moments of job launch and termination, thus allowing to employ abstract models with various levels of detail to estimate the time of execution. A simulated annealing algorithm is proposed for this problem. The paper provides the proof of asymptotic convergence of the algorithm and an experimental evaluation. The algorithm is also applied to a practical problem of scheduling in radiolocation systems.

Keywords: Systems design · Job shop scheduling · Scheduling algorithms · Reliability · Multiprocessor systems · Optimization problems · Real-time systems

1 Introduction

Real-time systems (RTS) often impose obligatory restrictions not only on the deadlines of the programs, but also on the reliability and other characteristics such as weight and volume. The co-design problem of finding the minimal necessary number of processors and scheduling the set of tasks on it arises in this relation. The limitations on the time of execution and the reliability of the RTS must be satisfied. This paper describes an algorithm of solving this problem. The algorithm can be tuned for solving instances of the problem by adjusting various settings. The algorithm permits to employ various techniques of computing the reliability of the RTS and various simulation methods for estimating the time of execution of a schedule. Thanks to this it can be used on different stages of designing the RTS. The program being scheduled changes over the course of designing the system with additional details introduced gradually, so the need to reschedule it and to define the hardware architecture more precisely may arise.

2 Problem Formulation

This paper considers only *homogeneous* hardware systems. Hence the system consists of a set of processors connected with a network device; all processors are identical, which means that they have equal reliability and the time of execution of any program

E. Pinson et al. (Eds.): ICORES 2014, CCIS 509, pp. 31–46, 2015.
DOI: 10.1007/978-3-319-17509-6_3

is equal on all processors. The structure of the network, on the other hand, is not defined strictly, allowing various models (bus, switch, etc).

The program to be scheduled is a set of interacting tasks. The program can be represented with its data flow graph $G = \{V, E\}$ where V is the set of vertices and E is the set of edges. Let M denote the set of available processors.

To improve reliability, two methods are used: processor redundancy and N-version programming.

Processor redundancy implies adding a new processor to the system and using it to run the same tasks as on some existing processor. In this case the system fails if both processors fail. Thc additional processor is used as hot spare, i.e. it receives the same data and performs the same operations as the primary processor, but sends data only if the primary one fails.

To use N-version programming (NVP, also known as multiversion programming), several versions (independent implementations) of a task are created. It is assumed that different versions written by different programmers will fail in different cases. The number of versions is always odd, and the execution of a task is deemed successful by majority vote, i.e. when more than a half of the versions produce the same output.

The reliability of the system depends on the following variables: $P(m_i)$ is the reliability of a processor, $Vers(v_i)$ is the set of available versions for the task v_i, $P(v_i)$ is the reliability of v_i counting all versions used. Formulae for $P(v_i)$ can be found in [3, 4, 8, 23]. The reliability of the whole system is calculated as the product of the reliability of its elements.

A schedule for the program is defined by task allocation, the correspondence of each task with one of the processors, and task order, the order of execution of the task on the processor.

If N-version programming is employed, the number of version must be specified for each instance of each task. Allocation and order are defined not for individual tasks, but for pairs "task - version".

Formally, a schedule of a system with processor redundancy and multiversion programming is defined as a pair (S, D) where S is a set of quadruples (v, k, m, n) where $v \in V, k \in Vers(v), m \in M, n \in \mathbb{N}$, so that

$$\forall v \in V \exists k \in Vers(v) : \exists s = (v_i, k_i, m_i, n_i) \in S : v_i = v, k_i = k;$$

$$\forall s_i = (v_i, k_i, m_i, n_i) \in S, \forall s_j = (v_j, k_j, m_j, n_j) \in S : (v_i = v_j \wedge k_i = k_j) \Rightarrow s_i = s_j;$$

$$\forall s_i = (v_i, k_i, m_i, n_i) \in S, \forall s_j = (v_j, k_j, m_j, n_j) \in S : (s_i \neq s_j \wedge m_i = m_j) \Rightarrow n_i \neq n_j.$$

D is a multiset of elements of the set of processors, M. The number of reserves of processor m is equal to the number of instances of m in D. Substantially m and n denote the placement of the task on a processor and the order of execution for each version of each task. The multiset D denotes the spare processors.

A schedule can be represented with a graph. The vertices of the graph are the elements of S. If the corresponding tasks are connected with an edge in the graph G, the same edge is added to the schedule graph. Additional edges are inserted for all pairs of tasks placed on the same processor right next to each other.

According to the definition, there can be only one instance of each version of each task in the schedule, all tasks on any processor have different numbers and the schedule must contain at least one version of each task. Besides these, one more limitation must be introduced to guarantee that the program can be executed completely. A schedule S is correct by definition if its graph has no cycles. $\bar{S}$ is the space of all correct schedules.

For every correct schedule the following functions are defined: $t(S)$ – time of execution of the whole program, $R(S)$ – reliability of the system, $M(S)$ – the number of processors used.

As mentioned before, the structure of the network is not fixed, so the time of execution depends on the actual model. Various models can be implemented (particularly, the algorithm was tested for bus, Ethernet switch and Fibre channel switch architectures), but all of them in the end have to build a time chart of the execution of the schedule. To calculate $t(S)$, it is necessary to define the start and end time of each task and each data transfer. $t(S)$ can be an analytic function, or it can be calculated with some algorithm, or it can even be estimated with simulation experiments with tools like the one described in [1 or 19]. If the X axis indicates time, different processors are represented with lines parallel to the X axis, the start and end times of all the tasks and transfers can be drawn in a chart like the one shown on Figs. 1 and 2 in Sect. 3.

Finally, the optimization problem can be formulated as follows. Given the program G, t_{dir}, the hard deadline of the program, and R_{dir}, the required reliability of the system, the schedule S that satisfies both constraints and requires the minimal number of processors is to be found:

$$\min_{S\in\bar{S}} M(S),$$

$$t(S) \leq t^{dir},$$

$$R(S) \geq R^{dir}.$$

Theorem 1. The problem stated above is NP-hard.

Proof. The NP-hardness can be proved by reducing problem (1) to the NP-hard subset sum problem: given the set of integers $a_1, \ldots a_n$, find out whether it can be split in two subsets with equal sums of its elements.

Let $B = \sum_{i=1}^{n} a_i$, $R_{dir} = 0$, $t_{dir} = B/2$. Graph G has n vertices and zero edges, $E = \varnothing$, so the tasks can be assigned to the processors in any order. The time of execution of each task v_i is defined as constant a_i. The time of execution of a task is defined in a natural way: if s_0 is assigned after $s_1, \ldots s_n$, then it is executed in the interval $(\sum_{i=1}^{n} a_i, \sum_{i=1}^{n} a_i + a_0)$.

If the subset sum problem has a solution consisting of two subsets, X and Y, then the tasks corresponding to X can be assigned on the first processor, and the rest can be assigned to the second processor. Obviously the time of execution will be $B/2$, the deadline will be met, and the number of processors is minimal, so the corresponding scheduling problem is solvable.

Similarly, if the subset sum problem has no solution, then for any of the possible divisions into two subsets the sum of one subset will exceed $B/2$, and thus the corresponding schedule will not meet the deadline.

This means that scheduling problem can be reduced to subset sum problem, and the reduction is obviously polynomial, because the only computation needed for the reduction is defining B which is a sum of n numbers. Therefore, the scheduling problem is NP-hard.

3 Proposed Problem Solution

3.1 Selecting the Method

The problem as formulated in Sect. 2 is unique, however, it is necessary to examine the solutions of similar problems. Out of all job shop scheduling problems we need to consider only those where the program is represented with a direct acyclic graph and the tasks cannot be interrupted. The definition of the schedule and the fault tolerance techniques can vary. Also we can ignore non-NP-hard scheduling problems, as their methods of solution are unlikely to be applicable to our problem. These limitations leave only the following possible methods of solution: exhaustive search, greedy strategies, simulated annealing and genetic strategies.

Exhaustive search is impractical in this case simply because of the sheer size of the solution space (the number of all transpositions of the tasks on all processors is more than $n!$). The target function (the number of processors) is discrete and can yield a limited set of integer values which makes using limited search methods such as branch and bound method impossible.

Greedy algorithms give an approximation of the optimal solution. The solution is constructed by scheduling tasks separately one after another according to a pre-defined strategy. For example, it is possible to select the position of the task so that the total execution time of all scheduled tasks is minimal. Such algorithm has polynomial complexity. This strategy can be called «do as soon as possible» strategy, it is discussed in [17]. More complex strategies, both reliability and cost/time driven are discussed in [18]. An approach that takes possible software and hardware faults into the account is discussed in [3]. Another solution is to do the exact opposite: first schedule all tasks on separate processors and then join processors while such operation is possible without breaking the deadlines [12].

The main drawback of greedy algorithms is potential low accuracy. There is no theoretical guarantee that the solution is close to the optimal, in fact, it is possible to artificially construct examples where a greedy strategy gives a solution infinitely distant from the optimal one. This drawback can be partially fixed by adding a random operation to the algorithm and running it multiple times, however, this way the main advantage that is low complexity is lost.

Simulated annealing algorithm [10] deals with a single solution on each step. It is mutated slightly to create a candidate solution. If the candidate is better, then it is accepted as the new approximation, otherwise it is accepted with a probability decreasing over time. So on the early steps the algorithm is likely to wander around the solution steps, and on the late steps the algorithm descends to the current local optimum. Simulated annealing does not guarantee that the optimal solution will be found, however, there are proofs that if the number of iterations is infinite, the algorithm converges in probability to the optimal solution [14]. Reference [21] formulates the principal steps needed to apply simulated annealing to job shop scheduling problem. It is necessary to define the solution space; define the neighborhood of each solution, in other words, introduce the elementary operations on the solution space; define the target function of the algorithm. Reference [16] gives experimental proofs of the efficiency of simulated annealing for job shop scheduling. This work also suggests an improvement over the standard algorithm: heuristics. In the classical algorithm, the candidate solution is chosen from the neighborhood randomly, however, knowing the structure of the schedules, it is possible to direct the search by giving priority to specific neighbors. Reference [9] also suggest the use of heuristics and gives an example of successful application of simulated annealing to scheduling.

The widely known genetic algorithms give an approximation of the optimal solution, and there is a hypothesis about the asymptotical convergence [5]. The first problem related to the application of genetic algorithm to scheduling is the encoding. If the tasks are independent, the schedule can be encoded simply by the list of processors where the corresponding tasks are assigned [15]. However, this is not viable for more complex models such as the one considered in this paper. For such cases, more sophisticated encoding is necessary, and the operations of crossover and mutations do not resemble the traditional operations with bit strings; schedules exchange whole parts that do not break the correctness conditions [6]. Reference [7] shows an example of an evolutionary strategy resembling the genetic algorithm applied to scheduling problems.

The main problem with genetic algorithms in regard to the discussed scheduling problem is low speed. As the algorithm has to allow using various models for time estimation, the time estimation can be complex and resource-consuming. It is impossible to avoid estimating time for all solutions of the population on each step. Therefore, if the population is substantially large, the algorithm can work very slowly as opposed to the simulated annealing algorithm that requires time estimation only once on each iteration.

Summing up the survey of the scheduling methods, we can conclude that simulated annealing is the preferable method both in terms of potential accuracy (asymptotic convergence can be proved) and speed (lower than greedy algorithm but substantially higher than genetic algorithm). The actual algorithm is discussed in detail in the next subsection.

3.2 Simulated Annealing Algorithm Description

The proposed algorithm of solution is based on simulated annealing [9]. Each iteration of the algorithm consists of the following steps:

Step 1. Set the initial solution $(S_0, D_0) \in \bar{S}$ which becomes the first approximation $((S, D) = (S_0, D_0))$.

Step 2. Set the current temperature to T_0 $(T = T_0)$.

Step 3. Apply the mutation operation to the current approximation (*S*, *D*) to get the new correct solution $(S', D') \in \bar{S}$. If this solution is the better than all previously visited solutions, it is recorded.
Step 4. Compute the value of the solution accuracy difference: $\Delta F = F(S', D') - F(S, D)$. If $\Delta F \le 0$ (the new solution is better), it is accepted as the current approximation $((S,D) = (S',D'))$. If $\Delta F > 0$ (the new solution is worse), it is accepted with the probability = $e^{-\Delta F/T}$.
Step 5. Repeat steps 3 and 4 a fixed number of times with constant temperature.
Step 6. If the stopping condition is satisfied, terminate the algorithm.
Step 7. Lower the temperature and return to step 3.

To construct a simulated annealing algorithm for the particular problem, the following problems have to be solved:

- Define the system of operations on schedules in order to implement mutation on step 3.
- Define the heuristic strategy that suggests which operation is applied on step 3.
- Select the temperature function on step 7.
- Define the function *F*(*S*, *D*) used to evaluate the solutions on step 4.
- Select the stopping condition for step 6.

The following operations on schedules are defined.

Add spare processor. In the schedule (S,D) a new element is added to the multiset *D*.

Delete spare processor. In the schedule (S,D) he element m is removed from the multiset *D*, if there is more than one instance of *m* in *D*.

Move vertex. This operation changes the order of tasks on a processor or moves a task on another processor. It has three parameters: the task to be moved, the processor where it is moved and the position on the target processor. The correctness of the resulting scheduled must always be checked during this operation.

Let *Trans*(*s*) be the set of tasks transitively depending on *s*: the set of all s_i, such that the graph *G* contains a chain (v, v_i).

The set *Succ*(*s*) can be constructed with the following method. Let $N_0 = Trans(s)$. If $N_i = \{s_1, s_2, \ldots, s_n\}$, and $s_{n+1}, \ldots, s_{n+k}$ satisfy $\forall l \in [1..k] : \exists i \in [1..n] : m_i \neq m \wedge m_i = m_{n+l} \wedge n_i < n_l$. Then $N_i = N_{i+1} \cup Trans(s_{n+1}) \cup Trans(s_{n+2}) \cup \ldots \cup Trans(s_{n+k})$. If $N_{i-1} = N_i$, then $N_i = Succ(s)$. *Succ*(*s*) is the set of tasks that depend on *s* indirectly.

Finally we can formulate the correctness condition of *Move vertex* operation. The task to move is $s_1 = (v_1, k_1, m_1, n_1)$, the target processor is m_2 and the target number is n_2, and the following condition must be true:

$$\forall s_i : m_i = m_2 : (n_i < n_2 \Rightarrow s_i \notin Succ(s_1)) \wedge (n_i \geq n_2 \Rightarrow s_1 \notin Succ(s_i)),$$

Then the operation requires the following substitution:

$$s'_1 = (v_1, k_1, m_2, n_2), \forall s_i : m_i = m_2 : n_i \geq n_2 \Rightarrow s'_i = (v_i, k_i, m_i, n_i + 1).$$

Add versions. Versions can be added only in pairs because the total number of versions must be odd if NVP is used. Two new versions are added on a new empty processor.

Delete versions. Versions are deleted in pairs. Two elements corresponding to two different versions of the same task are removed from the schedule.

Theorem 2. The system of operations is complete: if $(S_1, D_1), (S_2, D_2)$ are correct schedules, there exists a sequence of operations that transforms (S_1, D_1) to (S_2, D_2) such that all interim schedules are correct.

Proof. Applying any operation results in a correct schedule by definition of the operations. It is easy to see that each operation can be reversed [13], thus to prove the completeness of the system of operations it is enough to show how to transform both (S_1, D_1) and (S_2, D_2) to some schedule (S_0, D_0).

First let us enumerate all tasks. As the graph G has no cycles, for each task v it is possible to define *Level*(v). *Level*(v) = 1 if there are no edges terminating in v. *Level* (v) = n if all edges terminating in v start from vertices from levels below n. Assume that there are p_1 tasks at level 1, they can be numbered from 1 to p_1. Similarly, if there are p_2 tasks on level 2, they can be numbered from p_1 + 1 to $p_1 + p_2$. In the end all tasks will be numbered from 1 to N (where N is the total number of tasks).

The canonical schedule for program G is the schedule consisting of quadruples $s_i = (v_i, k_i, m_1, i)$, where $\{i\}$ are the numbers defined above. In this schedule all tasks are located on one processor, and due to the definition of the indices $\{i\}$ it has no cycles. Now let us show that any schedule can be transformed to the canonical schedule.

First all reserve processors and additional versions are deleted. After that the number of elements in the schedule will be equal to the number of tasks in graph G. Then an empty processor m_0 is selected, and the tasks are moved to it according to their respective numbers, each task is assigned the last position. This will be the canonical schedule, and now we need to prove, that all operations in this procedure were correct.

It can be proved with induction. The first operation is always correct, because the first task doesn't depend on any other (as it is on level 1), and since it is moved to the first position, no edges in the schedule graph terminate in it, hence cycles cannot appear.

Now assuming that tasks 1...p have been moved, let us examine the move of the task number p + 1. By definition, the operation is correct if

$$\forall s_i : m_i = m_2 : (n_i < n_2 \Rightarrow s_i \notin Succ(s_1)) \wedge (n_i \geq n_2 \Rightarrow s_1 \notin Succ(s_i)).$$

Since the task v_{p+1} becomes the last one on the new processor, the latter part of the equation is always true, so the condition is reduced to $\forall i$: $i \leq p \Rightarrow s_i \notin Succ(s_{p+1})$.

Let us analyze the set $Succ(s_{p+1})$. On the first iteration of its construction, it will contain quadruples from $Trans(s_{p+1})$. Due to the definition of the enumerations, none of the elements of $Trans(s_{p+1})$ can have a number lower than p + 1, because their level is higher. So, none of them is already assigned to processor m_0, so on the next iteration of constructing $Succ(s_{p+1})$ only tasks from $Trans(s_j)$, $j > p$ will be added. Accordingly, none of the elements of $Succ(s_{p+1})$ has a number lower than p, and it means that the correctness condition is satisfied.

Summing up, any two schedules (S_1, D_1) and (S_2, D_2) can be transformed to the canonical form with a sequence of correct operations. Using the reverse operations, the canonical form can be transformed to any of these tow schedules, Q.E.D.

The selection of the operation on each step of the algorithm is simple: if reliability requirements are not satisfied, either adding processors or adding versions is done with equal probability. Otherwise the operation is chosen from the remaining three operations, of course, if the operation is possible at all (i.e. for deleting versions some extra versions must already be present in the schedule). When the operation is selected, its respective parameters are chosen.

If the reliability of the system is lower that required, spare processors and versions should be added, otherwise they can be deleted. If the time of execution exceeds the deadline the possible solutions are deleting versions or moving vertices.

The selection of the operation is not deterministic so that the algorithm can avoid endless loops. Probability of selecting each operation, possibly zero, is defined for each of the four possible situations. These probabilities are given before the start of the algorithm as its settings.

Some operations cannot be applied in some cases. For example, if none of the processors have spare copies it is impossible to delete processors and if all versions are already used it is impossible to add more versions. Such cases can be detected before selecting the operation, so impossible operations are not considered.

When the operation is selected, its parameters have to be chosen according to the following rules.

Add versions. Among the tasks that have available versions one is selected randomly. Tasks with more versions already added to the schedule have lower probability of being selected.

Delete versions. The task is selected randomly. Tasks with more versions have higher probability of being selected.

Add spare processor. Similar to the addition of versions, processors with fewer spares have higher probability of being selected for this operation.

Delete spare processor. A spare of a random processor is deleted. The probability is proportional to the number of spare processors.

The probabilities for these four operations are set with the intention to keep balance between the reliability of all components of the system.

Move vertex. If $t < t_{dir}$ the main objective is to reduce the number of processors. The following operation is performed: the processor with the least tasks is selected and all tasks assigned to it are moved to other processors.

If $t > t_{dir}$ it is necessary to reduce the time of execution of the schedule. It can be achieved by reallocating some tasks, and we suggest three different heuristics to assist finding tasks that need to be moved: delay reduction, idle time reduction or mixed strategy.

Delay reduction strategy (shortened to S1). The idea of this strategy emerges from the assumption that if the time of the start of each task is equal to the length of the critical path to this task in graph G, the schedule is optimal. The length of the critical path is the sum of the lengths of all the tasks forming the path and it represents the earliest time when the execution of the task can begin.

For each element s it is possible to calculate the earliest time when s can start, i.e. when all the tasks that are origins of the edges terminating in s are completed. The difference between this time and the moment when the execution of s actually starts according to the current schedule is called the delay of task s. Since the cause of big delays is the execution of other tasks before the delayed one, the task *before* the task with the highest delay is selected for Move Vertex operation. If the operation is not accepted, on the next iteration the task before the task with the second highest delay is selected, and so on. If all tasks with non-zero delay have been tested, the task to move is selected randomly. The position (pair (m,n) from the triplet) is selected randomly among the positions where the task can be moved without breaking the correctness condition, and the selected task is moved to this position.

Figure 1 gives an example of delay reduction. Task 3 does not depend on task 4, so moving task 4 to the first processor reduces the delay of task 3, and the total time decreases accordingly.

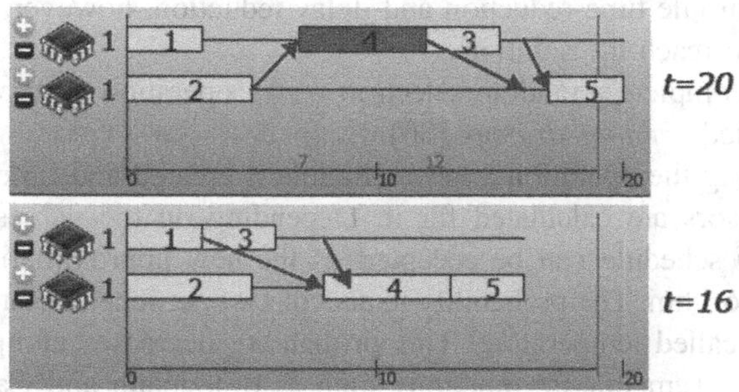

Fig. 1. Delay reduction strategy example.

Idle time reduction strategy (strategy S2). This strategy is based on the assumption that in the best schedule the total time when the processors are idle and no tasks are executed due to waiting for data transfer to end is minimal.

For each position (m,n) the idle time is defined as follows. If $n = 1$ then its idle time is the time between the beginning of the work and the start of the execution of the task in the position $(m,1)$. If the position (m,n) denotes the place after the end of the last task on the processor m, then its idle time is the time between the end of the execution of the last task on m and the end of the whole program. Otherwise, the idle time of the position (m,n) is the interval between the end of the task in $(m,n-1)$ and the beginning of the task in (m,n).

The task to move is selected randomly with higher probability assigned to the tasks executed later. Among all positions where it is possible to move the selected task, the position with the highest idle time is selected. If the operation is not accepted, the position with the second highest idle time is selected, and so on.

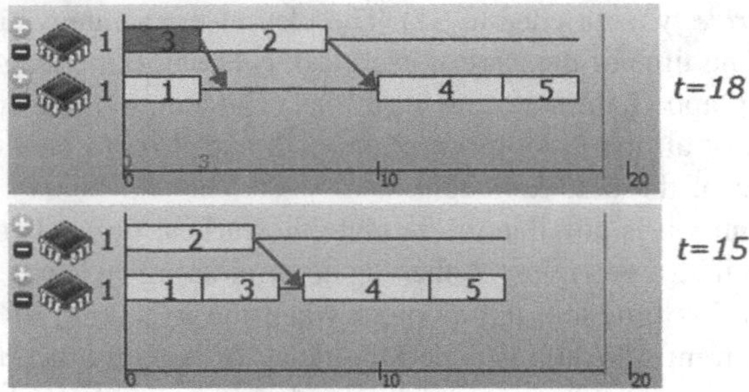

Fig. 2. Idle time reduction strategy example.

The idle time reduction strategy is illustrated in Fig. 2. The idle time between tasks 1 and 4 is large and thus moving task 3 allows reducing the total execution time.

Mixed strategy (strategy S3). As the name suggests, the mixed strategy is a combination of the two previous strategies. One of the two strategies is selected randomly on each iteration. The aim of this strategy is to find parts of the schedule where some processor is idle for a long period and to try moving a task with a big delay there, prioritizing earlier positions to reduce the delay as much as possible. This strategy has the benefits of both idle time reduction and delay reduction, however, more iterations may be required to reach the solution.

Additionally, completely random selection of the operation with a uniform distribution will be called *random strategy* (S0).

After performing the operation a new schedule is created and time, reliability and number of processors are calculated for it. Depending on the values of these three functions the new schedule can be accepted as the new approximation for the next iteration of the algorithm. The probability to accept a worse schedule on step 3 depends on the parameter called temperature. This probability decreases along with the temperature over time. Temperature functions such as Boltzmann and Cauchy laws [22] can be used as in most simulated annealing algorithms.

The correctness of the algorithm can be inferred from the fact that on all iterations the schedule is modified only by operations introduced in this section, and according to theorem 2, each operation leads to a correct schedule.

Theorem 3 (Asymptotic convergence). Assume that the temperature values decrease at logarithmic rate or slower: $t_k \geq \Gamma/log(k+k_0), \Gamma > 0, k_0 > 2$. Then the simulated annealing algorithm converges in probability to the stationary distribution where the probability to reach an optimal solution is $q_i = \frac{1}{|\Im|}\chi_{\Im}(i)$, where $\Im$ is the set of optimal solutions.

Proof. As shown in [14], the simulated annealing algorithm can be represented with an inhomogeneous Markov chain. The stationary distribution exists if the Markov chain is strongly ergodic. The necessary conditions of strong ergodicity are (1) weak ergodicity, (2) the matrix $P(k)^T$ has an eigenvalue equal to 1 for each k, (3) for its eigenvectors $q(k)$ the series $\sum_{k=1}^{\infty} q(k) - q(k+1)_1$ converges [21]. First we need to prove that the Markov chain is strongly ergodic. Condition (2) means that there exists an eigenvector

q such that $P(k)^T \cdot q = q$, or for each row of the matrix, $q_i = \sum_{j \in I} q_j \cdot P_{ji}$, which is exactly equivalent to the detailed balance equations. It is possible to check that

$$q_i = \frac{|\varepsilon(i)|}{\sum_j \left(|\varepsilon(j)| \cdot \frac{min\left(1, e^{\frac{f(i)-f(j)}{t_n}}\right)}{min\left(1, e^{\frac{f(j)-f(i)}{t_n}}\right)} \right)}$$

is the solution of this set of equations.

Let $min\left(1, e^{\frac{f(i)-f(j)}{t_n}}\right) = A_{ij}$. Notice that the following equation holds: $A_{ij} \cdot A_{jk} = A_{ik}$. Now it is easy to check the detailed balance equations.

$$\frac{|\varepsilon(i)|}{\sum_j (|\varepsilon(j)| \cdot \frac{A_{ij}}{A_{ji}})} G_{ij} \cdot A_{ij} = \frac{|\varepsilon(j)|}{\sum_k \left(|\varepsilon(k)| \cdot \frac{A_{jk}}{A_{kj}} \right)} G_{ji} \cdot A_{ji}$$

$$\frac{1}{\sum_j (|\varepsilon(j)| \cdot \frac{A_{ij}}{A_{ji}})} = \frac{1}{\sum_k \left(|\varepsilon(k)| \cdot \frac{A_{jk}}{A_{kj}} \right)} \cdot \frac{A_{ji}}{A_{ij}}$$

$$\frac{1}{\sum_j (|\varepsilon(j)| \cdot \frac{A_{ij}}{A_{ji}})} = \frac{1}{\sum_k \left(|\varepsilon(k)| \cdot \frac{A_{jk}}{A_{kj}} \cdot \frac{A_{ij}}{A_{ji}} \right)}$$

$$\frac{1}{\sum_j (|\varepsilon(j)| \cdot \frac{A_{ij}}{A_{ji}})} = \frac{1}{\sum_k \left(|\varepsilon(k)| \cdot \frac{A_{ik}}{A_{ki}} \right)}$$

The last equation is obviously correct.

To check condition (3) the following calculations can be performed.

$$\begin{aligned}
&\sum_{k=1}^{\infty} \| q(k) - q(k+1) \|_1 \\
&= \sum_{k=1}^{\infty} \sum_i |q_i(k) - q_i(k+1)| \\
&= \sum_{k=1}^{\infty} \sum_i \left| \frac{|\varepsilon(i)|}{\sum_j \left(|\varepsilon(j)| \cdot \frac{A_{ij}(k)}{A_{ji}(k)} \right)} - \frac{|\varepsilon(i)|}{\sum_j \left(|\varepsilon(j)| \cdot \frac{A_{ij}(k+1)}{A_{ji}(k+1)} \right)} \right| \\
&\leq C \cdot \sum_{k=1}^{\infty} \sum_i \left| \frac{1}{\sum_j (A_{ij}(k)/A_{ji}(k))} - \frac{1}{\sum_j (A_{ij}(k+1)/A_{ji}(k+1))} \right| \\
&= C \cdot \sum_i \left| \frac{1}{\sum_j (A_{ij}(1)/A_{ji}(1))} - \frac{1}{\sum_j (A_{ij}(k)/A_{ji}(k))} \right| .
\end{aligned}$$

Considering the conditions of the theorem, the last item is proportional to 1/*k*, and thus it converges to 0.

To check condition (1) it is possible to use the necessary condition of weak ergodicity [21]: prove that the series $\sum_{k=1}^{\infty}(1-\tau_1(P(k)^{N_k}))$ diverges. Let us find a lower bound for *P*(*k*).

$$P(k) \geq \left(\min_{i,j} G_{ij}\right) \cdot \exp\left(-\frac{\min(1, f(i)-f(j))}{t_k}\right) = C_1 e^{-C_2/t_k}$$

$$\sum_{k=1}^{\infty}(1-\tau_1(P(k)^{N_k})) \geq \sum_{k=1}^{\infty} C_1^{N_k} e^{-C_2 N_k/t_k}$$

Considering that $t_k \geq \frac{\Gamma}{\log(k+k_0)}$, we have a series like $C \cdot \sum_{k=1}^{\infty}\frac{1}{k}$, that diverges.

Finally it is necessary to find the limit of the vector *q* when the temperature approaches 0. Let us examine the limit of the expression in the denominator.

$$\lim_{n\to\infty}\sum_j\left(\frac{min\left(1, e^{\frac{f(i)-f(j)}{t_n}}\right)}{min\left(1, e^{\frac{f(j)-f(i)}{t_n}}\right)}\right)$$

If solution *i* is optimal, then the denominator is always equal to 1 regardless of *j*. The numerator will be equal to 1 is *j* is also an optimal solution, i.e. $f(i) = f(j)$, otherwise, if $f(i) < f(j)$, then $e^{\frac{f(i)-f(j)}{t_n}} \to 0$. So, the item in the sum converges to 1 if it corresponds to an optimal solution j, and converges to 0 otherwise. Therefore the limit is $|\Im|$, and $q_i = \frac{1}{|\Im|}$.

If solution *i* is not optimal, then in one of the denominators contains $f(j) - f(i) > 0$, and so the sequence $e^{\frac{f(i)-f(j)}{t_n}} \to \infty$, and the corresponding $q_i \to 0$.

Finally we can conclude that $q_i = \frac{1}{|\Im|}\chi_{\Im}(i)$, Q.E.D.

Theorem 5. The computational complexity of one iteration is O(N(N+E)), where N is the number of vertices of the program graph *G* and *E* is the number of its edges [24].

4 Experiments

The algorithm was tested both on artificial and practical examples. Artificial tests are necessary to examine the behavior of the algorithm on a wide range of examples. As the general convergence is theoretically proved, the aim of the experiments is to find the actual speed of the algorithm and to compare different strategies among each other.

Graph on Fig. 3 shows the results of the comparison of three strategies. We generated random program graphs with a pre-defined number of vertices and the number of edges proportional to the number of vertices. The number of vertices varies from 5 to 200 with step 5. For each example the algorithm was run 300 times, 100 times for each

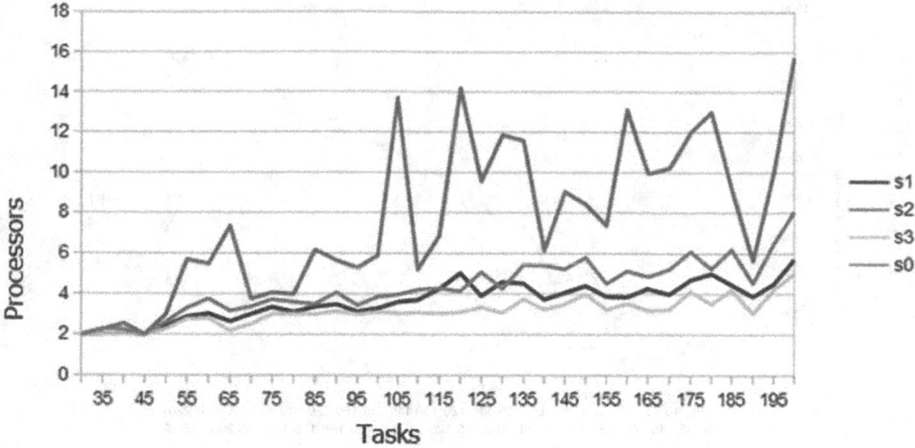

Fig. 3. Comparison of the strategies.

strategy, to make the results statistically important. The number of iterations of the algorithm was set fixed.

Figure 3 shows the average value of the target function (number of processors) depending on the number of vertices. The functions are not monotonous because of the random nature of the examples: it is not possible to guarantee that a solution with some number of processors exists, so a program of n tasks might require more processors than a program of $n + 5$ tasks. The idle time reduction strategy works worse than the other two, which is a sharp contrast with the previous version of the algorithm as shown in [24].

Let Res_i be the number of processors reached by the algorithm with strategy Si. The following statistical hypotheses [20] hold for the conducted sample of experiments.

- Strategies S1 and S3 work better than S2 and S0 $Res_1 > Res_2$, $Res_1 > Res_0$, $Res_3 > Res_2$, $Res_3 > Res_0$;
- $Res_3 - Res_1 \leq 1$;
- The results given by the algorithm are locally optimal.

Figure 4 shows the number of iterations required to reach the best result found for the corresponding problem. In each experiment, the algorithm conducted $10N$ operations where N is the number of tasks. However, after some point the continuing iteration stopped improving the result. Experiments show that the speed of the mixed strategy is practically equal to the speed of the delay reduction strategy, with S3 being slightly faster. Idle time reduction strategy is significantly faster, but it can be explained with the low quality of its results, hence fewer steps are needed to reach such results.

The practical problem we solved with the proposed algorithm is related to the design of radiolocation systems and is described in detail in [11, 25]. Briefly, the problem is to find the minimal number of processors needed to conduct the computation of the source of radio signals. The signals are received by an antenna array and then a special parallel method is used computes the results. The method is based on splitting the whole frequency diapason into L intervals and calculating the data for each

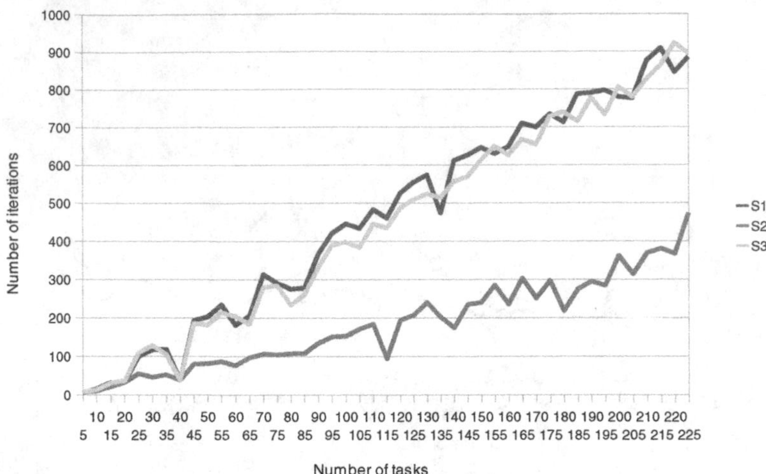

Fig. 4. Comparison of the speed of the strategies.

interval separately, preferably on parallel processors. Each of L threads is split into M subthreads as well.

In real systems the size of the array is a power of 2, usually between 256 and 1024 and the number of frequency intervals (L) is also power of 2, usually 32 or 64. M is a small number, usually between 2 and 5. As the vast majority of complex computations are done after splitting to frequency intervals, L is the main characteristic of the system that influences the overall performance. Therefore, the quality of the algorithm can be estimated by comparing the number of processors in the result with the default system configuration where $L * M$ processors are used. Figure 5 shows the quotient of these two numbers, depending on L, for radiolocation problem. Lower quotient means better result of the algorithm.

As we can see, the algorithm optimizes the multiprocessor system by at least 25 % in harder examples with many parallel tasks, and by more than a half in simpler cases.

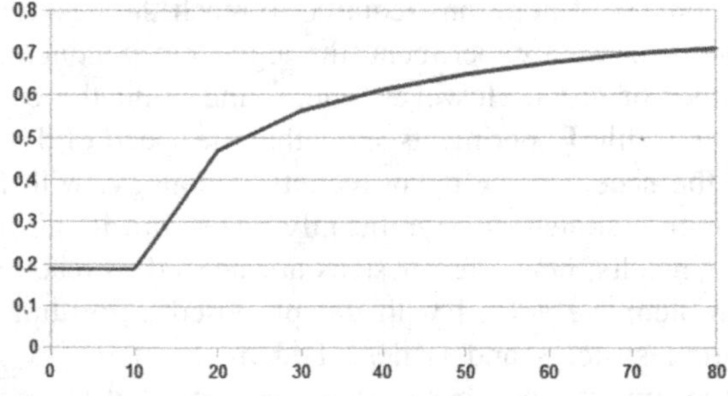

Fig. 5. Optimization rate (X axis shows the values of L, Y axis shows the optimization rate).

5 Conclusions

In this paper we formulate a combinatorial optimization problem arising from the problem of co-design of real-time systems. We suggest a heuristic algorithm based on simulated annealing, provide its description and prove the basic features, including asymptotic convergence.

Experimental evaluation of the different heuristic strategies within the discussed algorithm showed that one of the strategies was lacking compared to the other two. Mixed and delay reduction strategies have equal quality, while the mixed strategy converges slightly faster.

References

1. Antonenko, V.A., Chemeritsky, E.V., Glonina, A.B., Konnov, I.V., Pashkov, V.N., Podymov, V.V., Savenkov, K.O., Smeliansky, R.L., Vdovin, P.M., Volkanov, D.Y., Zakharov, V.A., Zorin D.A.: DYANA: an integrated development environment for simulation and verification of real-time avionics systems. In: European Conference for Aeronautics and Space Sciences (EUCASS), Munich (2013)
2. Avizienis, A., Laprie, J.C., Randell, B.: Dependability and its threats: a taxonomy. In: Building the Information Society, Proceedings IFIP 18th World Computer Congress, Toulouse, pp. 91–120 (2004)
3. Balashov, V.V., Balakhanov, V.A., Kostenko, V.A., Smeliansky, R.L., Kokarev, V.A., Shestov, P.E.: A technology for scheduling of data exchange over bus with centralized control in onboard avionics systems. Proc. Inst. Mech. Eng. Part G: J. Aerosp. Eng. **224**(9), 993–1004 (2010)
4. Eckhardt, D.E., Lee, L.D.: A theoretical basis for the analysis of multiversion software subject to coincident errors. IEEE Trans. Softw. Eng. **11**, 1511–1517 (1985)
5. Goldberg, D.E.: Genetic Algorithms in Search, Optimization, and Machine Learning. Addison Wesley, Reading (1989)
6. Hou, E.S., Hong, R., Ansari, N.: Efficient multiprocessor scheduling based on genetic algorithms. In: 16th Annual Conference of IEEE Industrial Electronics Society, IECON 1990, pp. 1239–1243. IEEE, November 1990
7. Jedrzejowicz, P., Czarnowski, I., Szreder, H., Skakowski, A.: Evolution-based scheduling of fault-tolerant programs on multiple processors. In: Rolim, J., et al. (eds.) Parallel and Distributed Processing. Lecture Notes in Computer Science, vol. 1586, pp. 210–219. Springer, Berlin (1999)
8. Laprie, J.-C., Arlat, J., Beounes, C., Kanoun, K.: Definition and analysis of hardware- and software-fault-tolerant architectures. Computer **23**, 39–51 (1990)
9. Kalashnikov, A.V., Kostenko, V.A.: A parallel algorithm of simulated annealing for multiprocessor scheduling. J. Comput. Syst. Sci. Int. **47**(3), 455–463 (2008)
10. Kirkpatrick Jr., S., Gelatt, D., Vecchi, M.P.: Optimization by simulated annealing. Science **220**(4598), 671–680 (1983)
11. Kostenko, V.A.: Design of computer systems for digital signal processing based on the concept of "open" architecture. Avtomatika i Telemekhanika **55**(12), 151–162 (1994)
12. Kostenko, V.A., Romanov, V.G., Smeliansky, R.L.: Algorithms of minimization of hardware resources. Artif. Intell. **2**, 383–388 (2000)

13. Kostenko, V.A.: Scheduling algorithms for real-time computing systems admitting simulation models. Program. Comput. Softw. **39**(5), 255–267 (2013)
14. Lundy, M., Mees, A.: Convergence of an annealing algorithm. Math. Program. **34**(1), 111–124 (1986)
15. Moore, M.: An accurate and efficient parallel genetic algorithm to schedule tasks on a cluster. In: 2003 Proceedings of the International Parallel and Distributed Processing Symposium, p. 5. IEEE (2003)
16. Orsila, H., Salminen, E., Hämäläinen, T.D.: Best practices for simulated annealing in multiprocessor task distribution problems. In: Simulated Annealing, pp. 321–342 (2008)
17. Qin, X., Jiang, H., Swanson, D.R.: An efficient fault-tolerant scheduling algorithm for real-time tasks with precedence constraints in heterogeneous systems. In: Proceedings of the International Conference on Parallel Processing 2002, pp. 360–368. IEEE (2002)
18. Qin, X., Jiang, H.: A dynamic and reliability-driven scheduling algorithm for parallel real-time jobs executing on heterogeneous clusters. J. Parallel Distrib. Comput. **65**(8), 885–900 (2005)
19. Smelyansky, R.L., Bakhmurov, A.G., Volkanov, D.Y., Chemeritskii, E.V.: Integrated environment for the analysis and design of distributed real-time embedded computing systems. Program. Comput. Softw. **39**(5), 242–254 (2013)
20. Sprinthall, R.C.: Basic Statistical Analysis. Allyn and Bacon, Boston (2006)
21. Van Laarhoven, P.J., Aarts, E.H., Lenstra, J.K.: Job shop scheduling by simulated annealing. Oper. Res. **40**(1), 113–125 (1992)
22. Wasserman, P.D.: Neural computing: theory and practice. Van Nostrand Reinhold Co., New York (1989)
23. Wattanapongsakorn, N., Levitan, S.P.: Reliability optimization models for embedded systems with multiple applications. IEEE Trans. Reliab. **53**, 406–416 (2004)
24. Zorin, D.A., Kostenko, V.A.: Algorithm for synthesizing a reliable real-time computing system architecture. J. Comput. Syst. Sci. Int. **51**(3), 410–417 (2012)
25. Zorinl, D.A.: Scheduling signal processing tasks for antenna arrays with simulated annealing. In: Proceedings of the 7th Spring/Summer Young Researchers' Colloquium on Software Engineering (SYRCoSE), Kazan, pp. 122–127 (2013)

Nash Equilibria for Multi-agent Network Flow with Controllable Capacities

Nadia Chaabane Fakhfakh[1,2(✉)], Cyril Briand[1,3], and Marie-José Huguet[1,2]

[1] CNRS, LAAS, 7 Avenue du Colonel Roche, 31400 Toulouse, France
nadia.chaabane@laas.fr

[2] INSA, LAAS, Univ. Toulouse, 31400 Toulouse, France

[3] UPS, LAAS, Univ. Toulouse, 31400 Toulouse, France

Abstract. In this work, a multi-agent network flow problem is addressed where a set of transportation-agents can control the capacities of a set of elementary routes A third-party agent, a customer, is interesting in maximizing the product flow transshipped from a source to a sink node through the network and offers a reward that is proportional to the flow value the transportation agents manage to provide. This problem can be viewed as a Multi-Agent Minimum-Cost Maximum-Flow Problem where the focus is put on finding stable strategies (i.e., Nash Equilibria) such that no transportation-agent has any incentive to modify its behavior. We show how such an equilibrium can be characterized by means of augmenting or decreasing paths in a reduced network. We also discuss the problem of finding a Nash Equilibrium that maximizes the flow and prove its NP-Hardness.

Keywords: Multi-agent network flow · Nash equilibria · Complexity · Min-Cost Max-Flow

1 Introduction

Multi-agent network games have become a promising interdisciplinary research area with important links to many application fields such as transportation networks, supply chain management, web services, production management, etc. [1,2]. In these applicative areas, decision processes often involve several agents, each one having its own autonomy, its own objectives and its own constraints. These actors, often referred to as agents, need to cooperate together to fulfill a global (social) goal, provided their own objective is also satisfied. This paper stands at the crossroads of two disciplines, namely multi-agent systems and social networks. A network flow that involves a set of agents, each one being in charge of a part of the network, is considered in this paper, where every agent is able to control the capacities of its arcs at a given cost. We address the problem of finding a Nash equilibrium that maximizes the flow transported through the network. A lot of features used in this work are inspired by the Multi-Agent Project Scheduling (MAPS), as presented in [3], especially the payment scheme:

E. Pinson et al. (Eds.): ICORES 2014, CCIS 509, pp. 47–62, 2015.
DOI: 10.1007/978-3-319-17509-6_4

the outcome of an agent depends on its own strategy and on the satisfaction of a customer, which depends on the flow circulating in the network. This paper mainly discusses the complexity of finding a Nash Equilibrium that maximizes the flow in the network.

To the best of our knowledge, the research presented here is an original way of presenting a transportation problem using multi-agent network flow with controllable arcs capacities. One important application is the expansion of transportation network capacity (railway, roads, pipelines, etc.) to meet current peak demand or to absorb future increase in the transportation demands. Therefore, it is possible to increase the capacity of the network using two solutions: either increase the capacity of one or many existing arcs or installing a new arc between two nodes. A natural problem in many network applications is where to increase arc capacities so that to increase the overall flow in the network at minimum cost. There exists substantial research on capacity expansion (or capacity planning) problems in different domains, such as manufacturing [4], electric utilities [5], telecommunications [6], inventory management [7], and transportation [8–10].

As regards to social networks, the prediction of agents' behavior is of interest. Several papers focus on games associated with various forms of networks, see [11] for an overview. In a recent work, Apt and Markakis (2011) studied the complexity of finding a Nash Equilibrium for the multi-agent social networks with multiple products, in which the agents, influenced by their neighbors, can choose one out of several alternatives [12]. In [13], a cooperative network flow game is considered, where an external party gives an additional payment to the coalition, which may stabilize the game if the payment is sufficiently high. They study the Cost of Stability (CoS) in threshold network flow games where each agent controls an edge in the network.

A decade ago, some researchers have paid attention to a particular multi-agent network problem: the Multi-Agent Project Scheduling problem (MAPS) that describes a project scheduling environment in which the activities of the project network are partitioned among a set of agents. In the seminal work of Evaristo and Van Fenema (1999) [14], a special framework for distributed projects is proposed, with costs and rewards shared among agents. In an earliest work [15], the authors considered a MAPS problem where each agent can control the duration of its activities at a given cost. The project activities and precedence constraints are classically modeled with an activity-on-arc graph. A reward is offered to the agents when they manage to finish the project earlier than expected, as proposed in [16]. It has been demonstrated in [3, 17] that finding a Nash equilibrium minimizing the project makespan is NP-hard in the strong sense. Moreover, using the concepts of an increasing and decreasing cut defined in [18] and the duality between maximum flow and minimum cut problems, Briand et al. (2012b) proposed an efficient integer linear program formulation for this problem [19].

The paper is organized as follows: Sect. 2 defines formally the Multi-Agent Minimum-Cost Maximum-Flow problem and introduces some important notations. Thereafter, Sect. 3 introduces the requirements for agents' strategies and presents some important properties. In Sect. 4, some useful particular cases are considered, namely the single agent, the general multi-agent and the special multi-agent cases. Section 5 focuses on the complexity of some decision problems. Finally, conclusions and future directions are drawn in Sect. 6.

2 Problem Statement and Notations

We focus on a Minimum-Cost Maximum-Flow problem in a Multi-Agent context. This problem will be further referred to as `MA-MCMF`. In this work, a major assumption is that arc capacities are controlled by some agents, called transportation-agents, each arc being assigned to a specific agent.

As in [16], we assume that a customer-agent gives a reward proportional to the flow that circulates inside the network. This reward is shared among transportation-agents according to some ratios collectively agreed during the network design phase [20]. Considering a network flow with limited arc capacities, this problem consists in sending a maximum amount of products (for the customer) from a source node to a sink node, at minimum cost (for the transportation-agents).

2.1 Problem Definition

The `MA-MCMF` problem is defined by a tuple $< G, \mathcal{A}, \underline{Q}, \overline{Q}, C, \pi, W >$ where:

- $G = (V, E)$ is a network flow. V is the set of nodes, $s, t \in V$ being the source and the sink nodes of the network flow G, respectively. E is the set of arcs, each one having its capacity and receiving a flow. An arc e from node i to node j is denoted by $e = (i, j)$.
- $\mathcal{A}$ is a set of m transportation-agents: $\mathcal{A} = \{A_1, \ldots, A_u, \ldots, A_m\}$. Arcs are distributed among the agents. An agent A_u owns a set of m_u arcs, denoted E_u. Each arc (i, j) belongs to exactly one transportation-agent (i.e., $E_u \cap E_v = \emptyset$ for each agent's pair $(A_u, A_v) \in \mathcal{A}^2$ such that $u \neq v$).
- $\underline{Q}$ (resp. $\overline{Q}$) represents the vector of normal (resp. maximum) capacity for each arc $(i, j) \in E$: $\underline{Q} = (\underline{q}_{i,j})_{(i,j)\in E}$ and $\overline{Q} = (\overline{q}_{i,j})_{(i,j)\in E}$.
- $C = (c_{i,j})_{(i,j)\in E}$ is the vector of costs where $c_{i,j}$ is the unitary cost incurred by agent A_u, for increasing $q_{i,j}$ by one unit. The vector C_u denotes the cost vector incurred when augmenting the capacity of its arcs.
- π refers to the reward given by the final customer. This reward is proportional to the flow that circulate from s to t.
- $W = \{w_u\}$ defines the sharing policy of rewards among the agents. The A_u reward for a gain of one unit of maximum flow equals $w_u \times \pi$.

In such a network game, each transportation-agent has to determine its *individual strategy*, i.e., the capacity $q_{i,j}$ of its own arcs, satisfying the constraints $q_{i,j} \in [\underline{q}_{i,j}, \overline{q}_{i,j}]$. The individual strategy of A_u is denoted $Q_u = (q_{i,j}), (i, j) \in E_u$. It represents the vector of capacities chosen by A_u for its arcs, with $\underline{Q} \leq Q_u \leq \overline{Q}$. A *strategy* S in the network flow is the vector of individual strategies of all agents: $S = (Q_1, \ldots, Q_m)$.

The price paid by transportation-agent A_u for its individual strategy Q_u equals:

$$P_u(Q_u) = C_u \times (Q_u - \underline{Q}_u) = \sum_{(i,j) \in E_u} c_{i,j} \times (q_{i,j} - \underline{q}_{i,j}) \tag{1}$$

Given a strategy S, $F(S)$ denotes the *maximum flow* that can circulate on the network flow given the current values of capacities. For each arc (i,j), the circulating flow $f_{i,j}$ is such that $\underline{q}_{i,j} \leq f_{i,j} \leq q_{i,j}$. The maximum flow $F(S)$ is equal to the sum of flow circulating in the forward arcs of source node (i.e., $F = \sum_{(s,j) \in E} f_{s,j}$). Let us remark that $F(S)$ can be computed in polynomial time using the well-known Ford-Fulckerson algorithm [22]. We denote by $\underline{F}$ the maximum flow when capacities $q_{i,j}$ are set to $\underline{q}_{i,j}$ for all transportation-agents (in other words, the largest possible flow at zero cost) and by $\overline{F}$ the maximum flow obtained when capacities $q_{i,j}$ are set to $\overline{q}_{i,j}$. Therefore, for any strategy S, it holds that $\underline{F} \leq F(S) \leq \overline{F}$.

With respect to the above payment scheme, the total reward given by the customer-agent for a circulating flow $F(S)$ under a strategy S is $\pi \times (F(S) - \underline{F})$.

The profit $Z_u(S)$ of transportation-agent A_u under strategy S is equal to the difference between its reward and spending:

$$Z_u(S) = w_u \times \pi \times (F(S) - \underline{F}) - P_u(Q_u) \tag{2}$$

$Z(S) = (Z_1(S), \ldots, Z_m(S))$ represents the overall profit vector.

The strategy profile S_{-u} denotes the strategies of the $(m-1)$ agents, but agent A_u, that is $S_{-u} = (Q_1, Q_2, .., Q_{u-1}, Q_{u+1}, .., Q_m)$. Therefore, a strategy where only one agent A_u modify its strategy is denoted by $S = (Q_u, S_{-u})$ and the profit of agent A_u resulting from such a strategy is denoted by $Z_u(Q_u, S_{-u})$.

Example of a MA-MCMF Problem. Let us consider a customer-agent willing to transport a flow of products from a given source node A to a given sink node D. Two transportation-agents A_1 and A_2 are involved in the transportation process. The customer-agent gives a reward $\pi = 120$ which is shared between agents following the sharing policy $w_1 = w_2 = \frac{1}{2}$. Figure 1 displays the network topology. The set of arcs of each transportation-agent are $E_1 = \{b = (A,C), c = (B,C), d = (B,D)\}$ and $E_2 = \{a = (A,B), e = (C,D)\}$, which are represented

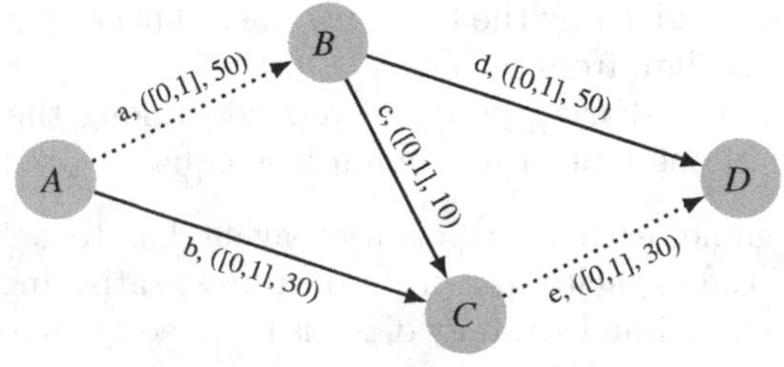

Fig. 1. Example of MA-MCMF problem.

with plain and dotted arcs, respectively. Each arc in the graph of Fig. 3 is valuated by the interval of normal and maximum capacities ($[\underline{q}_{i,j}, \overline{q}_{i,j}]$), and by the cost of increasing arc capacities ($c_{i,j}$). For instance, arc b of agent A_1 is valuated by the interval of capacities [0,1] and the cost 30. Transportation-agent A_1 can choose to open the route b from A to C with capacity $q_{A,C} = 1$ and transportation-agent A_2 can choose to put the capacity of the route e from C to D to capacity $q_{C,D} = 1$. With this strategy, the maximum flow of product than can be transported $F(S)$ is equal to 1 (from A to C and from C to D). The total reward given by the customer-agent equals 120 and the rewards for A_1 and A_2 are equal to 60. The cost for A_1 to open the route b for one unit of flow is 30 and the cost for A_2 to open the route e for one unit of flow is 30 too. Then, $Z_1 = Z_2 = 60 - 30 = 30$, which means that this strategy is profitable for both agents.

2.2 Mathematical Formulation

Each agent having the objective of maximizing its own profit, the problem can be formalized as the following multi-objective mathematical program where $Z_u(S)$ is computed according to Eq. (2):

$$\begin{array}{ll} Max\ (F, Z_1(S), Z_2(S), \ldots, Z_m(S)) & \\ s.t. & \\ (i) \quad f_{i,j} \leq q_{i,j},\ \forall (i,j) \in E & \\ (ii) \quad \sum_{(i,j)\in E} f_{i,j} - \sum_{(j,i)\in E} f_{j,i} = \begin{cases} 0\ \forall i \neq s,t \\ F\ ,\ i = s \\ -F\ ,\ i = t \end{cases} & \\ (iii) \quad \underline{q}_{i,j} \leq q_{i,j} \leq \overline{q}_{i,j},\ \forall (i,j) \in E & \\ \qquad f_{i,j} \geq 0,\ \forall (i,j) \in E & \end{array} \tag{3}$$

The mathematical formulation (5) aims at finding an overall strategy S that maximizes both the flow and the profit of all agents, each agent A_u deciding the arc capacity $q_{i,j}$, $\forall (i,j) \in E_u$. Constraints (i) represent the capacity constraints. Constraints (ii) impose the conservation of the flow.

3 Efficiency vs. Stability

A strategy is said *efficient* if it corresponds to a Pareto-optimal solution with respect to the above multi-objective program (5). The notion of Pareto optimality is concerned with social efficiency [21]. A Pareto strategy is preferred to any other strategy dominated by it.

Definition 1. *Pareto optimal strategy: A strategy S is Pareto-optimal if it is not dominated by any other strategy S'. In other words, it does not exist any strategy S' such that $Z_u(S') \geq Z_u(S)$ for all A_u, with at least one inequality being strict.*

The set of Pareto optimal strategies is denoted by S^P.

On the other hand, a strategy is *stable* if there is no incentive for any agent to modify its decision in order to improve its profit. The stability of a strategy ensures that agents can trust each other. It is connected to the notion of a *Nash equilibrium* in non-cooperative game (see [23–25]).

Definition 2. *Nash Equilibrium strategy: given a sharing reward policy w_u, a strategy $S = (Q_1, \ldots, Q_m)$ is a Nash Equilibrium if for any agent A_u with strategy Q'_u, the following equation holds:*

$$Z_u(Q_u, S_{-u}) \geq Z_u(Q'_u, S_{-u}), \;\; \forall Q'_u \neq Q_u \tag{4}$$

We refer to S^N as the set of Nash equilibria. Ideally, agents should choose a strategy which satisfy both Pareto optimality and Nash stability (i.e., $S \in S^N \bigcap S^P$). Nevertheless, since $S^N \bigcap S^P$ can be empty, such a strategy does not always exist. In this case, we are looking for a Nash equilibrium that is as efficient as possible with respect to the customer viewpoint. A Nash equilibrium that maximizes the flow circulating is indeed suitable both for maximizing the total reward and the customer satisfaction. The aim of this study is to find a stable strategy profile S^* (i.e., a Nash Equilibrium) that maximizes the flow circulating.

Let us also define the concept of a *poor* strategy. This concept will be useful for characterizing properly Nash equilibria.

Definition 3. *Poor strategy: A strategy $S = (Q_1, \ldots, Q_m)$ with flow $F(S)$ is a poor strategy if and only if it exists an agent A_u and an alternative strategy Q'_u such that $Z_u(S') > Z_u(S)$ and $F(S') = F(S)$, where $S' = (Q'_u, S_{-u})$.*

In other words, S is a poor strategy if and only if one agent is able to increase its profit by changing unilaterally its strategy (modifying the capacity of some of its arcs), without modifying the overall flow in the network, nor the profits of other agents. It is obvious that for any poor strategy S, $S \notin S^N \bigcup S^P$.

A poor strategy $S = (Q_u, S_{-u})$ can be easily transformed into a non-poor strategy $S' = (Q'_1, \ldots, Q'_m)$ by proceeding to an adaptation of the strategy Q_u of agent A_u while keeping strategy defined by S_{-u} fixed for the $m-1$ agents but agent A_u such that $S_{-u} = (Q_1, Q_2, \ldots, Q_{u-1}, Q_{u+1}, \ldots, Q_m)$.

Given $F(S) = F(S')$ and S_{-u}, a non-poor strategy S' can be the solution of the following linear program:

$$\begin{aligned}
&Max \textstyle\sum_{A_u \in \mathcal{A}} Z_u(S') = \sum_{A_u \in \mathcal{A}} [w_u \times \pi \times (F(S') - \underline{F}) - P_u(Q'_u)] \\
&s.t. \\
&(i) \quad Z_u(S') > Z_u(S), \, \forall A_u \in \mathcal{A} \\
&(ii) \quad F(S') = F(S) = \textstyle\sum_{(s,j)} f'_{s,j} \\
&(iii) \quad f'_{i,j} \leq q'_{i,j}, \, \forall (i,j) \in E \\
&(vi) \quad \textstyle\sum_{(i,j) \in E} f'_{i,j} - \sum_{(j,i) \in E} f'_{j,i} = \begin{cases} 0 \; \forall i \neq s,t \\ F \;,\; i = s \\ -F \;,\; i = t \end{cases} \\
&(v) \quad \underline{q}_{i,j} \leq q'_{i,j} \leq \overline{q}_{i,j}, \, \forall (i,j) \in E_u \\
&\qquad f'_{i,j} \geq 0, \, \forall (i,j) \in E
\end{aligned} \tag{5}$$

The mathematical program (5) is used both to verify if a strategy is poor and to ameliorate the strategy in order to remedy to its poorness. For the former concern, if a solution to (5) exists and is different from S, then the strategy S is poor. For the latter concern, the mathematical program (5) gives a non-poor strategy S' since it aims at maximizing the sum of profits of all the agents under the constraint that the flow remains constant and the profit of every agent in S' is at least greater or equal to the profit in S (i.e., $Z_u(S') > Z_u(S)$, $\forall A_u \in \mathcal{A}$). Therefore the following proposition holds.

Proposition 1. *Any solution of the mathematical program (5) is non poor solution.*

4 Case Analysis

For sake of simplicity, it is assumed throughout this section, that $\underline{q}_{i,j}=0$. Therefore, the initial minimum circulating flow at zero cost is equal to $\underline{F}=0$.

4.1 The Single-Agent Case

This section presents or recalls some basic properties related to classical network flow theory. In the single agent case (all the arcs belong to the same agent), a non-poor strategy S for a given flow $F(S)$ is a strategy that minimizes the overall cost. Such minimization problem is well-identified in the literature as the minimum-cost maximum-flow problem [26].

Let us define, in the following section, how the total flow can be either increased or decreased, at minimum cost, using augmenting or decreasing paths. These notions will be used in Sect. 4.2.

Increasing the Max-Flow. Given a flow $F(S)$ for strategy S, we are interested in increasing the flow value at minimum cost. For this purpose, we recall the well-known notion of an *augmenting path*, based on the concept of a residual graph $G_f(S)$, which is defined below.

Definition 4. *Residual graph: Given a network $G=(V,E)$ and a flow $F(S)$, the corresponding residual graph $G_f(S)=(V,E_r)$ for a given strategy S is defined as follows: each arc $(i,j)\in E$, having a maximum capacity $\overline{q}_{i,j}$ and a flow $f_{i,j}$ in G, is replaced by two arcs (i,j) and (j,i) in the residual graph. The arc (i,j) has cost $c_{i,j}$ and residual capacity $r_{i,j}=\overline{q}_{i,j}-f_{i,j}$ and the arc (j,i) has cost $c_{j,i}=-c_{i,j}$ and residual capacity $r_{j,i}=f_{i,j}$.*

Definition 5. *Augmenting path: An augmenting path is a path P in $G_f(S)$ from the source s to the sink t through which the flow can be increased.*

We refer to $\mathcal{P}$ as the set of augmenting paths. The greatest flow augmentation that can be achieved using $P\in\mathcal{P}$ is $r_p=min\{r_{ij}:(i,j)\in P\}$.

An augmenting path in $G_f(S)$ is made of forward arcs (having the same direction in G) and backward arcs (having the opposite direction than the ones in G). The set of forward and backward arcs are denoted P^+ and P^-, respectively.

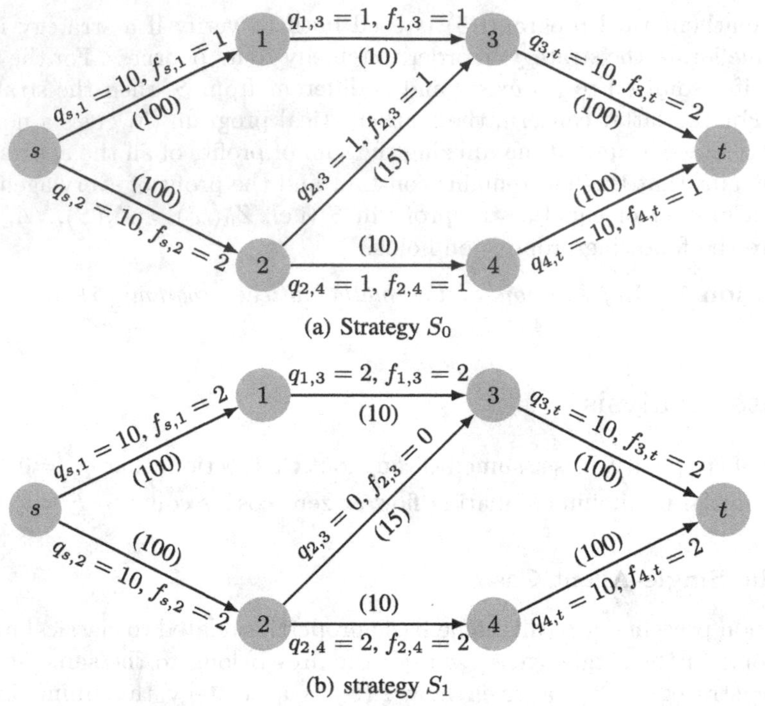

(a) Strategy S_0

(b) strategy S_1

Fig. 2. Example of single-agent network flow.

The cost of augmenting the flow by one unit using the augmenting path $P \in \mathcal{P}$ is denoted $cost(P)$. It is expressed as follows:

$$cost(P) = \sum_{(i,j) \in P+} c_{i,j} - \sum_{(i,j) \in P-} c_{i,j} \tag{6}$$

Decreasing the Max-Flow. When considering the problem of decreasing the flow at minimum cost in the network, we introduce the new concept of a *decreasing path.*

Definition 6. *Decreasing path: a decreasing path $\overline{P}$ is a path in $G_f(S)$ from the sink node t to the source node s through which the flow can be decreased.*

We refer to $\overline{\mathcal{P}}$ as the set of decreasing paths.

Similarly, a decreasing path in $G_f(S)$ is made of forward arcs (having the opposite direction than the one in $G_f(S)$) and backward arcs (having the same direction in $G_f(S)$). The set of forward and backward arcs are denoted $\overline{P}^+$ and $\overline{P}^-$, respectively. The profit of decreasing the flow along $\overline{P}$ can be expressed as follows:

$$profit(\overline{P}) = \sum_{(i,j) \in \overline{P}^+} c_{i,j} - \sum_{(i,j) \in \overline{P}^-} c_{i,j} \tag{7}$$

Example. Consider the network flow $G(V,E)$, displayed in Fig. 2, composed of seven arcs $E = \{a,b,c,d,e,f,g\}$. The set of vertex is $V = \{s,1,2,3,4,t\}$. Each arc in the graph is valuated by the interval of normal and maximum capacities, and by the cost of increasing arc capacities $([\underline{q}_{i,j}, \overline{q}_{i,j}], c_{i,j})$, where $[\underline{q}_{i,j}, \overline{q}_{i,j}]$ =[1,20], $\forall (i,j) \in E$. For instance, arc $(s,1)$ from node s to node 1 is valuated by the capacity interval [1,20] and the cost 100.

The initial strategy S_0 is described in Fig. 2(a) with flow equal to $F(S_0) = 3$. The best way to increase the flow in the network is to use the augmenting path having minimum cost, (i.e., $P = s - 1 - 3 - 2 - 4 - t$), to increase the flow on forward arcs by one unit and decrease the flow on backward arcs by one unit. With the obtained strategy $S_1 = (2,2,2,0,2,2,2)$ (see Fig. 2(b)), the maximum flow of product that can be transported $F(S_1)$ is equal to 4 and the cost incurred by the flow increase throughout the augmenting path P is equal to $cost(P) = 5$.

4.2 The Multi-agent Case

In the multi-agent context, any agent A_u can decrease (or increase) unilaterally its arc capacities to improve its profit Z_u. In this context, we introduce the concept of *profitability* of an augmenting or a decreasing path and provide a characterization of a Nash equilibrium strategy for the `MA-MCMF` problem.

Increasing the Max-Flow. Let us introduce the notion of a *profitable augmenting path.* In the multi-agent context, an augmenting path is composed by a set of forward and backward arcs $P = \{P^+, P^-\}$ such that by simultaneously increasing $q_{i,j}$ increased by one unit $\forall (i,j) \in P^+$ and decreasing by one unit $\forall (i,j) \in P^-$, it is possible to increase the overall flow by one unit.

The cost of an augmenting path for agent A_u, $cost_u(P)$ is expressed as follows:

$$cost_u(P) = \sum_{(i,j)\in P^+ \cap E_u} c_{i,j} - \sum_{(i,j)\in P^- \cap E_u} c_{i,j} \qquad (8)$$

Definition 7. *Profitable augmenting path. An augmenting path $P \in \mathcal{P}$ is said profitable for all agents if, for every agent A_u involved in P, $cost_u(P) < w_u \times \pi$.*

This means that through a profitable augmenting path, increasing the flow by one unit, is profitable for all the agents owning the arcs of the path (i.e., the profit of an agent A_u for increasing the flow by one unit verify $Z_u(S) = w_u \times \pi - cost_u(P) > 0$, where $cost_u(P)$ is the reduced cost).

Decreasing the Max-Flow. Now, the notion of *profitability* is introduced. In the multi-agent context, a decreasing path $\overline{P} = \{\overline{P}^+, \overline{P}^-\}$ is composed of forward and backward arcs. If $q_{i,j}$ is decreased by one unit, $\forall (i,j) \in \overline{P}^+$, and increased by one unit, $\forall (i,j) \in \overline{P}^-$, the overall flow is decreased by one unit.

Considering an agent A_u, the profit $profit_u(\overline{P})$ generated by decreasing capacity by one unit through a decreasing path is defined as follows:

$$profit_u(\overline{P}) = \sum_{(i,j)\in\overline{P}^+\cap E_u} c_{i,j} - \sum_{(i,j)\in\overline{P}^-\cap E_u} c_{i,j} \tag{9}$$

Definition 8. *Profitable decreasing path. A decreasing path $\overline{P} \in \overline{\mathcal{P}}$ is profitable if there is one agent A_u such that $profit_u(\overline{P}) > w_u \times \pi$.*

In other words, through a profitable decreasing path, decreasing the flow by *one* unit is profitable for one agent, to the detriments of the others.

In the multi-agent context, it is important to characterize strategies in which some agents can decrease or increase the overall flow. Therefore, it is important to find profitable augmenting paths in order to increase flow without generating decreasing paths that are profitable for some agent, hence preserving stability.

Proposition 2. *Nash Equilibrium.*
For a given non-poor strategy profile S, S is a Nash Equilibrium if and only if:

- $\forall A_u \in \mathcal{A}$, $\forall P \in \mathcal{P}$ *such that* $(i,j) \in E_u$
$$cost_u(P) > w_u \times \pi \tag{10}$$
- $\forall A_u \in \mathcal{A}$, $\forall \overline{P} \in \overline{\mathcal{P}}$
$$profit_u(\overline{P}) \leq w_u \times \pi \tag{11}$$

Proof. Consider a strategy S and a transportation-agent A_u. If S is poor, then S is not a Nash equilibrium. If S is non poor, A_u can only improve its situation by increasing or decreasing the flow. In the former case, for an additional unit of flow, A_u receives $w_u \times \pi$. Such a flow increase is profitable to A_u if and only if there is an augmenting path P such that $cost_u(P) < w_u \times \pi$, which contradicts equation (10). In the latter case, vice-versa, decreasing the flow by one unit is profitable if and only if there exists a decreasing path $\overline{P}$ such that $profit_u(\overline{P}) > w_u \times \pi$, which contradicts equation (11). Therefore, if and only if for no agent any of those conditions holds, no agent A_u can individually improve its profit, and S is a Nash equilibrium. □

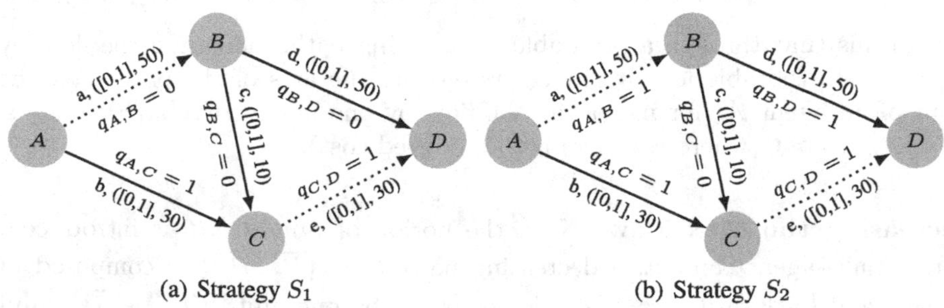

Fig. 3. Example of multi-agent network flow.

Example. Let us come back to the first example (cf. Sect. 2.1) to illustrate the notions of augmenting and decreasing path in the multi-agent case.

Consider an initial flow on the network equal to its minimum value $\underline{F} = 0$ corresponding to an initial strategy S_0. Increasing the flow is possible throughout the profitable augmenting path $P = (A-C-D)$, which leads to the strategy $S_1 = (0, 1, 0, 0, 1)$ (see Fig. 3(a)) with $F(S_1) = 1$ and $Z_1(S_1) = Z_2(S_1) = 60 - 30 = 30$ where the part of shared reward is $w_u \times \pi = 60$ and the cost of the path P is $cost_u(P) = 30$ for both agents. From this strategy, the flow can be increased along the profitable augmenting path $P'(= A - B - D)$, leading to the strategy $S_2 = (1, 1, 0, 1, 1)$ (see Fig. 3(b)) with $F(S_2) = 2$. The cost of the augmenting path for every agent is equal to $cost_u(P') = 50$ and the part of the shared reward for the additional unit of flow is equal to $w_u \times \pi = 60$. Therefore, the profit of both agents is equal to $Z_u(S_2) = Z_u(S_1) + (60 - 50) = 30 + 10 = 40$.

Note that, for the strategy S_2, there exists a profitable decreasing path $P" = (D-B-C-A)$ from sink node D to source node A which is profitable for agent A_1. In fact, A_1 can improve its own profit, by decreasing back the flow on b and d by one unit and increasing the flow on arc c by one unit. This leads to the strategy $S_3 = (1, 0, 1, 0, 1)$ (see Fig. 4) with $F(S_3) = 1$ and profits $Z_1(S_3) = 60 - 10 = 50$ and $Z_2(S_3) = 60-(50+30) = -20$, which is obviously bad for A_2. Therefore, although the strategy S_2 corresponds to a Pareto Optimum, which leads to a maximization of agent's profits, it is not a stable strategy. Strategy S_1 is a Nash Equilibrium but not Pareto Optimum. Therefore, in our example there is no a strategy which is both in S^N and S^P. The motivation of this paper is to search for a Nash-stable solution which is as efficient as possible, i.e., which maximizes $F(S)$.

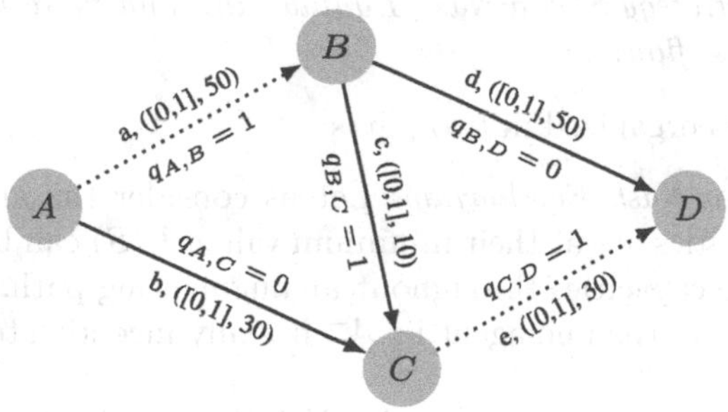

Fig. 4. Strategy S_3.

4.3 The Special Case $|E_u| = 1, \forall A_u$

In this section, we consider the special multi-agent case where each arc is managed by a specific agent. For this case, we show that finding a Nash equilibrium that maximizes the flow can be done in polynomial time. For sake of simplicity, we denote by u the unique arc of the agent A_u. In this context, increasing the flow by one unit brings to the agent A_u the reward $w_u \times \pi$ and since an agent manages only a single arc u then it is easy to compare the reward with the cost of increasing arc's capacity c_u of arc u. It is possible to divide the set of agents $\mathcal{A}$ into two subsets $\mathcal{A}^+$ and $\mathcal{A}^-$ as follows:

$$\mathcal{A}^+ = \{A_u, 1 \leq u \leq m, / c_u < w_u \times \pi\}$$

$$\mathcal{A}^- = \{A_u, 1 \leq u \leq m, / c_u \geq w_u \times \pi\}$$

Signification of each group of agents: On the one hand, for any agent belonging to the group $\mathcal{A}^+$, it is profitable to increase the capacity of its arc (i.e., $w_u \times \pi - c_u > 0$) if it increases the overall flow in the network (i.e., its arcs belong to an augmenting path). On the other hand, it is not profitable for any agent belonging to $\mathcal{A}^-$ to increase its arc capacity since $w_u \times \pi - c_u \leq 0$.

Consider the initial strategy $S = (Q_1, \ldots, Q_m)$ defined by:

$$Q_u = (\overline{q}_u), \forall A_u \in \mathcal{A}^+$$

$$Q_u = (\underline{q}_u), \forall A_u \in \mathcal{A}^-$$

We highlight that the strategy S can be poor since some arcs of the agents belonging to $\mathcal{A}^+$ can have an opened capacity greater than the value of the flow traversing them (i.e., $f_{i,j}(S) < q_{i,j}(S)$). Nevertheless, using LP formulation (5), finding a non-poor strategy $\hat{S}$ starting from S is easy. This leads to a non-poor strategy $\hat{S}$ with the same value of flow $F(\hat{S}) = F(S)$.

Notice that $\hat{S}$ may not be unique, since different non-poor strategies can be obtained.

We are going to prove now the following property:

Proposition 3. *Strategy $\hat{S}$ is a Nash Equilibrium, and there is no Nash Equilibrium with greatest flow.*

Proof. This proof is organized in two parts:

- *Proof that $\hat{S}$ is a Nash Equilibrium:* Let us consider the arcs of $A_u \in \mathcal{A}^-$. Since their capacities are at their minimum value, $F(\hat{S})$ can be increased only by increasing the capacities throughout an augmenting path. Since for agents in $\mathcal{A}^-$, $c_u > w_u \times \pi$, then no agent in $\mathcal{A}^-$ has any incentive to increase its arc capacity.

  Now let consider the agents $A_u \in \mathcal{A}^+$. If, in $\hat{S}$, $q_u = \overline{q}_u$ then A_u can improve its situation only by decreasing its arc capacity throughout a decreasing path. Since for agents in $\mathcal{A}^+$, $c_u \leq w_u \times \pi$, no agent A_u can take profit from decreasing back its arc capacity. If in $\hat{S}$, $q_u < \overline{q}_u$, one agent A_u might increase its arc capacity throughout an augmenting path such that all forward arcs belong to him (else it is not possible to increase the value of the flow). Since each agent owns exactly one arc, such a situation cannot occur.
  Finally, since no agent is able to improve its situation by itself, $\hat{S}$ is a Nash equilibrium.
- *Proof that $\hat{S}$ is the best Nash Equilibrium:* Suppose that there is a strategy S' such that $F(S') > F(\hat{S})$. This strategy requires that the capacity of at least one arc of $A_u \in \mathcal{A}^-$ has to be increased with respect to strategy $\hat{S}$. But since $c_u > w_u \times \pi$, $\forall A_u \in \mathcal{A}^-$, S' is not a Nash Equilibrium (see Proposition (2)).□

5 Problem Complexity

In this section, we discuss the complexity of finding a Nash equilibrium that maximizes the flow in the network.

5.1 Finding a Feasible Solution

Firstly, let us discuss the complexity of a simplified version of the considered problem in which we substitute the Nash Equilibrium constraint by a looser constraint stating that the profit of all agents has to be non-negative, i.e., $Z_u(S) \geq 0, \forall A_u \in \mathcal{A}$.

Proposition 4. *The multi-agent Min-Cost Max-Flow problem which aims at maximizing $F(S)$ under the constraints that agents have non-negative profits $Z_u(S) \geq 0$, with $q_{i,j} \in \mathbb{R}^+$, can be solved in polynomial time.*

Proof. This problem can be solved by the following linear mathematical problem where constraints (iii) impose that the profit of all agents has to be positive or null:

$$
\begin{aligned}
& Max\ F = \textstyle\sum_{(s,j)\in E} f_{s,j} \\
& s.t. \\
& (i) \quad f_{i,j} \leq q_{i,j},\ \forall (i,j) \in E \\
& (ii) \quad \textstyle\sum_{(i,j)\in E} f_{i,j} - \sum_{(i,j)\in E} f_{j,i} = \begin{cases} 0 \ \forall i \neq s,t \\ F\ ,\ i = s \\ -F\ ,\ i = t \end{cases} \\
& (iii) \quad w_u \times \pi \times (F - \underline{F}) - \textstyle\sum_{(i,j)\in E_u} c_{i,j} \times (q_{i,j} - \underline{q}_{i,j}) \geq 0,\ \forall A_u \in \mathcal{A} \\
& (iv) \quad \underline{q}_{i,j} \leq q_{i,j} \leq \overline{q}_{i,j},\ \forall (i,j) \in E \\
& \qquad\ \ f_{i,j} \geq 0,\ \forall (i,j) \in E
\end{aligned}
$$

Therefore, this problem can be solved using linear programming in polynomial time. □

5.2 Finding a Nash Equilibrium with Bounded Flow

We now consider the decision problem to determine if there exists a strategy which is a Nash equilibrium, with a flow greater than a given value. This problem can be defined as follows:

Nash-Equilibrium Bounded Flow (NEBF). Given a tuple $< G, \mathcal{A}, \underline{Q}, \overline{Q}, C, \pi, W >$ as defined in Sect. 2 and an integer φ, is it possible to find a Nash Equilibrium strategy profile S such that $F(S) > \varphi$?

Proposition 5. *Problem NEBF is strongly NP-complete.*

Proof. The NP-completeness of this problem can be proved using a reduction from the well-known 3-partition problem, which is known to be NP-complete in the strong sense [27]. First, `MA-MCMF` is in NP since, given a strategy S, $F(S)$ can be determined in polynomial time using classical Min-Cost Max-Flow algorithms. Let us recall the definition of the 3-partition problem.

3-Partition. Given a set $\zeta = \{a_0, \ldots, a_{K-1}\}$ of $K = 3k$ positive integers, such that $\sum_{l=0}^{K-1} a_l = k \times B$ and $a_l \in]B/4, B/2]$, is it possible to partition ζ into k subsets so that the sum of integers in each subset is equal to B?

An instance of the MA-MCMF problem with controllable capacities can be generated from an arbitrary instance of the 3-partition problem as follows.

From the 3-partition problem instance, we build up a network G with $k \times K$ arcs and $K+1$ nodes where the first one is source node $V_0 = s$ and the last one is the sink node $V_K = t$. An agent $A_u \in \mathcal{A} = \{A_1, \ldots, A_k\}$ owns K arcs.

The tail of an arc e_i is $V_{i \operatorname{div} K}$, its head is $V_{(i \operatorname{div} K)+1}$. Between nodes $V_{i \operatorname{div} K}$ and $V_{i \operatorname{div} K+1}$, there are k parallel arcs, indexed from i to $(i+K)$ step k, each of them belonging to a specific agent: arc e_i belongs to $A_{i \operatorname{div} K}$. The cost of arc e_i is $c_{e_i} = a_{i \operatorname{mod} K}$. In other words, to any positive integer $a_l \in \zeta$ is associated k parallel arcs with, same head and tail, maximum capacity $\overline{q}_{e_i} = 1$ and cost a_l. The total reward is set to $\pi = (B+\epsilon)k$, ϵ being an arbitrary small positive value. The sharing policy is defined by $w_u = 1/k$. Therefore, agent's unit reward is $w_u\pi = B+\epsilon$, identical for all agents. The objective is to determine whether it exists a Nash strategy such that $F(S) > 0$?

For illustration, the resulting network flow obtained from the 3-partition instance defined by $k = 3$, $\zeta = \{7, 8, 7, 7, 7, 8, 9, 10, 9\}$ and $B = 24$. We have $k = 3$ agents and $K * k = 27$ arcs is displayed in Fig. 5. Between nodes i and $i+1$, we find $k = 3$ arcs with cost a_{i+1}. The problem is to find, whether it exists, a Nash strategy such that the flow is strictly greater than 0. In that example, using the augmenting path with bold arcs allows to obtain a one-unit total flow, which is a Nash equilibrium since every agent does not pay more than its part of reward ($w_u\pi = B+\epsilon = 24+\epsilon$). But we remark that, any equivalent stable path is also a solution to the original 3-Partition problem.

Let us prove this last property in a general way. Consider the strategy $\underline{S}$ where all arcs have normal capacity, $q_{i,j} = 0$. The resulting flow obviously equals to $F(\underline{S}) = 0$. With respect to $\underline{S}$, we observe that an agent can increase the flow by the amount $\delta \in]0, 1]$, increasing the capacities of all its arcs by the same amount δ. However, doing so, the agent pays $kB\delta$ and only gains $(B+\epsilon)\delta$. Hence, the new strategy is not profitable and cannot be a Nash equilibrium. In order to obtain a Nash equilibrium, the total cost incurred by each agent for increasing its arc capacities must not exceed B, otherwise at least one agent will be interested in decreasing back its capacities (i.e., the residual graph cotains a profitable decreasing path).

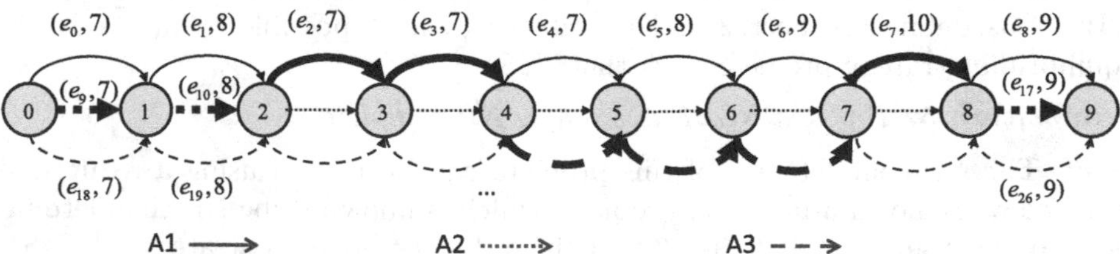

Fig. 5. Reduction from 3-PARTITION problem with $k = 3$.

Due to the topology of the network, in order to increase the flow, exactly $K = 3k$ arcs must be involved in an augmenting path. In any Nash equilibrium strategy with flow strictly greater than 0, the augmenting path having to be profitable for every agent, it must be made of exactly three arcs per agent. The total cost for every agent equals exactly B.

6 Conclusions

This paper presents a new game theory framework for a multi-agent network flow problem with controllable capacities. We consider that a final customer gives a reward, shared among agent, for any additional unit of flow circulating in the network. Each agent has the possibility to modify the capacities of its arcs at a given cost. We particularly point out the notions of efficiency and stability of a strategy and we introduce the notion of profitable augmenting or decreasing paths. We also prove that finding a Nash Equilibrium strategy with maximum flow is NP-hard in the strong sense.

Further works are ongoing to propose a linear mathematical model to find a Nash Equilibrium. Distributed heuristic able to find a Nash equilibrium are also under study.

Acknowledgements. This work was supported by the ANR project no. ANR-13-BS02-0006-01 named Athena.

References

1. Chen, B., Cheng, H.H.: A review of the applications of agent technology in traffic and transportation systems. IEEE Trans. Intell. Transp. Syst. **11**(2), 485–497 (2010)
2. Pechoucek, M., Marik, V.: Industrial deployment of multi-agent technologies: review and selected case studies. Auton. Agent. Multi-Agent Syst. **17**(3), 397–431 (2008)
3. Briand, C., Agnetis, A., Billaut, J.C.: The multiagent project scheduling problem: complexity of finding an optimal Nash equilibrium. In: Proceedings of PMS Conference, pp. 106–109 (2012)
4. Zhang, F., Roundy, R., Akanyildrim, M.C., Huh, W.T.: Optimal capacity expansion for multi-product, multi-machine manufacturing systems with stochastic demand. IIE Trans. **36**(1), 23–26 (2004)
5. Malcolm, S., Zenios, S.: Robust optimization for power systems capacity expansion under uncertainty. J. Oper. Res. Soc. **45**(9), 1040–1049 (1994)
6. Laguna, M.: Applying robust optimization to capacity expansion of one location in telecommunications with demand uncertainty. Manage. Sci. **44**(11), 101–110 (1998)
7. Hsu, V.N.: Dynamic capacity expansion problem with deferred expansion and age-dependent shortage cost. Manage. Sci. **4**(1), 44–54 (2002)
8. Magnanti, T., Wong, R.: Network design and transportation planning: models and algorithms. Transp. Sci. **18**(1), 1–55 (1984)

9. Fulkerson, D.R.: Increasing the capacity of a network: the parametric budget problem. Manage. Sci. **5**(4), 472–483 (1959)
10. Ordonez, F., Zhao, J.: Robust capacity expansion of network flow. Networks **50**(2), 136–145 (2007)
11. Tardos, E., Wexler, T.: Network formation games and the potential function method. In: Nisan, N., Roughgarden, T., Tardos, E., Vazirani, V.V. (eds.) Algorithmic Game Theory, pp. 487–516. Cambridge University Press, Cambridge (2007)
12. Apt, K.R., Markakis, E.: Diffusion in social networks with competing products. In: Persiano, G. (ed.) SAGT 2011. LNCS, vol. 6982, pp. 212–223. Springer, Heidelberg (2011)
13. Resnick, E., Bachrach, Y., Meir, R., Rosenschein, J.S.: The cost of stability in network flow games. In: Královič, R., Niwiński, D. (eds.) MFCS 2009. LNCS, vol. 5734, pp. 636–650. Springer, Heidelberg (2009)
14. Evaristo, R., Van Fenema, P.C.: A typology of project management: emergence and evolution of new forms. Int. J. Proj. Manage. **17**(5), 275–281 (1999)
15. De, P., Dunne, E.J., Ghosh, J.B., Wells, C.E.: Complexity of the discrete time-cost tradeoff problem for project networks. Eur. J. Oper. Res. **45**(2), 302–306 (1997)
16. Estevez, M.A.: Fernandez: a game theoretical approach to sharing penalties and rewards in projects. Eur. J. Oper. Res. **216**(3), 647–657 (2012)
17. Agnetis, A., Briand, C., Billaut, J.-C., Sucha, P.: Nash equilibria for the multiagent project scheduling problem with controllable processing times. J. Sched. **18**(1), 15–27 (2014)
18. Kamburowski, J.: On the minimum cost project schedule. Int. J. Manage. Sci. **22**(4), 401–407 (1994)
19. Briand, C., Sucha, P., Ngueveu, S.U.: Finding an optimal Nash equilibrium to the multi-agent project scheduling problem. J. Oper. Res (Submitted to)
20. Cachon, G.P., Lariviere, M.A.: Supply chain coordination with revenue-sharing contracts: strengths and limitations. Int. J. Manage. Sci. **51**(1), 30–44 (2005)
21. Ehrgott, M.: Multicriteria Optimization. Algorithmic Game Theory, pp. 487–516. Springer, New York (2005)
22. Ford, L.R., Fulkerson, D.R.: Flow in Networks. Princeton University Press, Princeton (1958)
23. Nash, J.: Equilibrium points in n-person games. Proc. Natl. Acad. Sci. **36**(1), 48–49 (1950)
24. Nash, J.: Non-cooperative games. Ann. Math. **54**(2), 286–295 (1951)
25. Shoham, Y., Leyton-Brown, K.: Multiagent Systems: Algorithmic, Game-Theoretic, and Logical Foundations. Cambridge University Press, New York (2009). ISBN 978-0-521-89943-7
26. Busacker, R.G., Gowen, P.J.: A Procedure for Determining a Family of Minimal-Cost Network Flow Patterns. Defense Technical Information Center, 15th edn. (1961)
27. Garey, M.R., Johnson, D.S.: Computers and Intractability: A Guide to the Theory of NP-Completeness. W.H. Freeman & Co., New York (1979)

A Fuzzy Model for Selecting Safeguards to Reduce Risks in Information Systems

E. Vicente, A. Mateos(✉), and A. Jiménez-Martin

Artificial Intelligence Department, Technical University of Madrid, Madrid, Spain
e.vicentecestero@upm.es, {amateos,ajimenez}@fi.upm.es

Abstract. Information systems can be represented by acyclic directed graphs where the nodes denote assets and the arcs connecting nodes represent the degree of dependency between assets. Threats are events that can trigger an incident in the organization, causing damage or intangible material loss to assets, and safeguards are measures for addressing threats. In this paper, we propose a fuzzy approach for selecting safeguards that minimizes costs while keeping the degree of dependency between support assets and terminal assets within acceptable levels. The approach is based on dynamic programming and uses the simulated annealing metaheuristic to solve optimization problems.

Keywords: Risk analysis · Fuzzy logic · Dynamic programming · Simulated annealing

1 Introduction

There are several risk analysis and management methodologies for information systems (IS) that conform to International Organization for Standardization (ISO) standards, specifically the ISO 27000 family of standards. Some examples of these methodologies are MAGERIT, by the Spanish Ministry of Public Administrations [5,11], by the Central Computing and Telecommunications Agency (UK); or NIST SP 800-30 [14], by the National Institute of Standard and Technology (USA).

These methodologies do not, however, consider uncertain valuations, but use precise values on different, usually percentage, scales. Boolean values are sometimes even used to indicate whether or not assets are dependent on each other regardless of the degree of such dependency. In no case is vague or imprecise information about the input parameters allowed. In our opinion, this is an important drawback of these methodologies.

In [15] we proposed an extension of the MAGERIT methodology based on classical fuzzy computational models. This methodology includes the following milestones:

1. *Identification and Valuation of Assets.* An asset is anything that is of value to the organization and therefore requires protection. A few data, information or business process assets often account for the total value of an organization's

E. Pinson et al. (Eds.): ICORES 2014, CCIS 509, pp. 63–78, 2015.
DOI: 10.1007/978-3-319-17509-6_5

assets. These assets are called *terminal assets.* Other assets (*support assets* such as hardware, software, personnel, facilities, ...) are valuable insofar as they are beneficial to the terminal assets, and they inherit the terminal asset value, according to the resulting benefit. Thus, support assets have no intrinsic value; they take their value from terminal assets.

The identified assets of the organization are then valued. Some assets may have a monetary value (how much money the organization would lose if this asset stopped working), whereas others require a qualitative assessment (if an asset stops working the losses would be very high, low, medium...). As mentioned above, the support assets inherit their values from terminal assets depending on how they influence each other. So, we have to determine the dependency relationships of the terminal assets with respect to support assets, and also dependency relationships between support assets.

2. *Threat Identification.* A *threat* is an event that can trigger an incident in the organization, causing damage or intangible material loss to assets. Threats may be of natural or human, accidental or deliberate origin. Some threats can affect more than one asset. In such cases, threats can cause different impacts depending on what assets are affected. A detailed list of threats is available in Annex C of ISO IEC 27005. MAGERIT suggests two threat assessment measures: *degradation*, the damage that the threat can cause to the asset, and *frequency*, how often the threat materializes.
3. *Identification and Valuation of Impact and Risk Indicators.* It is then necessary to qualitatively identify the consequences and establish impact and risk indicators for the valued assets and threats. The impact of a threat on an asset is the product of the asset value multiplied by the respective degradation. Risk is the product of the impact of the threat multiplied by the respective frequency.
4. *Selection of Safeguards.* Safeguards are measures for addressing threats. They can be procedures, personnel policies, technical solutions or physical security measures at the facilities. These safeguards can be *preventive*, if they reduce the frequency of threats; or *palliative*, if they reduce the degradation of assets caused by threats [11].

As described below, experts use a linguistic term scale (see Table 1) to represent asset values, their dependencies and the frequency and asset degradation associated with possible threats. Risk analysis computations are then based on the trapezoidal fuzzy numbers associated with linguistic terms.

However, direct assignment based on a rigid linguistic term scale is not always advisable since the expert has no say in the number of linguistic terms that the scale is to include and about the appearance of their associate trapezoidal fuzzy numbers. In that case we propose the use of the betting and lottery-based method for fuzzy probability elicitation described in [17]. Betting and lottery-based methods commonly used to assign probabilities can also be used to assign fuzzy probabilities [6,12]. In this section we briefly describe these methods and show how a fuzzy number representing the probability judgment can be extracted from experts.

Table 1. Linguistic term scale.

Term	Trapezoidal fuzzy number
Very low (VL)	(0, 0, 0, 0.05)
Low (L)	(0, 0.075, 0.125, 0.275)
Medium-low (ML)	(0.125,0.275, 0.325, 0.475)
Medium (M)	(0.325, 0.475, 0.525, 0.675)
Medium-high (MH)	(0.525, 0.675, 0.725, 0.875)
High (H)	(0.725, 0.875, 0.925, 1)
Very high (VH)	(0.925, 1, 1, 1)

Betting Method. For two selected monetary values $x > y$, the expert is given the option between either of the two following gambles:

- *b1*: If event A happens, then you win x\$. Otherwise, you lose y\$.
- *b2*: If event A does not happen, then you win y\$. Otherwise, you lose x\$.

If the expert has no preference for either bet, the respective expected utilities of both bets are equal, and it follows that $p(A) = x/(x + y)$. If the expert chooses one of the two gambles, then the expected utility of the selected gamble should be higher than for the rejected gamble. Then, the analyst has to update monetary values and offer the expert two new gambles. Thus, an interactive process is enacted until two alternative gambles are reached to which the expert is indifferent.

Lottery-based methods. For a given probability and monetary values x\$ and y\$, the expert is given the choice between the following lotteries:

- *l1*: If event A happens, then you win x\$. Otherwise, you lose y\$.
- *l2*: You win x\$ with probability p, or y\$ with probability $1 - p$.

If the expert has no preference for either of the lotteries, then the respective expected utilities are equal, and it follows that $p(A) = p$. Otherwise, the expert must readjust the value p, keeping the same monetary values. This again generates an interactive process, enacted until a couple of lotteries are reached to which the expert is indifferent.

The betting and lottery-based methods assume that the expert is able to provide a specific value for the probability of an event. However, a more realistic scenario is where experts have an imprecise and vague idea of that value. Consequently, experts will have an interval rather than a precise value in mind at the point when they are indifferent to either bet or lottery, that is, for the lottery-based method there will be an interval $[a, c]$ such that if $p = [a, c]$, then the expert has no preference for either lottery $l1$ or $l2$. Similarly, the betting method can result in an interval of indifference $[b, d]$.

Current protocols for probability elicitation like the above recommend the use of several methods to test the consistency of the expert and the existence of bias. In this regard, the development of betting and lottery-based methods meets this recommendation and establishes the following:

- If $[a,c] \cap [b,d] = \varnothing$, then the expert's probabilistic judgment was inconsistent.
- If any of the intervals is contained in the other $[a,c] \subseteq [b,d]$ (or $[b,d] \subseteq [a,c]$), then we assume that the trapezoidal fuzzy number (b,a,c,d) (or (a,b,d,c)) designates the expert probabilistic judgment.
- If $[a,c] \cap [b,d] \neq \varnothing$, is uncountable, and none of the intervals is contained in the other, then, assuming that $a \leq b \leq c \leq d, (a,b,c,d)$ designates the expert probabilistic judgment.

Thus, we consider the set of trapezoidal fuzzy numbers with support in [0,1], TF[0,1], i.e., $\widetilde{A} = (a,b,c,d)$ with $0 \leq a \leq b \leq c \leq d \leq 1$ and with a trapezoidal function in the vertices $(a,0),(b,1),(c,1),(d,0)$ [2–4,16].

Consequently, the following operators proposed in (Xu et al. 2010) accounting for trapezoidal fuzzy numbers will be used to make computations. Given $\widetilde{A}_1 = (a_1,b_1,c_1,d_1)$, $\widetilde{A}_2 = (a_2,b_2,c_2,d_2) \in TF[0,1]$, then $\widetilde{A}_1 \oplus \widetilde{A}_2 = (a_1+a_2-a_1a_2, b_1+b_2-b_1b_2, c_1+c_2-c_1c_2, d_1+d_2-d_1d_2)$ and $\widetilde{A}_1 \otimes \widetilde{A}_2 = (a_1a_2, b_1b_2, c_1c_2, d_1d_2)$. $\oplus$ and $\otimes$ are two internal composition laws in TF[0,1] that verify the commutative and associative properties and both have a neutral element.

The assets of an IS are elements of value to the organization and therefore require protection (servers, files, personnel, facilities, hardware, software, ...).

As cited before, these assets are interrelated [11], forming an acyclic graph, where just a few data, information items or business process assets often account for the total value of an organization's assets. These assets are called *terminal assets*. Other assets (*support assets*, such as hardware, software, personnel, facilities, ...) are valuable insofar as they are beneficial to the terminal assets. In other words, the support assets inherit their values from terminal assets depending on how they influence each other, i.e., depending on the probability of that any failure in an asset being transferred to the terminal assets.

In general, we say asset A_j *directly depends* on asset A_i, denoted by $A_i \to A_j$, if a failure in asset A_i causes a failure in the asset A_j with any given probability. This probability is usually referred to as the *degree of direct dependency* of A_j with respect to A_i. Note that in this fuzzy adaptation the degrees of direct dependency between assets will be represented by linguistic terms, which have associated trapezoidal fuzzy numbers. We denote these degrees of direct dependency by $\widetilde{d}(A_i, A_j)$.

These dependencies form a directed acyclic graph (to terminal assets), so that there may be intermediate assets between any asset A_i and a terminal asset A_k which can propagate a fault generated in A_i through to the terminal A_k. Our aim then is to compute the transmission probability between A_i and A_k. This probability is called *degree of indirect dependency* between A_i and A_k, which is denoted by $\widetilde{D}(A_i, A_k)$ and can be computed as follows [15].

We denote by $\mathbf{P} = \{P_1, \ldots, P_s\}$ the set of paths in the network connecting A_i with A_k. These paths are a sequence of arcs connecting a sequence of vertices, such that the start vertex and the last vertex are A_i and A_k, respectively. Then,

(*A*) If all assets, excluding A_i and A_k, in the paths in $\mathbf{P}$ are influenced by only one asset, then

$$\widetilde{D}(A_i, A_k) = \bigoplus_{j=1}^{s} \widetilde{D}(A_i, A_k | P_j) \tag{1}$$

where $\widetilde{D}(A_i, A_k|P_j) = \widetilde{d}(A_i, A_{j_1}) \otimes \ldots \otimes \widetilde{d}(A_{j_n}, A_k)$ and $P_j : (A_i \to A_{j_1} \to \ldots \to A_{j_n} \to A_k)$.

(*B*) Otherwise, we assume that the first r paths in $\mathbf{P}$ are formed by assets (excluding A_i and A_k) influenced by only one asset, and the remaining $s - r$ paths include at least one asset simultaneously influenced by two or more assets. Then, for the r first paths, we proceed as in *(A)*, and we denote by $\mathbf{S}$ the set including the $s - r$ remaining paths. We proceed with $\mathbf{S}$ as follows:

(i) Consider the set of non-terminal assets in $\mathbf{S}$ influenced by two or more assets, denoted by *I*, and the subset of *I* including assets uninfluenced by any other asset in *I*, denoted by *NI*.
(ii) We consider an asset A_r in *NI*. Then, we simplify the paths in $\mathbf{S}$ that include asset A_r making $A_i \to A_r \to \ldots \to A_k$, with $\widetilde{d}(A_i, A_r) = \widetilde{D}(A_i, A_r)$ (computed as in A).
(iii) Remove repeated paths from $\mathbf{S}$ and keep only one instance.
(iv) Build *I* and *NI* again from $\mathbf{S}$.
(v) If *NI* is not empty, go to (ii). Otherwise, the algorithm finishes.

Let us denote the resulting set of paths by $\mathbf{S} = \{P'_1, \ldots, P'_m\}$ with $m \leq s - r$. Then, the degree of dependency of A_k regarding A_i is

$$\widetilde{D}(A_i, A_k) = \bigoplus_{j=1}^{r} \widetilde{D}(A_i, A_k|P_j) \bigoplus_{l=1}^{m} \widetilde{D}(A_i, A_k|P'_l). \tag{2}$$

Once we have computed the degree of indirect dependency between all assets regarding the terminal assets, we can compute the accumulated values for non-terminal assets $\widetilde{v}_l$. These values usually have three components (ISO/IEC serie 27000):

1. *Availability.* How much damage would it cause if the asset is not available or cannot be used? This is a typical services inspection.
2. *Confidentiality.* How much damage would it cause if the asset is disclosed to someone it should not be? This is a typical data inspection.
3. *Integrity.* How much damage would it cause if the asset is damaged or corrupt? This a typical data inspection. Data can be manipulated, be wholly or partially false, or even missing.

Therefore,

$$\widetilde{v}_{i_{(l)}} = \sum_{k=1}^{n} ((\widetilde{D}(A_i, A_k) \otimes \widetilde{v}_{k_{(l)}}) \tag{3}$$

where l denotes the lth component.

Once assets have been valueted, the next step in the risk analysis methodology is to identify possible threats and compute the corresponding impact and risk indicators for the IS.

Threats are characterized by how often the threat materializes (*frequency*) $\widetilde{f}$ and by the *degradation* $\boldsymbol{D} = (\widetilde{d}_1, \widetilde{d}_2, \widetilde{d}_3)$ that the threat can cause to the three asset components. Note again that the frequency and degradation levels will be selected by the expert from the linguistic term scale and, consequently, a trapezoidal fuzzy number will be associated with each of them.

Then, the *impact* of a threat on an asset A_i is

$$\widetilde{I}_{i_{(l)}} = \widetilde{d}_l \otimes \widetilde{v}_{i_{(l)}}, \tag{4}$$

and the *risk* to the asset is

$$\widetilde{R}_{i_{(l)}} = \widetilde{I}_{i_{(l)}} \otimes \widetilde{f}. \tag{5}$$

The results of these operations will be fuzzy numbers belonging to TF[0, 1], which, generally, do not match up with the fuzzy numbers associated with the linguistic terms of the scale. Thus, a similarity function must be used to identify the most similar trapezoidal fuzzy number in the linguistic term scale to the fuzzy number output from computations.

Different similarity functions have been proposed by several authors [3,4,7,18,19]. In [16] a new similarity function was proposed on the basis of the geometric distance between both fuzzy numbers, the distance between their centroids and/or the ratio between the common area and the joint area under the membership functions.

Following the risk analysis and management methodologies for IS, Sect. 2 deals with the selection of safeguards that can be enforced to reduce the transmission probability of a failure throughout the IS. The aim is to minimize costs while keeping the risk at acceptable levels. To do this, we propose a mixed technique based on dynamic programming and metaheuristics, specifically, simulated annealing.

2 Selection of Preventive Safeguards

From Eqs. (3), (4) and (5) and the algorithm for computing degrees of indirect dependency, we can derive the risk for the IS in each component l given a threat with frequency $\widetilde{f}$ and degradation $\boldsymbol{D} = (\widetilde{d}_1, \widetilde{d}_2, \widetilde{d}_3)$ in the support asset $\widetilde{A}_i$ as $\widetilde{R}_{i_{(l)}} = \sum_{k=1}^{n} \widetilde{DD}(A_i, A_k) \otimes \widetilde{v}_{k_{(l)}} \otimes \widetilde{f} \otimes \widetilde{d}_l$, $\widetilde{v}_{k_{(l)}}$ being the value (constant) assigned to the terminal asset $\widetilde{A}_k$ in the component l.

Safeguards are measures for addressing threats. They can be procedures, such as incident management and documentation; personnel policies, such as training and awareness of employees operating on the IS; technical solutions, such as identification and authentication mechanisms based on biometrics; or physical security measures of the facilities, such as temperature control systems.

These safeguards can be *preventive*, if they reduce the frequency of threats; or *palliative*, if they reduce the degradation caused by threats on assets [11]. As the degree of dependence between two assets is the transmission probability of failures, a special type of preventive safeguard is that which reduces dependencies between support and terminal assets.

In this section we propose a method for reducing the degrees of dependency from all support assets to terminal assets minimizing the costs for the company.

As mentioned above, the probability of transmission of failure $\widetilde{D}(A_i, A_k)$ is the result of fuzzy operations with the probabilities of transmission of failure through intermediate assets linking the attacked support asset with other asset.

In each of these intermediate assets, safeguards can be enforced to reduce the probability of transmission of a failure. The effect induced for a safeguard in the probability of transmission of failures between two assets A_u and A_v can also be defined as a linguistic term, which is represented by a fuzzy number $\widetilde{e}^{u,v} \in TF[0,1]$, so that if the degree of direct dependency between the assets A_u and A_v is $\widetilde{d}(A_u, A_v)$, then, when we implement a safeguard with effect $\widetilde{e}^{u,v}$, the degree of direct dependency is reduced to $\widetilde{d}(A_u, A_v) \otimes (\widetilde{1} \ominus \widetilde{e}^{u,v})$, where $\ominus$ denotes the usual subtraction operation between trapezoidal fuzzy numbers, i.e., $(a_1, a_2, a_3, a_4) \ominus (b_1, b_2, b_3, b_4) = (a_1 - b_4, a_2 - b_3, a_3 - b_2, a_4 - b_1)$.

Note that $\ominus$ is not an internal composition law in TF[0, 1], however,

- $\widetilde{A}, \widetilde{B} \in TF[0,1] \Rightarrow \widetilde{A} \otimes (\widetilde{1} \ominus \widetilde{B}) \in TF[0,1]$,
- $\widetilde{A} \otimes (\widetilde{1} \ominus \widetilde{B}) \leq \widetilde{A}$ with the partial order of the trapezoidal fuzzy numbers (i.e., $\widetilde{A} \leq \widetilde{B} \Leftrightarrow a_1 \leq b_1, a_2 \leq b_2, a_3 \leq b_3, a_4 \leq b_4$) and
- $\widetilde{A} \otimes (\widetilde{1} \ominus \widetilde{B})$ decreases with $\widetilde{B}$.

We consider the set of safeguards that hinder the direct transmission of failure between A_u and A_v, $S^{u,v}$. Each safeguard $S_p^{u,v} \in S^{u,v}$ has a monetary cost $c_p^{u,v}$ and an effect $\widetilde{e}_p^{u,v}$ over $\widetilde{d}(A_u, A_v)$, which is reduced to $\widetilde{d}(A_u, A_v) \otimes (\widetilde{1} \ominus \widetilde{e}_p^{u,v})$.

The problem of keeping an acceptable level (low or very low) for the failure transmission probabilities among support and terminal assets with minimal costs can be represented as follows:

$$\begin{aligned} min \quad & \sum_{u,v}\sum_{p} c_p^{u,v} x_p^{u,v} \\ s.t. \quad & \widetilde{D}(A_i, A_k) \leq \widetilde{U}_{ik} \; \forall i, k, \\ & x_p^{u,v} \in \{0,1\}, \forall u, v, p \end{aligned}$$

where i and k in the first set of constraints refer to non-terminal and terminal assets, respectively, $\widetilde{U}_{ik}$ is a residual value accepted by the experts, $x_p^{u,v}$ are the decision variables ($x_p^{u,v} = 1$ means that safeguard $S_p^{u,v}$ is selected), and $\widetilde{D}(A_i, A_k)$ is reassessed replacing values $\widetilde{d}(A_u, A_v)$ by the affected values regarding the selected safeguards $\widetilde{d}(A_u, A_v) \otimes [\bigotimes_p (\widetilde{1} \ominus \widetilde{e}_p^{u,v})]$, where A_u and A_v are two consecutive assets connected by an arc in some path between A_i and A_k.

Note that the fact that the usual order in $TF[0,1]$ is a partial order constitutes a very restrictive constraint in our optimization problem, so we will use the concept of similarity function to relax this constraint.

If we define a threshold $\alpha \in [0,1]$ and a similarity function S, the constraint $\widetilde{D}(A_i, A_k) \leq \widetilde{U}_{ik} \; \forall i, k$ can be replaced by $S(\widetilde{D}(A_i, A_k), \widetilde{U}_{ik}) \geq \alpha$. Thus, the restrictiveness of the constraint increases proportionally to the threshold value and the feasible solution set will be composed of solutions that verify these softened/relaxed constraints.

Remember that indirect dependencies are recursively computed following the algorithm described in Sect. 1. Thus, the degree of dependency of the support assets further away from the terminals can be computed from the degree of dependency of the closest assets. Therefore, the problem can be solved in stages, and the principle of optimality in dynamic programming is verified: Given an optimal sequence of decisions, every subsequence is, in turn, optimal. Then we proceed as follows:

- Let L_0 be the set of terminal assets.
- Consider L_1 including support assets whose children belong to L_0 only (L_1 is not empty because the graph is acyclic). Identify safeguards that minimize costs keeping the degrees of dependency over their children at an acceptable level.
- Consider L_2 including support assets whose children belong to $L_0 \cup L_1$ only. Identify safeguards that minimize costs keeping the degrees of dependency over L_0 under an acceptable level. Note that the degrees of indirect dependency from the children of L_2 to terminal assets have already been computed in the previous stage, so we just need to identify the direct degree of dependency over assets in $L_0 \cup L_1$.
- Consider L_i including support assets whose children belong to $L_0 \cup L_1 \cup \ldots \cup L_{i-1}$ only. Identify safeguards that minimize costs keeping the degrees of dependency over L_0 under an acceptable level. Note that again we just need to identify the direct degree of dependency on assets of $L_0 \cup L_1 \cup \ldots \cup L_{i-1}$.

Simulated annealing [1,10] is applied in each step of the algorithm to derive the optimal selection of safeguards. It is a trajectorial metaheuristic which is named for and inspired by annealing in metallurgy.

An initial feasible solution is randomly generated. In each iteration a new solution y is randomly generated from the neighborhood of the current solution, $y \in N(x_i)$. If the new solution is better than the current one, then the algorithm moves to that solution ($x_{i+1} = y$), otherwise the movement to the worst solution is performed with certain probability. Note that accepting worse solutions allows for a more extensive search for the optimal solution and avoids trapping in local optima in early iterations. The probability of accepting a worse movement is a function of both the temperature factor and the change in the cost function. The initial value of temperature (T) is high, which leads to a diversified search, since practically all movements are allowed. As the temperature decreases, the probability of accepting a worse movement falls. If the temperature is zero, then only better movements will be accepted, which makes simulated annealing work like hill climbing.

The pseudocode of simulated annealing for a minimization problem is:

- Generate an initial feasible solution x_0. Do $x^* = x_0$, $f^* = f(x_0)$, $i = 0$. Select the initial temperature $t_0 = T$ (t_i temperature in the step i)
- Repeat until stopping criterion is satisfied:
  - Randomly generate $y \in N(x_i)$
    - If $f(y) - f(x_i) \leq 0$, then

  - $x_{i+1} = y$
  - If $f(x^*) > f(y)$, then $x^* = y, f^* = f(y)$
* Else
  - $p \sim U(0,1)$
  - If $p \leq e^{-(f(y)-f(x_i))/t_i}$, then $x_{i+1} = y$
  - Else $x_{i+1} = x_i$

- $i = i + 1$
- Update temperature

3 An Illustrative Example

Let us consider the IS shown in Fig. 1 with the direct degrees of dependency assessed by the experts considering the linguistic terms of Table 1, which has only one terminal asset, A_6.

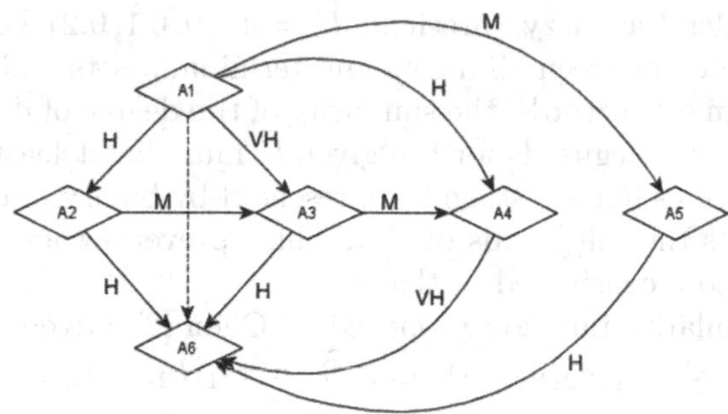

Fig. 1. Direct dependencies in the IS.

The set of paths in the analysis of the influence of A_1 over A_6 is $\mathbf{P} = \{P_1 : (A_1 \rightarrow A_2 \rightarrow A_6), P_2 : (A_1 \rightarrow A_2 \rightarrow A_3 \rightarrow A_6), P_3 : (A_1 \rightarrow A_2 \rightarrow A_3 \rightarrow A_4 \rightarrow A_6), P_4 : (A_1 \rightarrow A_3 \rightarrow A_6), P_5 : (A_1 \rightarrow A_3 \rightarrow A_4 \rightarrow A_6), P_6 : (A_1 \rightarrow A_4 \rightarrow A_6), P_7 : (A_1 \rightarrow A_5 \rightarrow A_6)\}$. Asset A_3 is influenced by A_1 and A_2, and A_4 is influenced by A_1 and A_3. Therefore, we proceed as in (B) of the algorithm described in Sect. 2, with $r = 2$ and $\mathbf{S} = \{P_2, P_3, P_4, P_5, P_6\}$:

(i) $I = \{A_3, A_4\}$ and $NI = \{A_3\}$.
(ii) Select A_3, then simplify paths P_2, P_3, P_4 and P_5 to $P_2' : (A_1 \rightarrow A_3 \rightarrow A_6)$, $P_3' : (A_1 \rightarrow A_3 \rightarrow A_4 \rightarrow A_6)$, $P_4' : (A_1 \rightarrow A_3 \rightarrow A_6)$ and $P_5' : (A_1 \rightarrow A_3 \rightarrow A_4 \rightarrow A_6)$, respectively, with $\widetilde{d}(A_1, A_3) = \widetilde{D}(A_1, A_3) = \left(\widetilde{d}(A_1, A_2) \otimes \widetilde{d}(A_2, A_3)\right) \oplus \widetilde{d}(A_1, A_3)$.
(iii) $\mathbf{S} = \{P_2', P_3', P_6\}$ since $P_2' = P_4'$ and $P_3' = P_5'$.
(iv) $I = \{A_4\}$ and $NI = \{A_4\}$.
(v) Go to (ii).
(ii) Select A_4, then simplify paths P_3' and P_6 to $P_3'' : (A_1 \rightarrow A_4 \rightarrow A_6)$, and $P_6' : (A_1 \rightarrow A_4 \rightarrow A_6)$, respectively, with $\widetilde{d}(A_1, A_4) = \widetilde{D}(A_1, A_4) = \left(\widetilde{d}(A_1, A_3) \otimes \widetilde{d}(A_3, A_4)\right) \oplus \widetilde{d}(A_1, A_4)$.

(iii) $\mathbf{S} = \{P_2', P_3''\}$ since $P_3'' \equiv P_6'$.
(iv) $I = \varnothing$ y $NI = \varnothing$.
(v) The algorithm finishes since $NI = \varnothing$.

Finally, $\mathbf{S} = \{P_2', P_3''\}$ and the degree of dependency of A_6 regarding A_1 is $\widetilde{D}(A_1, A_6) = \widetilde{D}(A_1, A_6|P_1) \oplus \widetilde{D}(A_1, A_6|P_7) \oplus \widetilde{D}(A_1, A_6|P_2') \oplus \widetilde{D}(A_1, A_6|P_3'') = (\widetilde{d}(A_1, A_2) \otimes \widetilde{d}(A_2, A_6)) \oplus (\widetilde{d}(A_1, A_5) \otimes \widetilde{d}(A_5, A_6)) \oplus (\widetilde{d}(A_1, A_3) \otimes \widetilde{d}(A_3, A_6)) \oplus (\widetilde{d}(A_1, A_4) \otimes \widetilde{d}(A_4, A_6))$.

The degree of dependency of A_6 regarding A_1 is $D(A_1, A_6) = (0.98, 0.99, 0.99, 1)$ if we consider the linguistic terms of Table 1 show in Fig. 1.

Let us consider a threat on asset A_1 with frequency $\widetilde{f} = M$ and degradation $\widetilde{d} = (H, H, H)$, then the risk to A_1 is $\widetilde{R}_{1_{(l)}} = (0.23, 0.415, 0.485, 0.675)$, $l = 1, 2, 3$.

We consider the asset network and the fuzzy direct dependencies shown in Fig. 1 corresponding to an IS. Besides, the set of available safeguards of failure transmission between support assets are shown in Table 2.

We also consider the fuzzy threshold $\widetilde{U} = (0, 0, 0.1, 0.2)$ below which the degree of dependency between all assets and terminal assets will be acceptable, and let $\alpha = 0.95$. In other words, the similarity of the degree of dependency after applying the selected safeguards for the given $\widetilde{U}$ must be at least 0.95.

The set of solutions in each stage is represented by binary matrices, in which each row represents the safeguards of S_p^{uv}, which prevents the failure transmission from asset u to v considered in that stage.

We use the similarity function proposed by Chen [2]: Given two trapezoidal fuzzy numbers $\widetilde{A} = (a_1, a_2, a_3, a_4)$ and $\widetilde{B} = (b_1, b_2, b_3, b_4)$, $S(\widetilde{A}, \widetilde{B}) = 1 - \sum_{i=1}^{4} \frac{|a_i - b_i|}{4}$. Although other similarity functions have been proposed in the literature [3,4,7,9,13,16,18,19], we have decided to use the geometric distance between both fuzzy numbers due to its low computational cost.

Dynamic programming is then executed as follows. First, note that $L_0 = \{A_6\}$, since the only terminal asset in the IS in Fig. 1 is A_6.

- Stage 1: $L_1 = \{A_4, A_5\}$. We adjust the degrees of dependency $\widetilde{D}(A_4, A_6) = \widetilde{d}(A_4, A_6) \otimes [\bigotimes_p^{10}(\widetilde{1} \ominus \widetilde{e}_p^{4,6} x_p^{4,6})] = VH \otimes [\bigotimes_p^{10}(\widetilde{1} \ominus \widetilde{e}_p^{4,6} x_p^{4,6})]$ and $\widetilde{D}(A_5, A_6) = \widetilde{d}(A_5, A_6) \otimes [\bigotimes_p^{15}(\widetilde{1} \ominus \widetilde{e}_p^{5,6} x_p^{5,6})] = H \otimes [\bigotimes_p^{15}(\widetilde{1} \ominus \widetilde{e}_p^{5,6} x_p^{5,6})]$, such that $S(\widetilde{D}(A_4, A_6), \widetilde{U}) \geq 0.95$ and $S(\widetilde{D}(A_5, A_6), \widetilde{U}) \geq 0.95$, $\widetilde{e}_p^{4,6}$ being the effect induced for the safeguard $S_p^{4,6}, p = 1, \ldots, 10$, $\widetilde{e}_p^{5,6}$ the effect induced for the safeguard $S_p^{5,6}, p = 1, \ldots, 15$, $x_p^{4,6} = 1$ or 0 if the safeguard $S_p^{4,6}, p = 1, \ldots, 10$, is selected or not, respectively, and $x_p^{5,6} = 1$ or 0 depending on whether or not the safeguard $S_p^{5,6}, p = 1, \ldots, 15$, minimizing the cost.
  As L_1 contains two elements, two optimization problems must be solved in this stage, associated with A_4 and A_5, respectively.
  Regarding asset A_4, solutions are represented by the vector $x^{4,6} = (x_1^{4,6}, \ldots, x_{10}^{4,6})$, see Table 2, where $x_p^{4,6} = 1$ if the safeguard $S_p^{4,6}$ is selected. The respective optimization problem to be solved using simulated annealing is:

$$\begin{aligned} min \quad & c_1^{4,6} x_1^{4,6} + \ldots + c_{10}^{4,6} x_{10}^{4,6} \\ s.t. \quad & S\left(\widetilde{D}(A_4, A_6), \widetilde{U}\right) \geq 0.95 \ . \\ & x_p^{4,6} \in \{0,1\}, p = 1, \ldots, 10 \end{aligned} \tag{6}$$

Table 2. Safeguards for A_1, A_2, A_3, A_4 and A_5.

Tag	Effect	Cost	Tag	Effect	Cost	Tag	Effect	Cost	Tag	Effect	Cost	Tag	Effect	Cost
$S_1^{1,2}$	L	100	$S_1^{1,3}$	MH	356	$S_1^{1,4}$	M	209	$S_1^{1,5}$	M	230	$S_1^{2,3}$	M	356
$S_2^{1,2}$	M	300	$S_2^{1,3}$	H	324	$S_2^{1,4}$	M	267	$S_2^{1,5}$	M	345	$S_2^{2,3}$	L	87
$S_3^{1,2}$	MH	550	$S_3^{1,3}$	L	110	$S_3^{1,4}$	MH	342	$S_3^{1,5}$	L	187	$S_3^{2,3}$	ML	267
$S_4^{1,2}$	M	430	$S_4^{1,3}$	ML	345	$S_4^{1,4}$	VH	789	$S_4^{1,5}$	M	321	$S_4^{2,3}$	M	320
$S_5^{1,2}$	ML	125	$S_5^{1,3}$	VL	87	$S_5^{1,4}$	M	234	$S_5^{1,5}$	MH	345	$S_5^{2,3}$	ML	156
$S_6^{1,2}$	L	240	$S_6^{1,3}$	MH	345	$S_6^{1,4}$	M	356	$S_6^{1,5}$	H	543	$S_6^{2,3}$	M	320
$S_7^{1,2}$	VL	100	$S_7^{1,3}$	M	200	$S_7^{1,4}$	M	276	$S_7^{1,5}$	MH	356	$S_7^{2,3}$	M	256
$S_8^{1,2}$	MH	324				$S_8^{1,4}$	M	200	$S_8^{1,5}$	M	206	$S_8^{2,3}$	M	300
$S_9^{1,2}$	VH	570				$S_9^{1,4}$	H	467	$S_9^{1,5}$	M	342	$S_9^{2,3}$	L	200
						$S_{10}^{1,4}$	H	342						
						$S_{11}^{1,4}$	L	127						
						$S_{12}^{1,4}$	M	207						
$S_1^{2,6}$	M	348	$S_1^{3,4}$	M	345	$S_1^{3,6}$	M	267	$S_1^{4,6}$	M	260	$S_1^{5,6}$	M	200
$S_2^{2,6}$	L	187	$S_2^{3,4}$	H	650	$S_2^{3,6}$	M	356	$S_2^{4,6}$	M	245	$S_2^{5,6}$	M	210
$S_3^{2,6}$	ML	254	$S_3^{3,4}$	M	200	$S_3^{3,6}$	M	378	$S_3^{4,6}$	ML	170	$S_3^{5,6}$	L	120
$S_4^{2,6}$	ML	367	$S_4^{3,4}$	M	367	$S_4^{3,6}$	M	324	$S_4^{4,6}$	M	256	$S_4^{5,6}$	ML	234
$S_5^{2,6}$	ML	567	$S_5^{3,4}$	M	388	$S_5^{3,6}$	M	345	$S_5^{4,6}$	M	367	$S_5^{5,6}$	M	267
$S_6^{2,6}$	M	390	$S_6^{3,4}$	H	453	$S_6^{3,6}$	M	231	$S_6^{4,6}$	M	289	$S_6^{5,6}$	MH	367
$S_7^{2,6}$	ML	256	$S_7^{3,4}$	L	189	$S_7^{3,6}$	MH	453	$S_7^{4,6}$	M	278	$S_7^{5,6}$	MH	366
$S_8^{2,6}$	M	307	$S_8^{3,4}$	L	256				$S_8^{4,6}$	M	345	$S_8^{5,6}$	M	254
$S_9^{2,6}$	L	235	$S_9^{3,4}$	M	345				$S_9^{4,6}$	M	240	$S_9^{5,6}$	ML	145
$S_{10}^{2,6}$	ML	124							$S_{10}^{4,6}$	MH	435	$S_{10}^{5,6}$	L	206
$S_{11}^{2,6}$	M	400										$S_{11}^{5,6}$	M	306
$S_{12}^{2,6}$	L	278										$S_{12}^{5,6}$	M	345
$S_{13}^{2,6}$	ML	260										$S_{13}^{5,6}$	M	280
												$S_{14}^{5,6}$	L	178
												$S_{15}^{5,6}$	MH	377

Table 3. Optimal solutions and costs for each asset.

Asset	Solution	Cost
A_5	$S_1^{5,6}$, $S_7^{5,6}$, $S_9^{5,6}$	711
A_4	$S_2^{4,6}$, $S_3^{4,6}$, $S_4^{4,6}$,$S_9^{4,6}$	911
A_3	$S_1^{3,6}$, $S_4^{3,6}$, $S_6^{3,6}$, $S_7^{3,6}$	1275
A_2	$S_7^{2,3}$, $S_1^{2,6}$, $S_5^{2,6}$, $S_7^{2,6}$,$S_{10}^{3,6}$	1551
A_1	$S_1^{1,2}$, $S_2^{1,3}$, $S_{10}^{1,4}$	1236
	Total cost	5684

The optimal solution and the associated costs are shown in the second row of Table 3, corresponding to vector $x^{4,6^*} = (0,1,1,1,0,0,0,0,1,0)$.
Regarding asset A_5, solutions are now represented by the vector $x^{5,6} = (x_1^{5,6}, x_2^{5,6}, \ldots, x_{15}^{5,6})$, see Table 2. The optimization problem to be solved is:

$$\begin{aligned} min \quad & c_1^{5,6} x_1^{5,6} + ... + c_{15}^{5,6} x_{15}^{5,6} \\ s.t. \quad & S\left(\widetilde{D}(A_5, A_6), \widetilde{U}\right) \geq 0.95\,. \\ & x_p^{5,6} \in \{0,1\}\,, p = 1, ..., 15 \end{aligned} \tag{7}$$

The optimal solution and the associated costs are shown in the first row of Table 3, corresponding to vector $x^{5,6^*} = (1,0,0,0,0,0,1,0,1,0,\ 0,0,0,0,0)$.

The new degrees of dependency after the application of the selected safeguards and the respective similarity to the fixed threshold, $\widetilde{U}$, are shown in the first two rows of Table 4.

The purpose of this paper is to describe how a mixture of dynamic programming techniques and metaheuristics can efficiently solve the problem and not to detail or compare the applied metaheuristic (simulated annealing) with others. However, we do think it is worthwhile to describe some parameters used in the implementation and to report a sensitivity analysis analyzing the effects caused by the changes to these parameters.

Table 4. New degrees of dependency after applying safeguards.

Asset	$\widetilde{D}(A_j, A_6)$	Similarity $\widetilde{U}$
A_5	$(0.015, 0.077, 0.114, 0.280)$	0.953
A_4	$(0.016, 0.072, 0.104, 0.269)$	0.959
A_3	$(0.008, 0.059, 0.096, 0.301)$	0.956
A_2	$(0.008, 0.057, 0.094, 0.316)$	0.953
A_1	(0.005,0.045,0.082,0.327)	0.951

- We randomly generate a sequence with binary values and check if the similarity constraint is verified to derive the initial solution. The length of the binary sequence depends on the problem (15 when dealing with $x^{5,6}$, 10 when dealing with $x^{4,6}\ldots$).
- The neighborhood of a solution is composed of any solutions that can be derived by changing the value of one of the binary elements of the solution, selected at random. If the resulting solution is not feasible, then it is discarded and another solution is generated in the neighborhood until a feasible solution is found.
- The initial temperature assures acceptance probabilities of worse solutions close to 0.9 in the initial iterations of the algorithm. The initial temperature is computed to obtain a high probability of acceptance ($\geq$0.9) of any neighbor of the initial solution, i.e., given the initial solution x_0, the minimum value T is computed such that $e^{-(f(y)-f(x_0))/T} \geq 0.9$, $\forall y \in N(x_0)$ and feasible, with

$f(y) - f(x_0) > 0$. In other words, $T = \max_{\substack{y \in N(x_0) \\ feasible}} \left\{ \frac{-(f(y)-f(x_0))}{ln(0.9)} \right\}$ because if we have $T \geq \frac{-(f(y)-f(x_0))}{ln(0.9)}$ $\forall y \in N(x_0)$ and feasible, with$f(y) - f(x_0) > 0$, then $ln(0.9) \leq \frac{-(f(y)-f(x_0))}{T}$ $\forall y \in N(x_0)$ and feasible, with $f(y) - f(x_0) > 0$, and since e^x is an increasing function, $0.9 \leq e^{\frac{-(f(y)-f(x_0))}{T}}$ $\forall y \in N(x_0)$ and feasible, with $f(y) - f(x_0) > 0$.
The pseudocode, starting from $x_0 = (x_0[1], \ldots, x_0[n])$, as follows:

* $y = x_0$, $T = 0$, $i = 1$.
* While $i \leq n$. Do $y[i] = 1 - y[i]$.
  · If y is a feasible solution then, if $\frac{-(f(y)-f(x_0))}{ln(0.9)} > T$, we have $T = \frac{-(f(y)-f(x_0))}{ln(0.9)}$
  · $y = x_0$, $i = i + 1$.

The solution x_0 has at most n feasible neighboring solutions. We have evaluated all neighboring solutions that are worse than the initial solution in those n steps. In the unfortunate event that the initial solution is the worst of its neighborhood, the initial value of the resulting T is null. Therefore we must start from another initial solution. This does not degrade the algorithm, because it can return to the neighborhood of the discarded solution at any time.
Thus the initial temperature that leads to the optimal solution over A_5 (for optimization problem (7)) is 3578. The temperature is maintained constant for $L = 20$ iterations and then it decreases after multiplying by 0.95, so that, after $h * L$ iterations, the temperature is $t_{h*L} = 0.95^h t_0$.

- The algorithm stops if f has not improved in the last 100 iterations.

Table 5 shows the best solutions reached after running the algorithm with different values for α to minimize $\widetilde{D}(A_5, A_6)$. Note that if the constraint is more restrictive, allowing only minor differences with the threshold $\widetilde{U}$, the set of safeguards for implementation will be larger. The same effect occurs when we use a more accurate (with a smaller support) threshold $\widetilde{U}$. Therefore, experts must choose lower or higher levels of acceptable accuracy regarding the dependency between assets, i.e., the accepted risk considering this fact.

Table 5. $\widetilde{D}(A_5, A_6)$ and associated costs for different α levels.

α	$\widetilde{D}(A_5, A_6)$	Similarity	Cost
0.8	(0.05,0.23,0.27,0.46)	0.81	554
0.9	(0.02,0.09,0.12,0.30)	0.93	653
0.95	(0.01,0.07,0.11,0.28)	0.95	711
0.98	(0.00,0.03,0.06, 0.20)	0.98	1021

- Stage 2: $L_2 = \{A_3\}$. The degrees of dependency $\widetilde{d}(A_3, A_6)$ and $\widetilde{d}(A_3, A_4)$ are adjusted by minimizing costs and incorporating the soft constraint $S(\widetilde{D}(A_3, A_6), \widetilde{U}) \geq 0.95$, where $\widetilde{D}(A_3, A_6) = [\widetilde{d}(A_3, A_6) \otimes (\bigotimes_{p=1}^{7} (\widetilde{1} \ominus \widetilde{e}_p^{3,6} x_p^{3,6}))] \oplus [\widetilde{d}(A_3, A_4) \otimes (\bigotimes_{p=1}^{9} (\widetilde{1} \ominus \widetilde{e}_p^{3,4} x_p^{3,4})) \otimes \widetilde{D}(A_4, A_6)]$. Note that $\widetilde{D}(A_4, A_6)$ was computed in Stage 1, $\widetilde{D}(A_4, A_6) = VH \otimes [(\widetilde{1} \ominus \widetilde{e}_2^{4,6}) \otimes (\widetilde{1} \ominus \widetilde{e}_3^{4,6}) \otimes (\widetilde{1} \ominus \widetilde{e}_4^{4,6}) \otimes (\widetilde{1} \ominus \widetilde{e}_9^{4,6})] = (0.016, 0.072, 0.104, 0.269)$. The optimization problem to be solved in this stage is

$$\begin{array}{ll} min\ c_1^{3,6} x_1^{3,6} + \ldots + c_7^{3,6} x_7^{3,6} + c_1^{3,4} x_1^{3,4} + \ldots + c_9^{3,4} x_9^{3,4} & \\ s.t. \qquad S(\widetilde{D}(A_3, A_6), \widetilde{U}) \geq 0.95 & . \\ \qquad x_p^{3,6}, x_q^{3,4} \in \{0,1\}, p = 1, \ldots, 7, q = 1, \ldots, 9 & \end{array}$$

The optimal solution and the associated cost is shown in the third row of Table 3, corresponding to vectors $x^{3,6^*} = (1,0,0,1,0,1,1)$ and $x^{3,4^*} = (0,0,0,0,0,0,0,0,0)$. The new degree of dependency after the application of the selected safeguards and the corresponding similarity to the fixed threshold, $\widetilde{U}$, are shown in the third row of Table 4.

- Stage 3: $L_3 = \{A_2\}$. The degrees of dependency $\widetilde{d}(A_2, A_3)$ and $\widetilde{d}(A_2, A_6)$ are adjusted minimizing costs and incorporating the soft constraint $S(\widetilde{D}(A_2, A_6), \widetilde{U}) \geq 0.95$, where $\widetilde{D}(A_2, A_6) = [\widetilde{d}(A_2, A_6) \otimes (\bigotimes_{p=1}^{13} (\widetilde{1} \ominus \widetilde{e}_p^{2,6} x_p^{2,6}))] \oplus [\widetilde{d}(A_2, A_3) \otimes (\bigotimes_{p=1}^{7} (\widetilde{1} \ominus \widetilde{e}_p^{2,3} x_p^{2,3})) \otimes \widetilde{D}(A_3, A_6)]$. Note that $\widetilde{D}(A_3, A_6)$ was computed in Stage 2, $\widetilde{D}(A_3, A_6) = [\widetilde{d}(A_3, A_6) \otimes (\widetilde{1} \ominus \widetilde{e}_1^{3,6}) \otimes (\widetilde{1} \ominus \widetilde{e}_4^{3,6}) \otimes (\widetilde{1} \ominus \widetilde{e}_6^{3,6}) \otimes (\widetilde{1} \ominus \widetilde{e}_7^{3,6})] \oplus [\widetilde{d}(A_3, A_4)) \otimes \widetilde{D}(A_4, A_6)] = (0.008, 0.059, 0.096, 0.301)$.

The optimal solution and the associated cost are shown in the fourth row of Table 3, corresponding to vectors $x^{2,3^*} = (0,0,0,0,0,0,1,0,0)$ and $x^{2,6^*} = (1,0,0,0,1,0,1,0,0,1,0,0,0)$. The new degree of dependency and similarity to $\widetilde{U}$, are shown in the fourth row of Table 4.

- Finally, $L_4 = \{A_1\}$. The degrees of dependency $\widetilde{d}(A_1, A_2)$, $\widetilde{d}(A_1, A_3)$, $\widetilde{d}(A_1, A_4)$ and $\widetilde{d}(A_1, A_5)$ are adjusted minimizing the cost and considering the soft constraint $S(\widetilde{D}(A_1, A_6), \widetilde{U}) \geq 0.95$, where $\widetilde{D}(A_1, A_6) = [\widetilde{d}(A_1, A_2) \otimes (\bigotimes_{p=1}^{9} (\widetilde{1} \ominus \widetilde{e}_p^{1,2} x_p^{1,2})) \otimes \widetilde{D}(A_2, A_6)] \oplus [\widetilde{d}(A_1, A_3) \otimes (\bigotimes_{p=1}^{7} (\widetilde{1} \ominus \widetilde{e}_p^{1,3} x_p^{1,3}) \otimes \widetilde{D}(A_3, A_6)] \oplus [\widetilde{d}(A_1, A_4) \otimes (\bigotimes_{p=1}^{12} (\widetilde{1} \ominus \widetilde{e}_p^{1,4} x_p^{1,4})) \otimes \widetilde{D}(A_4, A_6)] \oplus [\widetilde{d}(A_1, A_5) \otimes (\bigotimes_{p=1}^{9} (\widetilde{1} \ominus \widetilde{e}_p^{1,5} x_p^{1,5})) \otimes \widetilde{D}(A_5, A_6)]$. Note that $\widetilde{D}(A_2, A_6)$, $\widetilde{D}(A_3, A_6)$, $\widetilde{D}(A_4, A_6)$ and $\widetilde{D}(A_5, A_6)$ were computed in previous stages, $\widetilde{D}(A_2, A_6) = [\widetilde{d}(A_2, A_6) \otimes ((\widetilde{1} \ominus \widetilde{e}_1^{2,6}) \otimes (\widetilde{1} \ominus \widetilde{e}_5^{2,6}) \otimes (\widetilde{1} \ominus \widetilde{e}_7^{2,6}) \otimes (\widetilde{1} \ominus \widetilde{e}_{10}^{2,6}))] \oplus [\widetilde{d}(A_2, A_3) \otimes \left((\widetilde{1} \ominus \widetilde{e}_7^{2,3})\right) \otimes \widetilde{D}(A_3, A_6)] = (0.008, 0.057, 0.094, 0.316)$, $\widetilde{D}(A_3, A_6) = (0.008, 0.059, 0.096, 0.301)$, $\widetilde{D}(A_4, A_6) = (0.016, 0.072, 0.104,$

0.269) and $\widetilde{D}(A_5, A_6) = (0.01, 0.07, 0.11, 0.28)$. The optimal solution in this stage is shown in the last row of Tables 3 and 4.

After implementing the best safeguards, the risk caused by the previously considered threat over asset A_1 in each component is $\widetilde{R}_{1_{(l)}} = (0.001, 0.018, 0.039, 0.22)$, $l = 1, 2, 3$. The risks associated with this threat before and after implementation of safeguards are illustrated along with the risk threshold in Fig. 2.

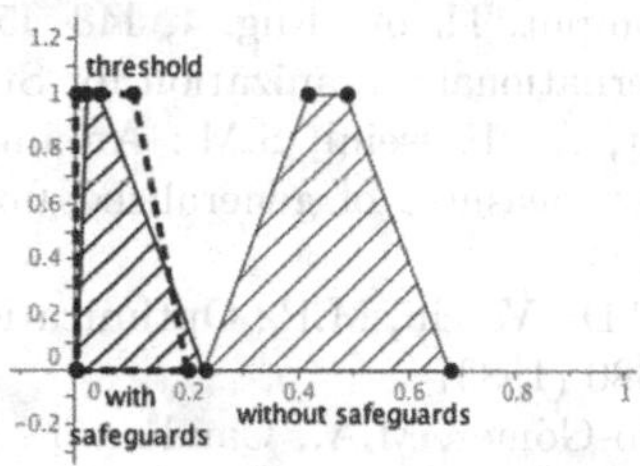

Fig. 2. Risk of A_1 before and after implementation of optimal safeguards.

4 Conclusions

We propose a model for selecting safeguards to reduce risks in information systems based on the reduction of the degree of dependency between support assets and terminal assets. As safeguards have associated costs, our aim is to select safeguards that minimize costs while keeping the risk with acceptable levels.

Although a metaheuristic could be used to solve this optimization problem, dynamic programming combined with simulated annealing was used because of the special structure of the constraint set. This leads to a more computationally efficient solution to the safeguard selection problem. Also the fuzzy environment allows experts to provide imprecise and vague failure propagation probabilities.

Another way to reduce system risk is to act on the probability of threats to each asset materializing or reducing the degradation of assets caused by threat materialization. This is a multiobjective problem (degradation has three components), which will be considered in future research.

Acknowledgements. The paper was supported by Madrid Government project S-2009/ESP-1685 and the Ministry of Science project MTM2011-28983-CO3-03.

References

1. Cerny, V.: Thermodynamical approach to the traveling salesman problem: an efficient simulation algorithm. J. Optim. Theory Appl. **45**, 41–51 (1985)
2. Chen, S.-M.: New methods for subjective mental workload assessment and fuzzy risk analysis. Cybern. Syst. **27**, 449–472 (1996)

3. Chen, S.-J., Chen, S.-M.: Fuzzy risk analysis based on similarity measures of generalized fuzzy numbers. IEEE Trans. Fuzzy Syst. **11**, 45–56 (2003)
4. Chen, S.-J., Chen, S.-M.: Fuzzy risk analysis based on the ranking of generalized trapezoidal fuzzy numbers. Appl. Intell. **26**, 1–11 (2007)
5. CCTA Risk Analysis and Management Method (CRAMM), Version 5.0. Central Computing and Telecommunications Agency (CCTA), London, 2003 (2009)
6. Finetti, B.: Foresight: its logical laws, its subjective sources. In: Kyburg, H.E., Smokler, H.E. (eds.) Studies in Subjective Probability. Wiley, New York (1964)
7. Gomathi, V.L., Sivaraman, G.: A novel similarity measure between generalized fuzzy numbers. Int. J. Comput. Theory Eng. **4**, 448–450 (2012)
8. ISO/IEC Serie 27000 International Organization for Standardization
9. Hejazi, S.R., Doostparast, A., Hosseini, S.M.: An improved fuzzy risk analysis based on a new similarity measures of generalized fuzzy numbers. Expert Syst. Appl. **38**, 9179–9185 (2011)
10. Kirkpatrick, S., Gelatt, C.D., Vecchi, M.P.: Optimization by simulated annealing. Science **220**(4598), 671–680 (1983)
11. López Crespo, F., Amutio-Gómez, M.A., Candau, J., Mañas, J.A.: Methodology for Information Systems Risk. Analysis and Management (MAGERIT Version 2). Book I, Book II and Book III. Ministerio de Administraciones Públicas, Madrid (2006)
12. Savage, L.J.: The Foundations of Statistics. Wiley, New York (1954)
13. Sridevi, B., Nadarajan, R.: Fuzzy similarity measure for generalized fuzzy numbers. Int. J. Open Probl. Comp. Sci. Math. **2**, 111–116 (2009)
14. Stoneburner, G., Gougen, A.: NIST 800–30 Risk Management. Guide for Information Technology Systems. National Institute of Standard and Technology, Gaithersburg (2002)
15. Vicente, E., Jiménez, A., Mateos, A.: A Fuzzy Approach to risk analysis in information systems. In: Proceedings of the 2nd International Conference on Operations Research and Enterprise Systems, pp. 130–133 (2013)
16. Vicente, E., Mateos, A., Jiménez, A.: A new similarity function for generalized trapezoidal fuzzy numbers. In: Rutkowski, L., Korytkowski, M., Scherer, R., Tadeusiewicz, R., Zadeh, L.A., Zurada, J.M. (eds.) ICAISC 2013, Part I. LNCS, vol. 7894, pp. 400–411. Springer, Heidelberg (2013)
17. Vicente, E., Jiménez, A., Mateos, A.: An interactive method of fuzzy probability elicitation in risk analysis. In: Intelligent Systems and Decision Making for Risk Analysis and Crisis Response, pp. 223–228. CRC Press, New York (2013)
18. Xu, Z., Shang, S., Qian, W., Shu, W.: A method for fuzzy risk analysis based on the new similarity of trapezoidal fuzzy numbers. Expert Syst. Appl. **37**, 1920–1927 (2010)
19. Zu, L., Xu, R.: Fuzzy risk analysis based on similarity measure of generalized fuzzy numbers. Springer, Berlin/Heidleberg (2012)

A Performance Improvement and Management Model for Small and Medium Sized Enterprises

Madani Alomar(✉) and Zbigniew J. Pasek

Industrial and Manufacturing Systems Engineering, University of Windsor, Windsor, Canada
{alomarm, zjpasek}@uwindsor.ca

Abstract. Manufacturers are faced with complex global challenges that contributed to significant changes in the business environment. These challenges drive business to continuously assess their performance and competitiveness. This paper proposes a model that will assist companies, particularly the small and medium-sized enterprises, assess their performance by prioritizing performance measures and selecting an adequate operations strategy under various market scenarios. The outlined model utilizes and integrates the Supply Chain Operations Reference framework and the Analytical Hierarchy Process approach to construct, link, and assess a multi-levels hierarchal structure. The model also assist small and medium-sized enterprises put more weight on supply chain attributes. The use and benefits of the proposed model are illustrated on a case of a family owned, medium-sized manufacturing enterprise.

Keywords: Small and medium sized enterprises · AHP · SCOR model · Supply chain strategy · Performance measurement · Expert choice

1 Introduction

Manufacturers today are faced with complex global challenges such as low cost competitors, fluctuating commodity prices, increasing customer expectations, and volatile economic conditions. The uncertainty associated with these factors has contributed on one hand to significant changes in the business environment resulting in tremendous growth and opportunities for new markets, and on the other hand in increased frequency and complexity of challenges that threaten the operations and survival of firms. These competitive pressures are driving manufacturing firms to continuously re-evaluate and adjust their competitive strategies, supply chains, and manufacturing technologies in order to improve performance, compete, and survive long term. Small and medium sized enterprises (SMEs) are much more vulnerable to these external pressures than larger companies, thus their responses often fall short, due to limited resources and capabilities (e.g., financial resources, managerial talent, and access to markets). Numerous studies have revealed that small businesses are extremely susceptible to failures; about 50 % of small businesses in Canada and 53 % in the United States fail to survive for more than five years [1]. Several research studies have linked the success of businesses to the type of performance measurement system (PMS) used by the firms and to the successful

E. Pinson et al. (Eds.): ICORES 2014, CCIS 509, pp. 79–94, 2015.
DOI: 10.1007/978-3-319-17509-6_6

design and implementation. Other researchers have considered strategic performance measurement system as a means to attain competitive advantage, continuous improvement and ability to successfully manage changes [2, 3]. Despite these results, several investigators found that many small enterprises predominantly emphasize financial index only [4–6] neglecting the others.

This paper proposes an approach methodology and a model that will assist SMEs in building a strategic and flexible performance measurement system that considers two types of supply chain strategies, and the supply chain performance attributes based on Supply Chain Operations Reference (SCOR) framework. The model relies on Analytical Hierarchy Process (AHP) approach to integrate various market scenarios, performance attributes and supply chain strategies into one comprehensive model. Unlike other previous works where the use of AHP and performance measures were mainly addressing the selection of best supplier, vendors, and markets or manufacturing departments, this work discusses the improvement of one enterprise performance under different market circumstances and the importance of different performance measures.

2 Performance Measurement Systems in SME

Performance measurement is at the core of a control and management system of an enterprise. It plays a key role in developing strategic plans and assessing organizational objectives. It is also important in assessing business ability to gain and sustain competitive advantage and directing corrective adjustments and actions as well [2]. Various researchers have linked the success of businesses to the type of performance measurement system used by them and to the successful design and implementation of the measurement system. Other researchers have considered strategic performance measurement system as means to attain competitive advantage, continuous improvement and ability to respond to internal and external changes [3]. In this sense, the performance measurement system is the instrument to support the decision-making either for launching, selecting actions or redefining objectives [7–9]. From a global perspective, performance measurement system as a multi-criteria instrument consists of a set of performance expressions or metrics [10].

The early generations of performance measurement models focused extensively on financial and accounting areas and completely ignored the operational and other non-financial issues. Currently, the new generation of performance measurement models makes a strong effort to be strategically oriented and to address other performance dimensions including combination of financial and non-financial areas [11]. Nevertheless, according to Tangen: “these new approaches have a good academic groundwork and are theoretically sound but they rarely help with the practical understanding of specific measures at an operational level”. This is considered a major obstacle in implementing multi-dimensional performance measurement system in small enterprises [12]. Other researchers have tied the failure of implementing existing performance measurement systems in small and medium sized enterprises to the following issues:

- Use of models or frameworks originally introduced for large enterprises, the one size fits all, leads to implementation failure [11].

- Improper use of well-known performance measurement models and frameworks [13].
- Informal approach to performance measurement models and frameworks (no rigorous plan or execution) [14].

Numbers of studies have revealed that many of the small and medium sized enterprises did not achieve the requirements of a strategic performance measurement system. For example, a study found that all companies under investigation had a surplus of financial measures, but their performance measurement systems were not derived from strategy, often unclear with complex or obsolete data, and historically focused on some outdated measures [4]. Another empirical survey conducted on 83 Danish enterprises [5] found that 50 % of these enterprises have either only one performance indicator such as cost or no performance indicators in place at all. An additional empirical study [6] revealed that majority of small and medium sized Canadian manufacturing firms continue using financial measures.

Despite the recommendations from industrial and academic experts, the proportion of firms that implement well-known performance measurement systems remains low. The results indicated that the types of performance measures used by the SMEs were rarely connected to strategy. The study also revealed that about 70 % of the companies failed to implement well-known strategic performance measurement models [6]. The majority of SMEs according to the previous studies use traditional management accounting systems. Nevertheless, the traditional management accounting systems and financial measures simply do not provide the richness of information that allows a company to remain competitive in today's market place [15] see also Table 1.

Table 1. Traditional versus no-traditional PMS.

Traditional performance measures	Non-traditional performance measures
Based on outdated traditional accounting system	Based on company strategy
Mainly financial measures	Mainly non-financial measures
Do not change over time	Change overtime as the needs change
Intended mainly for monitoring performance	Intended to improve performance
Not applicable for new advanced technology and methods, JIT, TQM	Applicable for new advances technology and methods: JIT, TQM
Ignoring continuous improvement	Support in achieving continuous improvement

It is necessary to understand that the metrics and the measures that are used in performance measurement system should have the power to capture the depth of organizational performance, the measures should reflect their clear relations with a range of levels of decision-making such as strategic, tactical, and operational, the metrics should reflect an acceptable balance between financial and non-financial measures, and the measurement system should ensure proper assignment of measures to the areas where they would be most suitable.

3 SME and the Challenges

Studies show that small and medium-sized enterprises are distinguished from larger firms by a number of key characteristics [4] such as personalized management with little delegation of authority, severe resource limitations in terms of skilled manpower, management and finance, and flexible structure, reactive or fire-fighting mentality, informal and dynamic strategies, dependency on small number of customers, limited markets, and high potential to innovativeness.

These characteristics are also viewed as challenges that influence the implementation of well-known performance measurement systems that are designed for larger firms in small and medium sized enterprises [16].

For example, the dynamic strategy of small business means that these businesses are more frequently revising their decisions than the larger firms. This greatly influences internal operations, and the relations with customers and suppliers. Such behaviour requires a better system of control with higher capability to control effectively and rapidly reflect these changes and their consequences on the internal operations as well as the external ones. These limitations of small manufacturing enterprises emphasize need for a performance measurement and control system that effectively reflects key business operations with fewer but critical measures that are written in form of an understandable structure, and flexible enough to fit specific needs of each individual enterprise and the changeable market conditions as well.

4 Analytical Hierarchy Process (AHP)

The Analytic Hierarchy Process (AHP), introduced in 1970 has become one of the most broadly used methods for multiple criteria decision-making (MCDM) [17]. It is a decision approach designed to assist in the solution of complex multiple criteria problems in a number of application areas. AHP is a problem-solving framework, flexible, organized method employed to represent the elements of a compound problem, hierarchically [18]. It has been considered to be an essential tool for both practitioners and academic researchers in organizing and analysing complex decisions [19]. AHP has been extensively used for selection process such as comparing the overall performance of manufacturing departments [20], manufacturing supply chain [21], benchmarking logistics performance [18], and vendor evaluation and selection [22]. More researchers are realizing that AHP is an effective technique and are applying it to several manufacturing areas [21]. AHP has several benefits [23]:

- It helps to decompose an unstructured problem into a rational decision hierarchy.
- Second, it can draw out more information from the experts or decision makers by employing the pair-wise comparison of individual groups of elements.
- Third, it sets the computations to assign weights to the elements.
- Fourth, it uses the consistency measure to validate the consistency of the rating from the experts and decision makers. The AHP procedure to solve a complex problem involves four steps [19]:

1. Breaking down the complexity of a problem into multiple levels and synthesizing the relations of the components are the underlying concepts of AHP
2. Pair-wise comparison aims to determine the relative importance of the elements in each level of the hierarchy. It starts from the second level and ends at the lowest. A set of comparison matrices of all elements in a level of the hierarchy with respect to an element of the immediately higher level are built so as to prioritize and convert individual comparative judgments into ratio scale measurements. The preferences are quantified by using a nine-point scale. The meaning of each scale measurement is explained in Table 2. Decision maker needs to express preference between each pair of the elements in terms of how much more one element is important than other element. Table 3 shows a matrix that expresses personal judgment and preferences.

Table 2. Comparison scale for the importance using AHP grading system.

Intensity of importance	Definition	Explanation
1	Equal importance	Two activities/factors contribute equally to the objective
3	Somewhat more important	Experience and judgment slightly favor one over the other
5	Strong importance	Experience and judgment strongly favor one over the other
7	Very strong importance	Experience and judgment very strongly favor one over the other. Its importance is demonstrated in practice
9	Absolutely extremely important	The evidence favoring one over the other is of the highest possible validity
2, 4, 6, 8	Intermediate values	When compromise is needed
Reciprocal	Opposite value	When activity I has one of the above numbers assigned to it with activity j, then j has the reciprocal value when compared to I.

Source: Saaty(2008).

Table 3. Pair-wise comparison for n number of elements at the same level.

	I1	I2	I3	*In*
I1	1	2	4	…
I2	0.5	1	…	…
I3	0.25	…	1	…
In	….	…	…	1

3. Relative weight calculation
   After the pair-wise comparison matrix is developed, a vector of priorities (i.e. eigenvector) in the matrix is calculated and is then normalized to sum to 1.0. This is done by dividing the elements of each column of the matrix by the sum of that column (i.e. normalizing the column). Then, obtain the eigenvector by

adding the elements in each resulting row to obtain a row sum, and dividing this sum by the number of elements in the row to obtain relative weight.

4. Consistency check

   A consistency ratio (CR) is used to measure the consistency in the pair-wise comparison. The purpose is to ensure that the judgments of decision makers are consistent. For example, when using AHP technique, a consistency ratio between factors and criteria can be obtained by the following equation:

$$\mathrm{CR} = \mathrm{CI}/\mathrm{RI} \tag{1}$$

Where:

CI: consistency index

RI: consistency ratio based on the value of n.

Checking consistency provides more information about the accuracy of the comparison and the decision alternatives selection. The final score of decision alternatives can be obtained by applying the following general equation:

$$Sk = \sum_{i=1}^{m} \sum_{j=1}^{ni} Wi\,wij\,rijk \tag{2}$$

Where:

Sk = overall decision of alternative k score

Wi = relative weight of criteria i

wij = relative weight of indicator j of criteria i

rijk = rating of decision alternative k and for indicator j of criteria i

ni = total number of indicators belong to criteria.

5 SCOR Performance Levels and Attributes

Supply Chain Council (SCC) is a global non-profit organization formed in 1996 to make and evolve a standard industry process reference model of the supply chain for the benefits of helping enterprises improve supply chain operations. SCC has established the supply chain framework- the (SCOR) process reference model for evaluating and comparing supply chain activities and related performance [24]. The SCOR model consists of standard supply chain processes, standard performance attributes and metrics, standard practices and standard job skills. It divides the supply chain attributes into two categories: internal and customer related attributes. The SCOR performance attributes such as: Supply Chain Reliability, Responsiveness, and Agility are considered as customer related attributes. Cost and Assets management are internal attributes. The SCOR performance section consists of two types of elements: Performance Attributes and Performance Metrics. A performance attribute is a combination of characteristics used to express a strategy. However, an attribute itself cannot be measured, it is used to set and identify strategic direction. The metrics that are assigned to each performance attribute measure the ability of the supply chain to achieve these attributes. Table 4 shows five performance attributes; two of them (the cost and assets management) are considered as internal-focused. Reliability, Responsiveness and Agility are considered as Customer-focused. Associated with the performance attributes are the level 1 strategic metrics. These level 1 metrics are the calculations by which an organization can measure how successful it is in achieving its desired position within the competitive market.

Table 4. SCOR performance attributes and definitions.

Performance Attribute		Definition
Internal	Costs: **CO**	The cost of operating the supply chain processes.
Internal	Assets management: **AM**	The ability to efficiently utilize assets
Customer	Reliability: **RL**	The ability to perform tasks and activates as planned or expected. It focuses on the outcomes of the processes
Customer	Responsiveness: **RS**	The speed at which tasks and activities are performed
Customer	Agility: **AG**	The ability to respond to external effects, i.e. demand and supply uncertainties.

For example, the performance attribute supply chain cost includes two types of costs: supply chain management cost and cost of goods sold. Reliability on the other hand involves only perfect order fulfilment. Each of level one strategic metric also divided to level 2 and 3 metrics, more information about SCOR performance attributes can be found at Supply Chain Council website [24]. However, the framework does not provide users and practitioners with any guidelines on how to use or where to start the evaluation that requires another tool that simplify such a complex framework.

6 The Approach

Since business conditions became more unpredictable and unstable, manufacturing firms are required to adjust their operations strategies in order to meet these changes. The evaluation of the alternative supply chain strategies; effective or responsive requires that the performance of the strategies on agility, reliability, responsiveness, cost, to be reevaluated, re-prioritized, quantified and aggregated to capture the new business goals. However, this process is not a straightforward task, since the performance and strategy evaluation process depends on many factors that by nature are interconnected and require a specific level of skill and qualifications that mostly do not exist in many SMEs. Successful performance measuring systems have to satisfy and completely fulfil the following points:

- The metrics used in performance measurement systems should have the power to capture and represent the organizational performance.
- The measures need to convey clear connections with a range of levels of decision-making such as strategic and operational.

- The metrics should also need to reflect an acceptable balance between non-financial and financial measures.
- A measurement system that ensures a suitable allocation of metrics to the areas where they would be most appropriate.

Therefore, the framework outlined in this paper helps SMEs construct and build a strategic performance measurement system which involves the two types of supply chain strategies: Efficient and Responsive, and supply chain performance attributes based on SCOR model. The framework utilizes AHP approach to integrate SCOR performance attributes, and the two types of supply chain strategies into one comprehensive model, (Fig. 1). The supply chain model is use for several reasons. First, SMEs need to think and act relying on a wider range of measures that covers financial and non-financial issues. Secondly, this effort aims at bridging the gap between supply chain models and SMEs. For example, a study revealed that there is a poor fit between supply chain management and the small and medium-sized enterprises. The authors attributed this poor fit to variety of reasons such as improper implementation of supply chain management by the small and medium-sized enterprises, and due to the lack of use of supply chain management to complement strategic focus [25]. The Expert Choice software was used to assist us in building the hierarchal structure of the company's overall goal, market scenarios, performance attributes and supply chain strategies. Expert Choice is intuitive, graphically based and structured in a user friendly fashion so as to be valuable for conceptual and analytical thinkers, novices and category. Expert Choice software is intended to help decision-makers and the software users overcome the limits of the human mind to synthesize qualitative and quantitative inputs from multiple stakeholders. The Expert Choice software [26]:

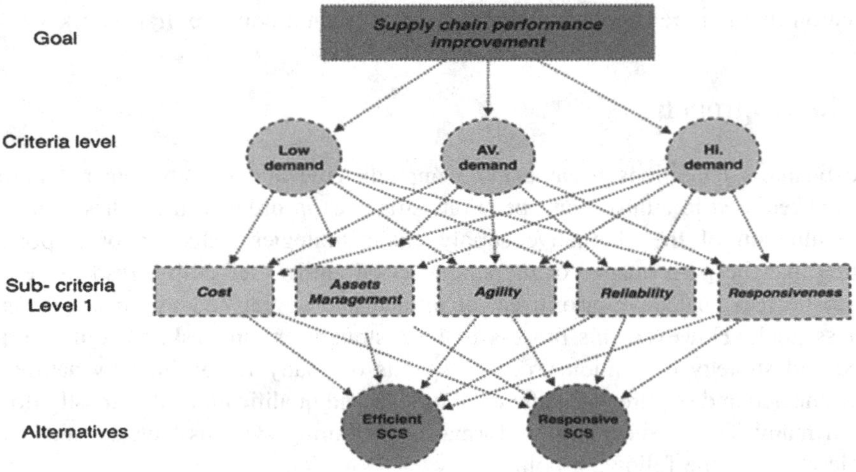

Fig. 1. The four levels structure of the model.

- Conveys structure and measurement to the planning and budgeting process
- Aids you determine strategic priorities and optimally allocates business resources
- Converses priorities and builds consensus
- Documents and justifies strategic decisions
- Enables you to move forward quickly and confidently

The AHP and Expert Choice software engage decision makers in structuring a decision into smaller parts, proceeding from the goal to objectives to sub-objectives down to the alternative courses of action. Decision makers then make simple pairwise comparison judgments throughout the hierarchy to arrive at overall priorities for the alternatives. The decision problem may involve social, political, technical, and economic factors [26]. The model is illustrated in the next section on a case of a medium-sized manufacturing enterprise. As shown in Fig. 1, two key supply chain strategies are considered at the last level that represents the available alternatives that the decision maker has to choose from based on market conditions, business environment and company's overall goal. The third level, the attributes level, includes: Cost, Assets management as internal or let us say financial attributes and Agility, Reliability, and Responsiveness as customer or non-financial performance and strategy attributes. Notice that the SCOR attributes bring financial and non-financial measures together to achieve an important part of the non-traditional performance system requirements. The second level or the scenario level shows various market conditions: low demand, average demand and high demand. Each and every business encounters one or more of these market conditions, but the question of how, when, and why one supply chain strategy is chosen over the other and on what basis usually remains fairly open. Some of these issues will be highlighted in the next section through the presented case study.

7 Case Study

A family-owned medium-size manufacturing firm, call it company x, specialized in production of plastic pipes and fittings products. The company strategy is to produce and deliver high quality products to its customers at the agreed delivery time and method. Most of its customers are large firms, mega project contractors and government agencies. Although the company operates in a highly competitive market, the plastic pipes and fittings market, its product prices are almost the highest compared to similar products on the market. Based on the information collected about the company policy and operations, the Expert Choice software was used to translate and build the four level hierarchal structures: the goal, scenarios, criteria, and alternatives levels.

The evaluation of these alternative strategies is carried out level-by-level, starting from top down towards the lower levels. The process begins on level two by assessing likelihood of occurrence of scenarios of different market demands during the planning period. The evaluation process of different scenario according to company x is shown in Table 5.

The results of the second level evaluation process show that the possibility of high demand scenario occurrence is relatively higher than the other ones, Fig. 2 above. The second step evaluates the relative effects of each criterion "attribute" on performance

Table 5. Pair wise comparison at level 2.

	Low	Av.	High
Low	1	4	3
Av.	1/4	1	2
High	1/3	1/2	1

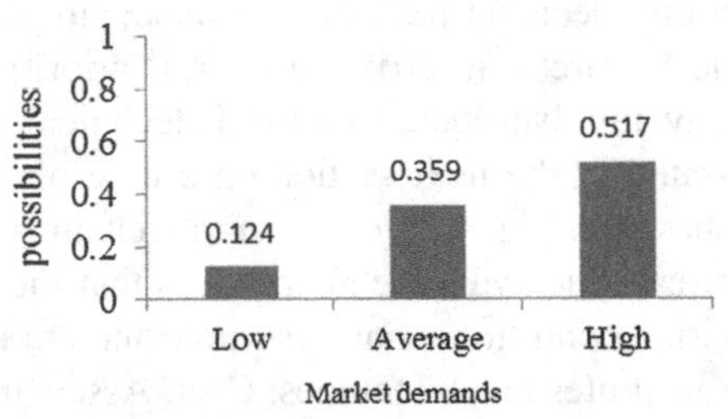

Fig. 2. The likelihood of different scenarios.

under a specific scenario. For example, what would be the relative effect of cost (CO), assets management (AM), agility (AG), reliability (RL), and responsiveness (RS) on performance if demand is low?, see Table 6. Notices that the relative effects of each performance attribute or criterion may vary depending on market conditions or product types.

Table 6. The pair wise comparison of performance attributes under low market demand.

	CO	AM	AG	RL	RS
CO	1	3	4	3	4
AM	0.33	1	3	2	2
AG	0.25	0.33	1	3	4
RL	0.33	0.50	0.33	1	1
RS	0.25	0.50	0.25	1	1

The results obtained from the evaluation process of performance attributes are shown in Fig. 3. In order to complete the level calculations one needs two more comparison processes for average and high market demand. The third step addresses the performance of each strategy on each performance criterion. Finally, the overall performance of each strategy can be calculated through the composition process by using Expert Choice. The performance of the two alternatives: efficient and responsive supply chain strategy is shown in Fig. 4.

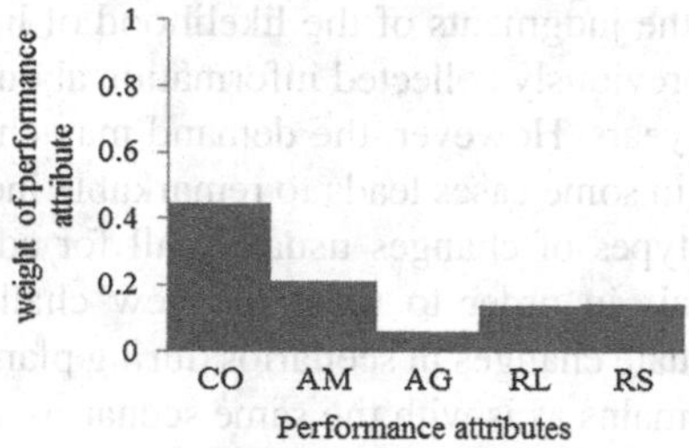

Fig. 3. Weights of performance attributes under low market demand.

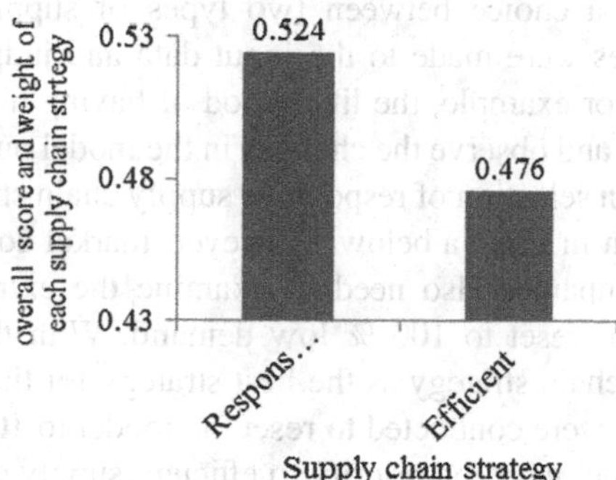

Fig. 4. Overall weight of the two alternatives.

8 Results and Discussion

The proposed framework was used to develop a model for a specific medium-sized manufacturing company. Notice that the company expectations of having high demand for the plastic pipes and fittings is about 52 %, 36 % for average demand and 12 % for low demand during the planning period. With high market demand, customers usually pay less attention to products prices and manufacturers without difficulty cover fixed and other related costs in mass production environment. This means that the company must place more emphasis on customer-related attributes as a major performance success factors. Within the planning period, the evaluation process clearly shows that focus on responsiveness is the most appropriate strategy that company x needs to adopt since the possibility of having high demand is relatively higher than the others. However, maintaining forever the same performance measures or supply chain will not help in rapidly changing business environment.

As the external environment changes frequently and rapidly, the group of performance attributes and measures in use by businesses must also change to reflect the changes in internal and/or external environment. Generally speaking, the changes in the performance measurement system can be done by adding, eliminating, replacing, or simply by reprioritizing performance measures and metrics. For example, a performance measure such as, for example supply chain responsiveness which initially has high priority may move down to low priority in other circumstances or because of changes in the internal and external business environment.

In the case presented, the judgments of the likelihood of having high, average and low demand are based on previously collected information about the market demands of company x in the last few years. However, the demand may change at any time during the planning period which in some cases leads to remarkable increase or decrease of the real market needs. These types of changes usually call for adjustments in businesses strategies, policies, or goals in order to meet the new challenges. For this reason, sensitivity analyses to evaluate changes in scenarios during planning period of company x were used. The model remains as is with the same scenarios of market demands: low, average, and high in the second level. The third level has five supply chain performance attributes: cost, assets management, agility, reliability, and responsiveness. And finally in fourth level provides a choice between two types of supply chains: efficient or responsive. Some changes were made to the input data and judgments of level 2, the market scenarios level. For example, the likelihood of having high demand was set to 100 % in order to capture and observe the changes in the model outputs. The 100 % high demand market resulted in selection of responsive supply chain strategy with about 0.66 priority weights as shown in Fig. 5a below. However, market conditions and demands always change, thus companies also need to examine the extremes of the markets. Therefore, the model was reset to 100 % low demand. With this setting, the model chooses efficient supply chain strategy as the best strategy for the low demand market, see Fig. 5b. Similar steps were conducted to reset the model to 100 % average demand. With this setting, the model gave the priority to efficient supply chain strategy but with less weight compared to 100 % low demand scenario, Fig. 5c.

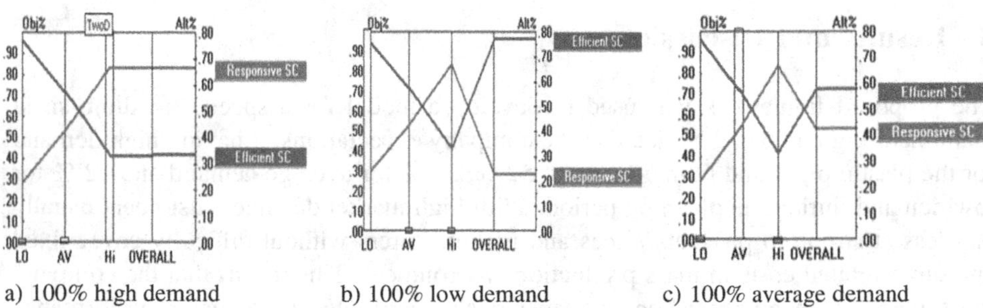

a) 100% high demand b) 100% low demand c) 100% average demand

Fig. 5. Various demand scenarios.

Table 7 shows the results of different scenarios generated using sensitivity analysis using Expert choice. In general, when the probability of the occurrence of low or average demand is 100 %, the performance of efficient supply chain strategy will be better than the performance of responsive supply chain strategy. When the probability of high demand is certain, likelihood of 100 %, responsive supply chain strategy should give better performance than efficient supply chain strategy. For company x, the market demand can be divided to three intervals or classes: low, average, and high. In addition, the company sets the limits for each one as shown in Table 8 below. Based on these intervals and the forecasted demand for the planning period, the coming 18 months, the company has to adopt both strategies but in different time periods as shown in Fig. 6.

Table 7. Different scenarios call for differing supply chain strategies.

Prob. Low	Prob. AV.	Prob. Hi	Priority efficient	Priority responsiveness	Strategy to adopt
1.00	0.00	0.00	0.768	0.232	Efficient
0.00	1.00	0.00	0.573	0.427	Efficient
0.00	0.00	1.00	0.333	0.667	Responsive

Table 8. Demand categories for company x.

Demand	Low	Average	High
Weight (tons)	0–2499	2500–4999	5000–8000

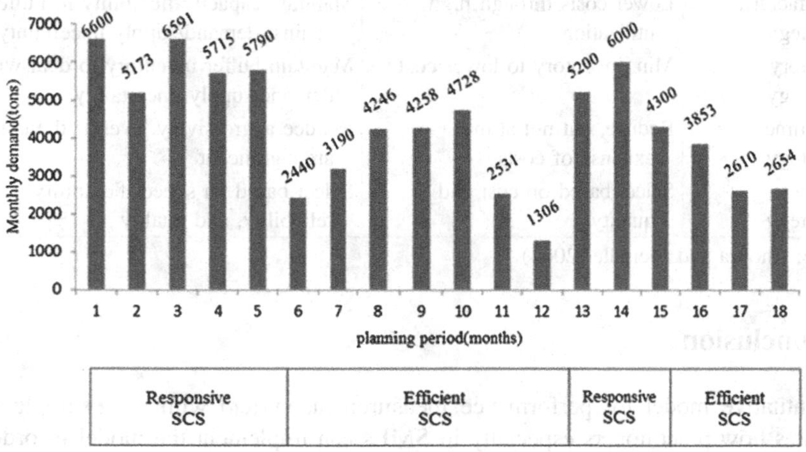

Fig. 6. Forecasted market demand of company x and the selection of the supply chain strategy.

The company needs to adopt responsive supply chain strategy for the first five months within the planning period and go back to efficient supply chain for the rest of the year.

The Fisher's framework suggests that there are two types of products, functional and innovative products [27]. Based on this classification, he suggested two types of supply chain strategies that fit each product type. For instance, he recommended efficient supply chain strategy for functional products, and a responsive supply chain for innovative types of products. Although efficient supply chain strategy performs well with functional products, i.e. plastic pipes and fittings, our case shows that there are few months within the planning period that require some degree of responsiveness in order to meet customer orders, particularly orders for government projects.

Nevertheless, implementation of the model requires users to be aware of the difference between the two strategies. For instance, in the presented case the company needs to minimize inventory to lower the cost during low demand time. It also needs to select material suppliers based on cost as a main factor while trying to reduce manufacturing costs and lower the margins. On the other hand (during high demand period), the company has to reduce lead time, put higher price margins, respond quickly to demand and select suppliers based on flexibility, speed and reliability. Table 9 shows the general differences and a comparison between the two strategies [28].

Table 9. Characteristics of efficient and responsive supply chain strategies.

	Efficient supply chain	Responsive supply chain
Primary goal	Supply demand at lowest cost	Respond quickly to demand
Product design strategy	Max. Performance at a min. product cost	Create modularity to allow postponement of product differentiation
Pricing strategy	Lower margins because price is a prime custom driver	Higher margins because price is not a prime customer driver
Manufacturing strategy	Lower costs through high utilization	Maintain capacity flexibility to buffer against demand/supply uncertainty
Inventory strategy	Min. inventory to lower cost	Maintain buffer inventory to deal with demand/supply uncertainty
Lead time strategy	Reduce, but not at the expense of costs	Reduce aggressively, even if the cost are significant
Supplier strategy	Select based on cost and quality	Select based on speed, flexibility, reliability, and quality

Source: Chopra and Meindle (2004)

9 Conclusion

A quantitative model for performance measurement system with the example used illustrates how practitioners especially in SMEs can implement the model in order to improve business performance. Using SCOR model helped in identifying a set of financial and non-financial performance measures that are generally used to evaluate supply chain performance in large firms. The use of AHP approach was useful in structuring the model to four levels: Overall goal, Scenarios, Criteria, and Alternatives. The use of Expert Choice software facilitated an excellent environment in structuring the model hierarchically, carrying out evaluation level by level, and making final alternatives evaluation and selection. Limited sensitivity analyses were performed in order to sense the difference when changes occur in the internal or external environment through our model. We witnessed through the case that the link between product type and supply chain strategy type works very well which proofs previous suggestions. We also observed that adding market demands with three different scenarios into the model provides us with different results for one market scenario, which suggests that there are two key players in strategy selection and that are the product type and the market demand. The authors of this paper believe that the outlined model achieves important directions of non-traditional performance measurement system such as: flexibility, easy to use, up to date, comprehensive, involves financial and non-financial measures, and based on business strategy as well. Unlike previous implementations of AHP and performance measures model, the proposed model introduced a new approach that SMEs can use to evaluate their internal needs and external requirements by combining the two approaches correctly. The proposed model also effectively engages users, mainly SMEs, to the world of supply chain management and operations. However, we need to implement this methodology and model in different business environments to insure the ease of use of the model and to collect more data and information about obstacles, limitations and ultimately the impact on business profit.

References

1. SME Research and Statistics. http://ic.gc.ca/eic/site/061.nsf/eng/h_02689. Accessed 2013
2. Holban, I.: Strategic performance measurement system and SMEs competitive advantage. In: International Conference on Economics and Administration, University of Bucharest, Faculty of administration and businesses (2009)
3. Cocca, A., Alberti, M.: "SME's three step pyramid: a new performance measurement framework for SMEs". In: 16th International Annual EUROMA Conference: Implementation–Realizing Operations Management Knowledge, Göteborg (2009)
4. Hudson, M., Lean, J., Smart, P.: Improving control through effective performance measurement in SMEs. Prod. Plan. Control **12**(8), 804–813 (2001)
5. Hvolby, H., Thorstenson, A.: Indicators for performance measurement in small and medium-sized enterprises. Proc. Inst. Mech. Eng. Part B J. Eng. Manuf. **215**(8), 1143–1143 (2001)
6. Gosselin, M.: An empirical study of performance measurement in manufacturing firms. Int. J. Prod. Perform. Manage. **54**(5/6), 419–437 (2005)
7. Bititci, U.: Modelling of performance measurement systems in manufacturing enterprises. J. Prod. Econ. **42**(2), 137–147 (1995)
8. Globerson, S.: Issues in developing a performance criteria system for an organization. Int. J. Prod. Res. **23**(4), 639–646 (1985)
9. Neely, A.: The performance measurement revolution: why now and what next? Int. J. Oper. Prod. Manage. **19**(2), 205–228 (1999)
10. Melnyk, S., Stewart, D., Swink, M.: Metrics and performance measurement in operations management: dealing with the metrics maze. J. Oper. Manage. **22**(3), 209–218 (2004)
11. Taticchi, P., Tonelli, F., Cagnazzo, L.: Performance measurement and management: a literature review and a research agenda. Meas. Bus. Excell. **14**(1), 4–18 (2010)
12. Tangen, S.: Performance measurement: from philosophy to practice. Int. J. Prod. Perform. Manage. **53**(8), 726–737 (2004)
13. Tenhunen, J., Rantanen, H., Ukko, J.: SME oriented implementation of performance measurement system. Lappeenranta University of Technology, Department of Industrial engineering and management, Lahti, Finland (2001)
14. Chennell, A., Dransfield, S., Field, J., Fisher, N., Saunders, I., Shaw, D.: "OPM: a system for organizational performance measurement". In: Performance Measurement in Past, Present and Future, pp. 96–103. Cranfield University, Cranfield, Cranfield (2005)
15. Dixon, J., Nanni, A., Vollmann, T.: "The new performance challenge: measuring operations for world-class competition". In: Business One, Dow Jones-Irwin, Homewood, IL (1990)
16. Garengo, P., Biazzo, S., Bititci, U.: Performance measurement systems in SMEs: a review for a research agenda. Int. J. Manage. Rev. **7**(1), 25–47 (2005)

17. Saaty, T.: Decision making with the analytic hierarchy process. Int. J. Serv. Sci. **1**(1), 83–98 (2008)
18. Chen, K., Huang, M., Chang, P.: Performance evaluation on manufacturing times. Int. J. Adv. Manuf. Technol. **31**(3–4), 335–341 (2006)
19. Cheng, E., Li, H.: Analytic hierarchy process: an approach to determine measures for business performance. Meas. Bus. Excell. **5**(3), 30–37 (2001)
20. Rangone, A.: An analytical hierarchy process framework for comparing the overall performance of manufacturing departments. Int. J. Oper. Prod. Manage. **16**(8), 104–119 (1996)
21. Wang, G., Huang, S., Dismukes, J.: Manufacturing supply chain design and evaluation. Int. J. Adv. Manuf. Technol. **25**(1–2), 93–100 (2005)
22. Haq, A., Kannan, G.: Fuzzy analytical hierarchy process for evaluating and selecting a vendor in a supply chain model. Int. J. Adv. Manuf. Technol. **29**(7–8), 826–835 (2006)
23. Cheng, E., Li, H., Ho, D.: Analytic hierarchy process: a defective tool when used improperly. Meas. Bus. Excell. **6**(4), 33–37 (2002)
24. Supply Chain Council. http://supply-chain.org/scor. Accessed 2013
25. Arend, R., Wisner, J.: Small business and supply chain management: is there a fit? J. Bus. Venture **20**, 403–436 (2005)
26. Collaboration and Decision Support Software for Groups & Organizations. http://expertchoice.com. Accessed 2013
27. Fisher, M.: What is the right supply chain for your product? Harvard Bus. Rev. **75**, 105–117 (1997)
28. Chopra, S., Meindl, P.: Supply Chain Management. Pearson Prentice-Hall, Upper Saddle River (2004)

Adjacency Variables Formulation for the Minimum Linear Arrangement Problem

Serigne Gueye[1(✉)], Sophie Michel[2], and Mahdi Moeini[3,4]

[1] Laboratoire d'Informatique d'Avignon (LIA), 339, chemin des Meinajaries, Agroparc, B.P. 91228, 84911 Avignon Cedex 9, France
serigne.gueye@univ-avignon.fr

[2] Laboratoire de Mathmatiques Appliques du Havre (LMAH), 25 rue Philippe Lebon, B.P. 540, Le Havre Cedex, France

[3] Centre de Recherche en Informatique de Lens (CRIL), Rue Jean Souvraz, SP 18, 62307 Lens Cedex, France

[4] Chair of Business Information Systems and Operations Research (BISOR), Technical University of Kaiserslautern, Postfach 3049, Erwin-Schroedinger-Str., Room 42-426, D-67653 Kaiserslautern, Germany
moeini@wiwi.uni-kl.de

Abstract. We present a new integer linear formulation using $O(n^2)$ variables, called **adjacency** variables, to solve the Minimum Linear Arrangement problem (MinLA). We give a couple of valid equalities and inequalities for this formulation, some of them deriving from on a new general partitioning approach that is not limited to our formulation. We numerically tested the lower bound provided by the linear relaxation using instances of the matrix market library. Our results are compare with the best known lower bounds, in terms of quality, as well computing times.

Keywords: Minimum linear arrangement problem · Integer programming · Graph partitioning · Cutting plane algorithms

1 Introduction

The Minimum Linear Arrangement (MinLA) is a combinatorial optimization problem, originally proposed by Harper [10] in 1961. It consists in arranging the nodes of a graph on a line in such a way to minimize the sum of the distances between the adjacent nodes. In the literature, the MinLA is known under different names such as the optimal linear ordering, the edge sum problem, or the minimum-1-sum (see e.g., [2,4,5,12,13,17,18]). Formally, let $G = (V, E)$ be a simple graph, where V (with $|V| = n$, and $|E| = m$) is the set of nodes and E denotes the set of edges. An arrangement is defined as a one-to-one function $\varphi : V \longrightarrow \{1, 2, ..., n\}$. MinLA consists in finding an arrangement φ minimizing the following sum

E. Pinson et al. (Eds.): ICORES 2014, CCIS 509, pp. 95–107, 2015.
DOI: 10.1007/978-3-319-17509-6_7

$$\sum_{uv \in E} |\varphi(u) - \varphi(v)|.$$

MinLA has many practical applications in design of VLSI layouts, graph drawing, or single machine job scheduling ([13,17,18]). It is known to be strongly NP-hard in its general form [8]. However there are polynomial time algorithms for some particular graphs such as trees, outerplanar graphs, and certain Halin graphs [5].

MinLA can be viewed as a particular case of the well-known Quadratic Assignment Problem (QAP) [14]. Consequently, the standard Koopmans and Beckmann's formulation may be used for modeling MinLA.

Let x_{ik} be binary variables defined as follows:

$$x_{ik} = \begin{cases} 1 \text{ if the node } i \text{ is assigned to the location } k \text{ of the line,} \\ 0 \text{ otherwise.} \end{cases}$$

Using these variables, the problem consisting of

$$\begin{array}{lll} (Q) \ : \ \text{Min} & \displaystyle\sum_{(i,j)\in E} \sum_{k=1}^{n} \sum_{l=1}^{n} |k-l| x_{ik} x_{jl} & \\ \text{such that} & \sum_{k=1}^{n} x_{ik} = 1, & \forall\, i = 1, ..., n, \\ & \sum_{i=1}^{n} x_{ik} = 1, & \forall\, k = 1, ..., n, \\ & x_{ik} \in \{0,1\} & \forall\, i,k = 1, ..., n, \end{array}$$

belongs to the class of 0–1 quadratic problems.

Let A be the adjacency matrix of G. The objective function of (Q) may be written as follows:

$$\sum_{i=1}^{n} \sum_{j=i+1}^{n} \sum_{k=1}^{n} \sum_{l=1}^{n} A_{ij} |k-l| x_{ik} x_{jl}.$$

Because of the hardness of finding optimal values to such formulation, lower bounding and heuristic algorithms are usually applied to get good approximations ([5,17,18]). We focus in this paper on the family of lower bounding techniques that uses integer linear programming techniques, in which our contribution may be also classified.

In a sequence of articles, Liu and Vanelli [16], Amaral et al. [2] and Caprara et al. [4,5] have proposed some integer linear programs. All uses a similar kind of variables called **"distance variables"**. For any couple (i,j), a distance variable D_{ij} (between i and j) is defined as follows

$$D_{ij} = \sum_{k=1}^{n} \sum_{l=1}^{n} |k-l| x_{ik} x_{jl}, \ \forall\, 1 \le i < j \le n. \tag{1}$$

Note that, for all fixed locations k_0 and l_0, taking $x_{ik_0} = 1$ and $x_{jl_0} = 1$ implies $D_{ij} = |k_0 - l_0|$. Thus, D_{ij} represents the distance between entities i and j, which depends on their respective locations on the line.
With these variables, (Q) becomes:

$$(LP_D) : \text{Min} \quad \sum_{i=1}^{n} \sum_{j=i+1}^{n} A_{ij} D_{ij} \tag{2}$$

such that

$$\sum_{i=1}^{n} x_{ik} = 1, \qquad \forall\, i = 1, ..., n, \tag{3}$$

$$\sum_{k=1}^{n} x_{ik} = 1, \qquad \forall\, k = 1, ..., n, \tag{4}$$

$$D_{ij} \geq |k - l|(x_{ik} + x_{jl} - 1), \forall\, i, j, k, l = 1, ..., n,\ i < j,\ k \neq l, \tag{5}$$

$$x_{ik} \in \{0, 1\}, \qquad \forall\, i, k = 1, ..., n, \tag{6}$$

$$D_{ij} \geq 0, \qquad \forall\, i, j = 1, ..., n,\ i \neq j. \tag{7}$$

The drawback of this formulation is that its linear relaxation admits the trivial solution $x_{ik} = 1/n$ for all i, k and $D_{ij} = 0$ for all $(i, j) \in E$, yielding a lower bound of zero. For this reason, it is necessary to analyze the polytope corresponding to the convex hull of the feasible integer points in order to derive additional valid inequalities. This kind of studies have been performed in [2,5,16].

Caprara et al. [6], as well as Amaral [1] and Gueye et al. [9] have reformulated a closed problem, called the Single Ray Facility Layout Problem, using the concept of **"betweeness variables"**. A betweeness variable t_{ilj} is defined as a binary variable that is equal to 1 if and only if the entity l is located between the entities i and j. Adding suitable cuts for the polytope induced by these types of variables lead to very good lower bounds for the minimum linear arrangement problem.

The goal of this paper is to propose a new integer linear formulation using $O(n^2)$ variables. We give some valid inequalities and equalities of the associated polytope. Some of them exploit a partitioning process of the graph G, and may be applied, on minor changes, for the distance variables formulation. We test our method with some instances of *the matrix market library*[1] [3].

The structure of the paper is as follows. In Sect. 2, the new formulation is given. In Sects. 3 and 4 valid constraints. They are derived from a lifting process as well as the property of the node degrees of G. To make the formulation stronger, new type of cuts, called **partitioning cuts**, and cutting planes algorithms are introduced in Sect. 5. The computational experiments are reported in Sect. 6 and the last Section includes some conclusions.

2 A Linear Formulation with $O(n^2)$ Variables

Let us consider the following **adjacency variables**

$$F_{kl} := \sum_{i=1}^{n} \sum_{j=1}^{n} A_{ij} x_{ik} x_{jl} \in \{0, 1\}. \tag{8}$$

[1] http://math.nist.gov/MatrixMarket/.

Remark 1. We note that F_{kl} is equal to 1 if the two entities located in k and l are linked by an edge in G, and 0 otherwise. As a consequence, one may see that the graph whose the adjacency matrix is represented by $F = \{F_{kl}\}$ is isomorphic to G.

If we apply this definition on (Q), the formulation becomes:

$$(LP_F) : \text{Min} \quad \sum_{k=1}^{n} \sum_{l=k+1}^{n} |k-l| F_{kl} \tag{9}$$

$$\text{such that} \quad \sum_{i=1}^{n} x_{ik} = 1, \quad \forall\, i = 1, ..., n, \tag{10}$$

$$\sum_{k=1}^{n} x_{ik} = 1, \quad \forall\, k = 1, ..., n, \tag{11}$$

$$F_{kl} \geq A_{ij}(x_{ik} + x_{jl} - 1), \forall\, i, j, k, l = 1, ..., n,\ k < l, \tag{12}$$

$$x_{ik} \in \{0, 1\}, \quad \forall\, i, k = 1, ..., n, \tag{13}$$

$$F_{kl} \geq 0, \quad \forall\, k, l = 1, ..., n,\ k < l. \tag{14}$$

Because of the definition of F, one may see that we transform the initial problem in a new one that consists in finding an optimal isomorphic graph for G. For any feasible solution, we can easily verify that the constraints (12) imply that F_{kl} is greater than A_{ij}. Since we are minimizing and because $F_{kl} \geq 0$, F_{kl} will be precisely equal to 1 if i and j are adjacent in G (i.e. $A_{ij} = 1$) and 0 otherwise.

Our linear model has $O(n^2)$ variables; but there are two similar drawbacks as for the distance variables formulation. In one hand, the formulation has a zero bound obtained by taking $x_{ik} = \frac{1}{n}$ and $F_{kl} = 0$. On the other hand, there are $O(n^4)$ number of constraints (12) that should be reduced. In the Sect. 3, we replace the constraints (12) by $O(n^3)$ equivalent ones. We introduce subsequently, in Sect. 4, some valid equalities related to the node degrees of G. The Sect. 5 deals with a new family of cuts based on a partitioning of the graph G.

3 Lifting Valid Inequalities

In order to reduce the number of constraints, we present the following theorem that shows how we can produce $O(n^3)$ inequalities equivalent to the constraints (12).

Theorem 1. *Let $k, l, i_0 \in \{1, 2, \cdots, n\}$. The following inequalities are valid for (LP_F):*

$$F_{kl} \geq \sum_{j=1}^{n} A_{i_0 j} x_{jl} + \sum_{\substack{k'=1 \\ k' \neq k}}^{n} \alpha_{k'} x_{i_0 k'}, \tag{15}$$

where $\alpha_{k'} := \min(A_{i'j} - A_{i_0 j})$, such that $i' \neq i_0, j$ and $j \neq i_0$ and $k' \neq k$.

Proof. Let us consider the subdomain where $x_{i_0,k'} = 0$ for all $k' \neq k$, implying that $x_{i_0,k} = 1$.

Taking into account the definition of F (see (8)), it follows that, for any l

$$F_{kl} \geq \sum_{j=1}^{n} A_{i_0 j} x_{jl}.$$

By using lifting techniques, assuming that $x_{i_0,k'} = 1$, we can find for any $k' \neq k$ the best coefficient, $\alpha_{k'}$, for which

$$F_{kl} \geq \sum_{j=1}^{n} A_{i_0 j} x_{jl} + \alpha_{k'}.$$

The corresponding value is given by: $\alpha_{k'} = \min(A_{i'j} - A_{i_0 j})$ such that $i' \neq i_0, j$ and $j \neq i_0$ and $k' \neq k$.
We obtain the valid inequalities by repeating the same lifting procedure for all $k' \neq k$. ■

Replacing the constraints (12) by (15) yields an equivalent formulation while reducing the constraint complexity by a factor n (i.e., $O(n^3)$ instead of $O(n^4)$). In the next section, we present some additional sets of valid equalities that strengthen our formulation.

4 Degree Valid Equalities

According to the definition of F_{kl}, we note that

$$\sum_{l=1}^{n} F_{kl}$$

is equal to the degree of any node having the label k. Since the sum of the degree of a simple graph with $|E| = m$ edges is $2m$ we have the following valid equalities.

Theorem 2. *We have*

$$\sum_{k=1}^{n} \sum_{l=1}^{n} F_{kl} = 2m,$$

where m is the number of edges of the simple graph G. ■

A stronger version of these inequalities is given through the following theorem.

Theorem 3. *For any connected graph G, the following equalities are valid*

$$\sum_{k=1}^{n} F_{kl} = \sum_{j=1}^{n} (\textit{degree of node } j)\ x_{jl}, \quad l \in \{1, \cdots, n\}. \tag{16}$$

Proof. For any value of l, we have

$$\sum_{k=1}^{n} F_{kl} = \sum_{i=1}^{n}\sum_{j=1}^{n} A_{ij}x_{jl} = \sum_{j=1}^{n}\sum_{i=1}^{n} A_{ij}x_{jl} = \sum_{j=1}^{n} x_{jl}\left(\sum_{i=1}^{n} A_{ij}\right)$$

Since *the degree of the node* (j) is equal to $\sum_{i=1}^{n} A_{ij}$, consequently,

$$\sum_{k=1}^{n} F_{kl} = \sum_{j=1}^{n}(\text{degree of node } j)\ x_{jl},$$

where l belongs to $\{1, \cdots, n\}$. ■

5 Partitioning Cuts

We present in this section the partitioning cuts based on the definition of F_{kl} and on the Remark 1. To properly define these cuts, we need to use weighted graphs.

So, let G_A be the **completed weighted graph** defines in the node set $V = \{1, 2, ..., n\}$ of G, and where each edge (k, l), with $1 \leq k < l \leq n$, is valuated by A_{kl}, the element (k, l) of the adjacency matrix A of G. Let $F = \{F_{kl}\}_{1 \leq k,l \leq n}$ be a square symmetric matrix whose each element corresponds to a **value** of the **variable** F_{kl}, and $\mathcal{D}(LP_F)$ be the set containing all feasible binary matrices F for LP_F. Remark that each element of $\mathcal{D}(LP_F)$ derived from a permutation of lines and columns of the adjacency matrix A. The permutation being indicated by the binary variables x_{ik}. For any element F of $\mathcal{D}(LP_F)$, the simple complete weighted graph where the node set is V, and each edge (k, l) are weighted by the value F_{kl} is noticed G_F. Because of the Remark 1, one may see that any quantitative property verified by G_A must be also verified by G_F. For instance, the optimal value of any **graph partitioning problem** on G_A must be the same than in G_F.

For a given p, a **node partition** of G_A in p subsets V_1, V_2,...,V_p, of cardinality n_1, n_2,...,n_p is such that:

- $V_i \cap V_j = \emptyset, \forall\ i, j \in \{1, 2, ..., p\},\ i \neq j$ (i.e. the sets are disjoints),
- $V_1 \cup V_2 \cup ... \cup V_p = V$,
- $|V_i| = n_i\ \forall\ i \in \{1, 2, ..., p\}$.

The **cut set** (or multicut), corresponding to the node partition, is defined as the set of weighted edges whose the nodes belongs to two different subsets. It is noticed $(V_1, V_2, ..., V_p)$. The **capacity** of the cut $(V_1, V_2, ..., V_p)$ is the sum of its edge weights. It will be noticed $c_A(V_1, V_2, ..., V_p)$. **The graph partitioning problem** consists in finding a node partition minimizing $c_A(V_1, V_2, ..., V_p)$. Several researchers have studied and developped good heuristics or exact methods to solve this problem. The special case $p = 2$ corresponds to the well-known graph bipartitioning problem.

Remark that since A is the adjacency matrix of G, $c_A(V_1, V_2, ..., V_p)$ may be also written as:

$$\sum_{(k,l)\in(V_1,V_2,...,V_p)} A_{kl} = c_A(V_1, V_2, ..., V_p).$$

Since for any $F \in \mathcal{D}(LP_F)$, all G_A properties must be respected by G_F, we have the following lemma.

Lemma 1. *Let p be any integer number, and $V_1,...,V_p$ an optimal partition of G_A in p subsets of sizes n_1, n_2,...,n_p. For any partition $(W_1, W_2, ..., W_p)$ in p subsets of sizes n_1, n_2,...,n_p we necessary have:*

$$\boxed{c_F(W_1, W_2, ..., W_p) = \sum_{(k,l)\in(W_1,W_2,...,W_p)} F_{kl} \geq c_A(V_1, V_2, ..., V_p)}.$$

■

The lemma gives a necessary condition for a F to be feasible, in particular optimal. As a consequence, given $\overline{F}_{kl}$ $(1 \leq k < l \leq n)$ an optimal solution of the linear relaxation of LP_F, if, for a partition $W_1, W_2, ..., W_p$, we have

$$c_{\overline{F}}(W_1, W_2, ..., W_p) < c_A(V_1, V_2, ..., V_p)$$

then,

$$\sum_{(k,l)\in(W_1,W_2,...,W_p)} F_{kl} \geq c_A(V_1, V_2, ..., V_p),$$

defines a cut that may be added on the formulation.

5.1 Separation Problem

To implement such a type of cuts, we need to solve the corresponding separation problem. Given an optimal solution $\overline{F}_{kl}$ $(1 \leq k < l \leq n)$ of the linear relaxation, it consists in finding an optimal partition of $G_{\overline{F}}$ giving the value $c_{\overline{F}}(W_1, W_2, ..., W_p)$. We also need to know $c_A(V_1, V_2, ..., V_p)$. Each value requires to solve a graph partitioning problem, the first one in $G_{\overline{F}}$, and the second one in G_A. Unfortunately, the graph partitioning problem is NP-hard and therefore finding an optimal solution is likely to be a difficult task. However, the problem has been intensively studied. It is currently possible to derive quickly some good lower bounds of the optimal value, using semidefinite programming or linearization techniques. Similarly, many metaheuristic schemes such as the ones used in the software Chaco [3] allow to find good upper bounds. Fortunately, these lower and upper bounds are already sufficient for our purposes as explained below.

If we notice $\overline{c}_A(V_1, V_2, ..., V_p)$ a lower bound of the optimal value $c_A(V_1, V_2, ..., V_p)$, and $\overline{W}_1, \overline{W}_2, ..., \overline{W}_p$ the best partition of $G_{\overline{F}}$ found by any metaheuristic scheme, we also have

$$\sum_{(k,l)\in(\overline{W}_1,\overline{W}_2,...,\overline{W}_p)} F_{kl} \geq c_A(V_1, V_2, ..., V_p) \geq \overline{c}_A(V_1, V_2, ..., V_p)$$

Thus, if $c_{\overline{F}}(\overline{W}_1, \overline{W}_2, ..., \overline{W}_p) < \overline{c}_A(V_1, V_2, ..., V_p)$, then the corresponding cut can be added on the formulation.

5.2 Cutting Planes Algorithm

We can then deduce the following cutting plane Algorithm 1.

Algorithm 1. Level p Partitioning Cutting Plane Algorithm.

cut : global boolean variables that indicates if some cuts have been found; p : fixed number of partitions;
cut = false;
for *all possible cardinalities* $n_1, n_2, ..., n_p$ *of the* p *subsets* **do**
    Solve the linear relaxation of (LP_F) to obtain the values $\overline{F}_{kl}$;
    Using any metaheuristic, solve the graph partitioning problem in $G_{\overline{F}}$ to obtain $(\overline{W}_1, \overline{W}_2, ..., \overline{W}_p)$;
    if $c_{\overline{F}}(\overline{W}_1, \overline{W}_2, ..., \overline{W}_p) < \overline{c}_A(V_1, V_2, ..., V_p)$ **then**
        add $\sum_{(k,l)\in(\overline{W}_1, \overline{W}_2, ..., \overline{W}_p)} F_{kl} \geq \overline{c}_A(V_1, V_2, ..., V_p)$;
        cut = true ;

The Level p Partitioning Cutting Plane Algorithm can be then embedded in a loop on p as described in the Global Algorithm 2. In this algorithm, for different values of p starting from 2 until a maximal values, some cuts are generated. And as long as some cuts are found the entire process is restarted.

Algorithm 2. Global Algorithm.

cut = true ;
p_{max} : maximum value for p ;
while *cut == true* **do**
    for $p = 2, ..., p_{max}$ **do**
        Level p Partitioning Cutting Plane Algorithm;

In the Algorithm 1, it may be observed that the number of iterations of the loop grows polynomially with p. Indeed, $p = 2$ implies $O(n)$ iterations, $p = 3$ implies $O(n^2)$ and in general we have $O(n^{p-1})$ iterations.

Furthermore, the values $\overline{c}_A(V_1, V_2, ..., V_p)$ used in the Algorithm 1 are the lower bounds of the graph partitioning problem on the graph G_A. For all $p \in \{1, 2, ..., p_{max}\}$, and all possible cardinalities of the subsets $V_1, V_2, ..., V_p$, these values may be precomputed **one time** before the global algorithm. We show in the Subsect. 5.3 how to obtain these bounds. At the opposed, deriving $c_{\overline{F}}(\overline{W}_1, \overline{W}_2, ..., \overline{W}_p)$, makes necessary to run, at each iteration of the Algorithm 1, a metaheuristic scheme.

5.3 Solving the Graph Partitioning Problem

The graph partitioning problems that give $\overline{c}_A(V_1, V_2, ..., V_p)$ can be modeled using a distance variables formulation.

Let us recall that we want to find a partition of the node set of G_A in p subsets V_1, V_2,...,V_p of size n_1, n_2,...,n_p, in such a way that the sum of the edge weight with extremities in two different partitions is minimized.

Let us consider the binary variables

$$y_{ik} = \begin{cases} 1 \text{ if the node } i \text{ of } G \text{ is put in the partition } k, \\ 0 \text{ otherwise.} \end{cases}$$

where $i \in \{1, 2, ..., n\}$ and $k \in \{1, 2, ..., p\}$.

For any $k, l \in \{1, 2, ..., p\}$ let δ_{kl} be equal to 1 if $k \neq l$ and 0 otherwise.

The graph partitioning problem can be formulated as follows:

$$\text{Min} \sum_{i=1}^{n} \sum_{k=1}^{p} \sum_{j=i+1}^{n} \sum_{l=1}^{p} A_{ij} \delta_{kl} y_{ik} y_{jl} \tag{17}$$

$$\text{such that} \quad \sum_{i=1}^{n} y_{ik} = n_k, \qquad \forall\, k = 1, ..., p, \tag{18}$$

$$\sum_{k=1}^{n} y_{ik} = 1, \qquad \forall\, i = 1, ..., n, \tag{19}$$

$$y_{ik} \in \{0, 1\}, \qquad \forall\, i = 1, ..., n; k = 1, ..., p. \tag{20}$$

As for the minimum linear arrangement problem, this is also a variant of a Quadratic Assignment Problem. The objective function represents the number of edges (i, j) with endpoints in two differents partitions (k and l). The first constraints determines the size of each partition, and the second one imposes that each node must be in one and only one partition.

Let us define the distance variables as follows:

$$D_{ij} = \sum_{k=1}^{p} \sum_{l=1}^{p} \delta_{kl} x_{ik} x_{jl}, \ \forall\, 1 \leq i < j \leq n. \tag{21}$$

It can be seen that, for all known locations k_0 and l_0, taking $y_{ik_0} = 1$ and $y_{jl_0} = 1$ implies $D_{ij} = \delta_{k_0 l_0}$ which in turn will be equal to 1 if $k_0 \neq l_0$ and 0 otherwise. In other words, D_{ij} is 1 if and only if i and j are assigned to two different partitions, or similarly if (i, j) is **cut** by the partition.

According to Deza and Laurent [7], the vector $D = \{D_{ij}\}_{1 \leq i < j \leq n}$ corresponds to the definition of a **multicut**. It is known that multicuts define a distance on the node set of G, and as a particular consequence respect the triangular inequalities. Indeed, it is easy to see that for any nodes i, j, h with $1 \leq i < j < h \leq n$

$$D_{ij} \leq D_{ih} + D_{jh},$$

since if $D_{ij} = 1$ it means that i and j are in two different partitions, implying that:

- either h are in the partition of i with $D_{jh} = 1$,
- or h is in the partition of j with $D_{ih} = 1$,
- or h is nor in the partition of i, nor in the partition of j with $D_{jh} = D_{ih} = 1$.

This leads to the following lemma.

Theorem 4. *Let i, j, h satisfy $1 \leq i < j < h \leq n$. We have the following triangular inequalities:*

$$D_{ij} \leq D_{ih} + D_{jh}, \tag{22}$$

$$D_{ih} \leq D_{ij} + D_{jh}, \tag{23}$$

$$D_{jh} \leq D_{ij} + D_{ih}. \tag{24}$$

In addition to the triangular inequalities, we can be added the following one that exploit for each i the sum of the distance.

Theorem 5. *Let $\delta_k = \sum_{\substack{l=1 \\ l \neq k}}^{p} n_l, \forall\, k = 1, 2, ..., n$. The following equalities is verified:*

$$\sum_{j=1}^{n} D_{ij} = \sum_{k=1}^{p} \delta_k y_{ik}, \forall\, i = 1, 2, ..., n\ . \tag{25}$$

Proof. We know that $D_{ij} = \sum_{k=1}^{p} \sum_{l=1}^{p} \delta_{kl} y_{ik} y_{jl}, \forall\, i, j.$.

Thus $\sum_{j=1}^{n} D_{ij} = \sum_{k=1}^{p} \left[\sum_{l=1}^{p} \delta_{kl} \sum_{j=1}^{n} x_{jl} \right] y_{ik}$. It follows with constraint (19) that

$$\sum_{j=1}^{n} D_{ij} = \sum_{k=1}^{p} \left[\sum_{l=1}^{p} \delta_{kl} n_l \right] y_{ik} = \sum_{k=1}^{n} \delta_k y_{ik}.$$

■

Using Theorems 4 and 5 we have the following linear program.

$$(LP_{GP}) : \text{Min} \qquad \sum_{i=1}^{n} \sum_{j=i+1}^{n} A_{ij} D_{ij}$$

such that (18)–(20), (22)–(25)

A closed formulation has been studied by Lisser and Rendl [15] with the difference that the set of constraints (25) do not appear in the corresponding paper.

To compute $\overline{c}_A(V_1, V_2, ..., V_p)$, for each number of partitions and each cardinalities of the subsets, somes problems similar to (LP_{GP}) are generated at the beginning of the global algorithm. Then, for each, we solve the linear relaxation in which the $O(n)$ ((18), (19), (25)) constraints are introduced directly while the $O(n^3)$ triangular inequalities ((22)–(24)) are generated iteratively by a cutting plane algorithm.

6 Numerical Experiments

The goal of this Section is to evaluate the lower bound quality corresponding to the linear relaxation of (LP_F) **strengthened with, degree valid equalities (16), and partitioning cuts**. We note LB this bound. The lifting inequalities are necessary to reach a feasible optimal integer solution since they make the formulation (LP_F) equivalent to (Q). But our numerical experiments show that they play a minor role in the quality of the lower bound. For this reason, the results do not included these inequalities.

We compare our results with the currently published, best known lower bounds given in Schwarz PhD thesis [18]. We use the benchmark instances that come from the matrix market library [3]. All experiments are conducted on a DELL laptop equipped with Intel Core i3 CPU of 2.40 GHz and 3456 MB of memory. All linear programs are solved with IBM Ilog Cplex 12.5.1. It includes the programs need for LB as well as for the formulations (LP_{GP}).

In the global Algorithm 2, one may see, that the number of possible cardinalities, for the subsets of p partitions ($p \in \{2, ..., p_{max}\}$) increases with p. Thus, it may becomes very difficult, or intractable, to consider big value for p. In our experiments, we limit ourselves to $p_{max} = 3$.

To find good heuristic solutions of the graph partitioning problems on the graphs $G_{\overline{F}}$ (line 6 of Algorithm 1), we use the free software package **Chaco 2.2** [11]. Chaco is an open source code, under license of the Sandia National Laboratories, designed to find heuristic solutions of partitioning problems of any number of partitions, and whatever is the size of each partition. It has been written in C and applies several heuristic methods. More precisely, we use the multilevel Kernighan-Lin heuristic of this software. At each iteration, we generate the graph $G_{\overline{F}}$. Then we give it to a Chaco C function that solve the problem and return the best partition found $(\overline{W}_1, \overline{W}_2, ..., \overline{W}_p)$. The results of our computational experiments are summerized in Table 1.

In this table, "*Prob*" is the name of the problem, $d(\%)$ is the density of the graph G, and LB our lower bound. We distinguish the time to solve the graph paritioning problems on G_A, noticed $t_{(LP_{GP})}$, to the time to perform the global algorithm notices t_{LB}. We indicate in the column LB^* the best lower bounds of the litterature, the time to obtain it (t_{LB^*}), and the best upper bounds of the litterature UB. Gap_{LB} (resp. Gap_{LB^*}) is the relative deviation from UB of the bound LB (resp. LB^*).

The best bounds of the litterature are, in terms of gaps, better than ours but the scheme we propose is able to generate for some instances good bounds faster. For instance, if we consider the time $t_{LP_{GP}}$ and t_{LB}, we reach some bounds closed

Table 1. Numerical results.

Prob	n	$d(\%)$	LB	$t_{LP_{GP}}$	t_{LB}	LB^*	t_{LB^*}	UB	Gap_{LB}	Gap_{LB^*}
bcspwr01.mtx	39	6	76,5	25,5	1,4	106	5,7	106	28	0
bcspwr02.mtx	49	5	116	203,5	4	161	14,9	161	28	0
can24.mtx	24	25	181	0,8	1,6	210	3,3	210	14	0
can61.mtx	61	14	943	919,4	28,2	1137	1125,1	1137	17	0
can62.mtx	62	4	143	685,4	5,2	210	49,6	210	32	0
can73.mtx	73	6	969	4484,7	8,8	971	2016,8	1100	12	12
curtis54.mtx	54	9	296	232,7	9,9	454	69,6	454	35	0
dwt59.mtx	59	6	172,2	936,2	9,5	289	39,4	289	65	0
dwt66.mtx	66	6	190	1145	3	192	34,2	192	1	0
dwt72.mtx	72	3	133,5	2140,1	19,2	167	38,4	167	20	0
ibm32.mtx	32	18	426	6,2	1,4	485	1241,3	485	12	0
will57.mtx	57	8	214	648,5	3	335	50,3	335	36	0

to 17 % of the optimal value in *ibm*32 and *can*61 with a reduced processing time. The CPU time of the instance *ibm*32 is particularly short.

Due to the fact that the method to compute the lower bounds of the graph partitioning problems are independent of the partitioning cuts algorithm, one may focused only on the time t_{LB}. Since, applying other mathematical programming techniques, such as in the paper of Lisser and Rendl [15], may reduced this time. So, in the case we consider only the time t_{LB} the computing time appears even more interesting in comparison to t_{LB^*}.

7 Conclusion and Perspectives

We have proposed a new formulation based on some variables expressing the adjacency between locations. This formulation has been strenghthened by valid inequalities and equalities, in particular by partitioning cuts for which a cutting plane algorithm have been designed. Notice that the partitioning cuts and the cutting plane algorithm are not limited to the adjacency variables formulation. It may be applied similarly to the distances variables to enhance the corresponding bounds. The numerical results are currently modest with some relative deviations between 1 %, for the better case, to 65 %, for the worst one.

As perspectives, it is necessary to decrease the processing time need to solve the graph partitioning problems on G_A. This can be done by a more carefull analysis of the important graph partitioning problems to solve. Actually for each value p, all possible partitioning problems with p partitions are generated and solved. It is also necessary to better described the polytope induced by the adjacency variables to make our formulation more competitive in terms of bound quality. Finally, we will also investigate the application of partitioning cuts to distance variables formulation for closed problems such as the Quadratic Assignment Problem.

Acknowledgements. This work was financially supported by the region of Haute-Normandie (France) and the European Union. This support is gratefully acknowledged.

References

1. Amaral, A.: A new lower bound for the single row facility layout problem. Discrete App. Math. **157**, 183–190 (2009)
2. Amaral, A., Caprara, A., Letchford, A., Gonzalez, J.: A new lower bound for the minimum linear arrangement of a graph. Electron. Notes Discrete Math. **30**, 87–92 (2008)
3. Mouawad, A.E., Nishimura, N., Raman, V., Simjour, N., Suzuki, A.: On the parameterized complexity of reconfiguration problems. In: Gutin, G., Szeider, S. (eds.) IPEC 2013. LNCS, vol. 8246, pp. 281–294. Springer, Heidelberg (2013)
4. Caprara, A., Gonzalez, J.: Laying out sparse graphs with provably minimum bandwidth. INFORMS J. Comput. **17**(3), 356–373 (2005)
5. Caprara, A., Letchford, A., Gonzalez, J.: Decorous lower bounds for minimum linear arrangement. INFORMS J. Comput. **23** (2010)
6. Caprara, A., Oswald, M., Reinelt, G., Schwarz, R., Traversi, E.: Optimal linear arrangements using betweeness variables. Math Program. Comput. **3**, 261–280 (2011)
7. Deza, M., Laurent, M.: Geometry of cuts and metrics. In: Algorithms and Combinatorics, vol. 15. Springer, Heidelberg (1997)
8. Garey, M., Johnson, D., Stockmeyer, L.: Some simplified np-complete graph problems. Theoret. Comput. Sci. **1**, 237–267 (1976)
9. Gueye, S., Michel, S., Yassine, A.: A 0–1 linear programming formulation for the berth assignment problem. In: IEEE LOGISTIQUA 2011, pp. 50–54 (2011)
10. Harper, L.H.: Optimal assignment of numbers to vertices. J. SIAM **12**, 131–135 (1961)
11. Hendrickson, B., Leland, R.: The chaco user's guide, version 2.0. Technical report. SAND95-2344, Sandia National Laboratories (1995)
12. Horton, S.: The optimal linear arrangement problem: algorithms and approximation. Ph.D. Thesis, Georgia Institute of Technology (1997)
13. Hungerlaender, P., Rendl, F.: A computational study and survey of methods for the single-row facility layout problem. Technical report (2012)
14. Koopmans, T.C., Beckmann, M.J.: Assignment problems and the location of economic activities. Econometrica **25**, 53–76 (1957)
15. Lisser, A., Rendl, F.: Graph partitioning using linear and semidefinite programming. Math. Program., Ser. B **95**, 91–101 (2003)
16. Liu, W., Vanelli, A.: Generating lower bounds for the linear arrangement problem. Discrete Appl. Math. **59**, 137–151 (1995)
17. Petit, J.: Experiments on the minimum linear arrangement problem. Technical report (1999)
18. Schwarz, R.: A branch-and-cut algorithm with betweenness variables for the Linear arrangement problem. Ph.D. Thesis, Universitaet Heidelberg (2010)

A Vessel Scheduling Problem with Special Cases

Selim Bora[1](✉), Endre Boros[2], Lei Lei[2], W. Art Chovalitwongse[3], Gino J. Lim[4], and Hamid R. Parsaei[1]

[1] Texas A&M University at Qatar, Doha, Qatar
selim.bora@qatar.tamu.edu
[2] Rutgers University, New Brunswick, NJ, USA
[3] University of Washington, Washington, USA
[4] University of Houston, Houston, TX, USA

Abstract. We study the inventory and distribution operations encountered in oil and petrochemical industry. We show some special cases for the NP-complete problem, and propose polynomial time solution methods. We propose two approaches for the main problem. One of them makes use of the minimum cost flow formulation of the same problem under some assumptions, and the other one uses Benders Decomposition. In addition, we propose another problem and its formulation which involves time-windows for delivery, for which the same approaches can be applied. However, methodology or the results for the latter problem are not given.

Keywords: Vessel scheduling · Bender's decomposition · Minimum cost network flow

1 Introduction

This study is focused on the demand-supply coordination problem encountered in petrochemical industry such as oil and gasoline. The cost to produce and deliver gasoline products to the market consists of three major components: the transportation cost of crude oil to refiners, the operation cost of refinery processing, and the cost of marketing and distribution. An oil company typically operates many tens of refineries, with several million barrels of crude oil per day and several billion dollars on crude transportation per year. As the retail gasoline prices continue to rapidly elevate around the world, effectively coordinating the demand and supply of gasoline products has therefore become even more crucial to oil companies. Particularly in this study, the company uses its own and chartered vessels to distribute the gasoline products to discharging/demand locations. Each discharging location carries its own inventories and serves as a depot of distribution for the local market. Since vessels are expensive in both variable and fixed costs, any inefficiency in the supply process could result in a substantial operating cost. The distribution scheduling problem encountered in this process is very complicated due to the involvement of heterogeneous vessels (e.g., in terms of their loading capacities, discharging and berthing times, and operating costs) and the fact that each vessel has multi-level of loading capacities

E. Pinson et al. (Eds.): ICORES 2014, CCIS 509, pp. 108–125, 2015.
DOI: 10.1007/978-3-319-17509-6_8

such that a load beyond the normal/base capacity will result in an extra overload cost. Practical issues faced include which vessel should deliver to which depot in which time period, whether a particular vessel trip should carry an extra load and by how much, and what should be the ending inventory at a depot in a particular period, etc. Due to high distribution cost of gasoline products, an effectively distribution schedule could help the company to further improve the profit of its supply chain and to strengthen its competitive advantage in the market place.

Our goal is to minimize the operating costs related to shipping and handling of goods. The fleet size is not fixed, nor an initial amount is set, so one of the tasks we have at hand is to determine the number of vessels that will be used within the planning horizon. Shipping costs can be divided into two categories:(1) The fixed cost related to either purchase or lease of a vessel, (2) the overloading cost which is incurred if the vessels carry above a certain capacity. There are two more costs that we need to watch out for. Each shipment made to a port may incur a holding or penalty cost based on the demand. If the demand is not met on time, it cannot be satisfied at a later time period, and therefore we need to pay penalty for each unit. Also, if the port is forced to hold some inventory, then a holding cost is charged. In addition to all these cost factors, we also need to consider the fact that each vessel is available for a certain amount of time within a period, and therefore even if a vessel has enough capacity, it may not have enough time to visit all the ports we desire.

Optimally solving distribution operations scheduling problem is not an easy task. Previous work related to industrial shipping varies a lot. Here, we focus on the existing results that are closely related to our work. A large summary of works related to various types of vessel scheduling and routing problem can be found in the literature survey by [8]. Two more recent surveys can be found more specifically in the area of combined inventory management and routing [1] and on fleet composition and routing [11]. The most recent literature survey is by [7], which takes a look at the publications in the last decade, and list possible research areas that could be pursued in this area.

Reference [16] presents an algorithm which combines the linear programming technique with that of dynamic programming to improve the solution to linear model for fleet planning. Even though their approach is similar, the problem they are dealing with requires demand satisfaction and initial fleet is already given, and the decision is to whether add new vessels to the existing fleet or not.

Reference [5] presented a better formulation to the original fleet deployment problem proposed by [14]. In this formulation, just like we do, there is a single loading port, finite number of customer ports, and a finite planning horizon. However, they require the demand to be met, and the fleet size is constant. The costs incurred are due to routes chosen, shipping cargoes, and unloading time. They show that this formulation is better for computational efficiency.

Reference [4] present a study regarding fleet size and design of optimal liner routes for a container shipping company. The problem is solved by generating a number of candidate routes for the different ships first, and then, the problem is formulated and solved as a linear programming model, where the columns represent the candidate routes. They extend this model to a mixed integer programming model that also considers investment alternatives to expanding fleet capacity. Reference [2] also present a model for determining the optimal number of ships and fleet deployment plan.

On the other hand, [13] were the first ones to propose dynamic programming application to ship fleet management. The problem they dealt with was to determine the sequence in which the currently owned ships should be sold and the extent to which charter ships should be taken on. They tackle the problem in two stages. The first stage determines a good priority ordering for selling the ships regardless of the rate at which charter ships are taken on. The second stage uses dynamic programming to determine an optimal level of chartering given the priority replacement order. This first stage priority ordering essentially reduces the dynamic programming calculation from a problem with as many as states as number of ships in fleet to a 1 state variable problem which is computationally manageable by dynamic programming methods. Several authors use benchmark instances to compare the results of different strategies and heuristics. Reference [10] define 20 test instances with 12100 nodes for the standard fleet size and mix vehicle routing problem. Reference [15] deals with trucks that vary in capacity and age are utilized over space and time to meet customer demand. Operational decisions (including demand allocation and empty truck repositioning) and tactical decisions (including asset procurements and sales) are explicitly examined in a linear programming model to determine the optimal fleet size and mix. The method uses a time-space network, common to fleet-management problems, but also includes capital cost decisions, wherein assets of different ages carry different costs, as is common to replacement analysis problems. A two-phase solution approach is developed to solve large-scale instances of the problem. Phase I allocates customer demand among assets through Benders decomposition with a demand-shifting algorithm assuring feasibility in each subproblem. Phase II uses the initial bounds and dual variables from Phase I and further improves the solution convergence through the use of Lagrangian relaxation.

A network optimization approach has been proposed by [3], where they formulate a multi-commodity capacitated distribution-planning problem as a nonlinear mixed integer programming model, and solve it as a generalized assignment problem within an algorithm for the overall distribution/routing problem based on a Bender's type decomposition.

Reference [12] proposes an approach to a bi-directional flow problem where each iteration starts with a given planning horizon, which is then partitioned into three planning intervals, where each interval consists of consecutive time periods in the given planning horizon. Afterwards, some constraint relaxations are applied to the problem in which all the forward demand and all the backward demand of the time periods in the third planning interval are consolidated into a single forward demand and a single backward demand, which is an idea we use in one of our approaches.

Reference [6] focuses on minimizing total tardiness, rather than the operating costs, and the routes for vessels are observed under three different cases, one of them being arbitrary, just like in our problem. Later on, they talk about the other problems in the literature and how their approach is related to them.

2 Problem Definition

This paper, brings together some of the ideas that were proposed in the literature before. We are given a fleet $|V|$ of container vessels, $v \in \mathbf{V}$ that distributes the goods from a main distribution center to a number of customer ports over a $|T|$-period planning horizon. Each vessel has two loading capacities: the regular loading capacity u_v^0, and the maximum loading capacity u_v^{max} so that carrying a load beyond u_v^0 will impose an over loading charge $g_v^0/unit$ and carrying a load beyond u_v^{max} violates the feasibility. In addition to this limitation, for every vessel there is total available time τ_v which is used up by the berthing time $b_{v,p}$ at ports which vary depending on vessel type. There are $|p|$ customer ports on the network, each port $p \in \mathbf{P}$ has a demand, $d_{p,t} \geq 0$ in period $t \in \mathbf{T}$. For every port, unsatisfied demand are penalized at $p_{p,t}/unit$ based on the unsatisfied demand and no backlogging is allowed. On the other hand, end of period inventory incurs a holding cost of $h_n/unit$. Let c_v^f denote the fixed cost if the vessel is being dispatched in a period. The problem is finding a feasible vessel dispatching schedule to minimize the total shortage and overage penalty plus the vessel overloading and fixed cost. The minimum cost flow network formulation proposed guarantees optimality when the number of vessels dispatched in every period is known. To define our problem more formally, we define the following set of variables:

- $S_{p,t} \in \mathbb{Z}_+$: amount of shortage at port p in period t
- $Q_{v,p,t} \in \mathbb{Z}_+$: amount of supply delivered to port p in period t via vessel v's regular capacity
- $O_{v,p,t} \in \mathbb{Z}_+$: amount of supply delivered to port p in period t via vessel v's overloading capacity
- $I_{p,t} \in \mathbb{Z}_+$: ending inventory at port p in period t
- $Y_{v,p,t} \in \{0,1\}$: $Y_{v,n,t} = 1$ if vessel v delivers to port p in period t
- $Z_{v,t} \in \{0,1\}$: $Z_{v,t} = 1$ if vessel v is dispatched in period t.

Based on this, the constraints to the problem will include the following:

A vessel must not be carrying anything if it's not dispatched, nor visiting ports:

$$Q_{v,p,t} + O_{v,p,t} \leq u_v^{max} Y_{v,p,t} \quad \forall\, v \in \mathbf{V},\ p \in \mathbf{P},\ t \in \mathbf{T}. \tag{1a}$$

$$Y_{v,p,t} \leq Z_{v,t} \quad \forall\, v \in \mathbf{V},\ p \in \mathbf{P},\ t \in \mathbf{T}. \tag{1b}$$

Vessels dispatched must not be used over their time and regular/maximum capacity:

$$\sum_{p \in \mathbf{P}} b_{v,p} Y_{v,p,t} \leq \tau_v \quad \forall\, v \in \mathbf{V},\ t \in \mathbf{T}. \tag{2a}$$

$$\sum_{p \in \mathbf{P}} Q_{v,p,t} \leq u_v^0 \quad \forall\, v \in \mathbf{V},\ t \in \mathbf{T}. \tag{2b}$$

$$\sum_{p \in \mathbf{P}} (Q_{v,p,t} + O_{v,p,t}) \leq u_v^{max} \quad \forall\, v \in \mathbf{V},\ t \in \mathbf{T}. \tag{2c}$$

The last group of constraints is to help to formulate our objective, which is a compositions of all expenses (penalties, etc.).

Vessel dispatching costs:

$$c^D = \sum_{v \in \mathbf{V}} \sum_{t \in \mathbf{T}} c_v^f Z_{v,t}. \tag{3a}$$

Early arrival penalties:

$$c^H = \sum_{p \in \mathbf{P}} \sum_{t \in \mathbf{T}} h_p I_{p,t}. \tag{3b}$$

Unsatisfied demands' penalties:

$$c^U = \sum_{p \in \mathbf{P}} \sum_{t \in \mathbf{T}} p_{p,t} S_{p,t}. \tag{3c}$$

Overloading penalties:

$$c^O = \sum_{v \in \mathbf{V}} \sum_{p \in \mathbf{P}} \sum_{t \in \mathbf{T}} g_v^0 O_{v,p,t}. \tag{3d}$$

Then our problem is to minimize $c^D + c^H + c^U + c^O$, subject to the constraints (1)–(3) and the sign and type restrictions in the definitions of the decision variables.

If the dispatching information is already available, i.e. $|V_1|$ vessels for t = 1, $|V_2|$ for t = 2, ..., $|V_T|$ for t=T, then there becomes no need for the binary variables. In addition, define new variables, $x_{v,k,n,t}$ and $r_{v,k,n,t}$, which are the normal and over flows shipped by vessel v dispatched in period k for port n to satisfy the demand on period t. Based on this definition, the following can be established:

$$Q_{v,p,t} = \sum_{k=t}^{T} x_{v,t,p,k} \quad \forall\, v \in \mathbf{V},\ p \in \mathbf{P},\ t \in \mathbf{T}. \tag{4a}$$

$$O_{v,p,t} = \sum_{k=t}^{T} r_{v,t,p,k} \quad \forall\, v \in \mathbf{V},\ p \in \mathbf{P},\ t \in \mathbf{T}. \tag{4b}$$

$$S_{p,t} = d_{p,t} - \sum_{k \in \mathbf{T}} \sum_{v \in \mathbf{V}_k} (x_{v,k,p,t} + r_{v,k,p,t}) \quad \forall\, p \in \mathbf{P},\ t \in \mathbf{T}. \tag{4c}$$

$$I_{p,t} = \sum_{k \in \mathbf{T}} \sum_{w=t+1}^{T} \sum_{v \in \mathbf{V}_k} (x_{v,k,p,w} + r_{v,k,p,w}) \quad \forall\, p \in \mathbf{P},\ t \in \mathbf{T}. \tag{4d}$$

Based on the above assumptions and definitions, we get the following model. Objective function is the same except that the last part is now a constant based on vessel dispatching information, i.e. $Z_{v,t}$ values are known. Constraints (5b) and (5c) assure that normal and over capacity are not exceeded, where as constraint (5d) prevents shipments for a specific demand to be more than the demand itself, therefore making the first part of the objective function always nonnegative.

$$min \sum_{p \in \mathbf{P}} \sum_{t \in \mathbf{T}} p_{p,t}(d_{p,t} - \sum_{k=1}^{T} \sum_{v \in \mathbf{V}_k} (x_{v,k,p,t} + r_{v,k,p,t}))$$

$$+ \sum_{v \in \mathbf{V}} \sum_{p \in \mathbf{P}} \sum_{t \in \mathbf{T}} g_v^0 \sum_{k=t}^{T} r_{v,t,p,k} + \sum_{v \in \mathbf{V}} \sum_{t \in \mathbf{T}} c_v^f Z_{v,t}$$

$$+ \sum_{p \in \mathbf{P}} \sum_{t \in \mathbf{T}} h_p \sum_{k=1}^{t} \sum_{w=t+1}^{T} \sum_{v \in \mathbf{V}_k} (x_{v,k,p,w} + r_{v,k,p,w}). \tag{5a}$$

$$s.t. \sum_{p \in \mathbf{P}} b_{v,p} Y_{v,p,t} \leq \tau_v \quad \forall\, v \in \mathbf{V},\ t \in \mathbf{T}. \tag{5b}$$

$$\sum_{k=t}^{T} \sum_{n \in N} (x_{v,t,n,k} + r_{v,t,n,k}) \leq u_v^{max} \quad \forall\, v \in \mathbf{V},\ t \in \mathbf{T}. \tag{5c}$$

$$\sum_{k=t}^{T} x_{v,t,n,k} + r_{v,t,n,k} \leq u_v^{max} Y_{v,n,t} \quad \forall\, v \in \mathbf{V},\ p \in \mathbf{P},\ t \in \mathbf{T}. \tag{5d}$$

$$\sum_{n \in N} \sum_{k=t}^{T} x_{v,t,n,k} \leq u_v^0 \quad \forall\, v \in \mathbf{V},\ t \in \mathbf{T}. \tag{5e}$$

$$\sum_{k=1}^{t} \sum_{v \in V_k} (x_{v,k,n,t} + r_{v,k,n,t}) \quad \forall\, p \in \mathbf{P},\ t \in \mathbf{T}. \tag{5f}$$

$$Y_{v,n,t} \leq Z_{v,t} \quad \forall\, v \in \mathbf{V},\ p \in \mathbf{P},\ t \in \mathbf{T}. \tag{5g}$$

Lemma 21. *The above problem can be reformulated without the berthing time constraint and solved as a minimum cost flow problem by assuming the knowledge of the number of vessels dispatched in each time period.*

Proof. First, we construct a dummy source node S, and a dummy sink node F. Associate to each vessel $v \in \mathbf{V}_k$, 2 nodes $(v, k)^P$ and $(v, k)^O$, one for normal and other for over capacity. These nodes are connected to the source node with 0 and g_v^0 costs, a lower bound of 0 and an upper bound u_v^0 and $u_v^{max} - u_v^0$ respectively. Add another set of $|P|$ nodes (p_p) for case of shortage at each port with 0 costs, 0 lower bounds and no upper bounds. Next, take care of the ports by adding $|P| * |T|$ nodes denoted (p, t) for each port n at every period t. The arcs between nodes corresponding to vessels and ports incur a holding cost of $h_p(t - k)$, has a lower bound of 0 and no upper bound. Also, there will be arcs between shortage nodes, (p_p), and ports, (p, t), where the shortage costs $p_{p,t}$ will be charged. Finally, add arcs between ports and the sink, with a lower and upper bound of $d_{p,t}$ and no cost. This network will have $2|V||T| + |P| + |P||T|$ many nodes, and $|V||P||T^2| + |P^2|$ many arcs, making minimum cost flow approach practical for problems of reasonable size. An example network is shown in Fig. 1. □

Lemma 22. *The objective function values and constraints for both problems above are the same, assuming we guessed the right number of vessels.*

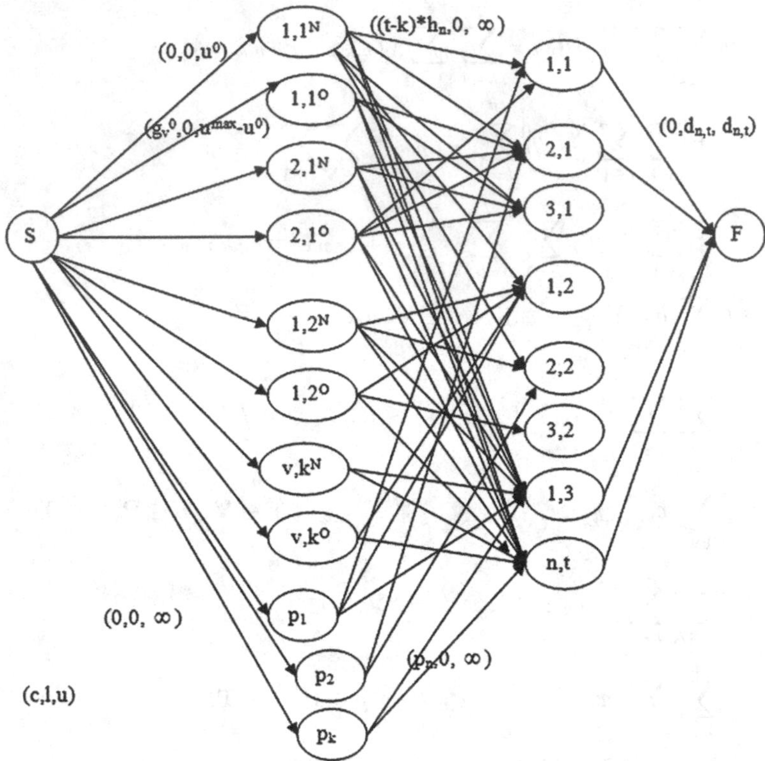

Fig. 1. An example of the network for $|V_1| = 2$, $|V_2| = 1$, $|V_3| = 2$, $|N| = 3$, and $|T| = 3$.

Proof. First of all, the fixed cost due to vessels for both problems will be the same. Next, assume $x^*_{v,k,n,t}$ and $r^*_{v,k,n,t}$ are the optimal flow vectors corresponding to the minimum cost flow problem. Then, using the equalities corresponding the variables of two problems, the objective function value of the original problem becomes:

$$\begin{aligned}
&\sum_{p \in \mathbf{P}} \sum_{t \in \mathbf{T}} p_{p,t}(d_{p,t} - \sum_{k \in \mathbf{T}} \sum_{v \in \mathbf{V}_k} (x_{v,k,p,t} + r_{v,k,p,t})) \\
&\quad + \sum_{v \in \mathbf{V}} \sum_{p \in \mathbf{P}} \sum_{t \in \mathbf{T}} g_v^0 \sum_{k=t}^{T} r_{v,t,p,k} + \sum_{v \in \mathbf{V}} \sum_{t \in \mathbf{T}} c_v^f Z_{v,t} \\
&\quad + \sum_{p \in \mathbf{P}} \sum_{t \in \mathbf{T}} h_p \sum_{k \in \mathbf{T}} \sum_{w=t+1}^{T} \sum_{v \in \mathbf{V}_k} (x_{v,k,p,w} + r_{v,k,p,w}) \qquad (6) \\
&= \sum_{p \in \mathbf{P}} \sum_{t \in \mathbf{T}} p_{p,t} S^*_{p,t} + \sum_{v \in \mathbf{V}} \sum_{p \in \mathbf{P}} \sum_{t \in \mathbf{T}} g_v^0 O^*_{v,k,p,t} \\
&\quad + \sum_{v \in \mathbf{V}} \sum_{t \in \mathbf{T}} c_v^f Z^*_{v,t} + \sum_{p \in \mathbf{P}} \sum_{t \in \mathbf{T}} h_p I^*_{p,t}
\end{aligned}$$

The first 2 lines of this equality (6) and the objective function of the minimum cost flow are exactly the same, which only leaves us with the inventory part. The w index is for shipments that are on a future date than current period t, and the k index is taking into account all shipments that have been made up to period t. Therefore, a shipment made on period k for period t will appear in the summation $(t-k)$ many times, allowing us to replace index w with t, remove the summation regarding w, and charge the holding cost as many times as necessary. This shows that both objective function values are the same.

As far as the constraints are concerned, first realize that in the original problem, (2a) is no longer required while berthing times are large enough. Similarly, (1a) and (1b) were associated with the fact that dispatching information was not available, so now, they could be dropped as well. Equation (2b) in the original problem is the same constraint as (5d) in the reduced problem, and they are both concerned with normal capacity of a vessel. Equations (5c) and (5d) of the reduced problem, added together, imply the same restriction on maximum vessel capacity as (2c) of the original problem. On the other hand, the flow balance constraint in the original problem is taken care of by two means: (1) the new index k for the variables, tells us when shipment was made, so we now whether a shipment is held at inventory or used immediately, (2) in the reduced problem, shipment for a specific demand will not be more than the demand itself, therefore shortage never becomes negative according to the relation between $S_{n,t}$ and $x_{v,k,n,t}$, $r_{v,k,n,t}$. □

3 Special Cases

Figure 2 is a list of special cases deduced from the general problem, for which we propose efficient solution approaches. Case 7 makes use of the method proposed by [9], used for solving knapsack problems with divisible item sizes.

Based on our minimum cost network flow approach, we propose the following two heuristics for no berthing time case:

3.1 Backward Heuristic

1. Divide the planning horizon into two groups, primary and secondary, for each port, the new demand is equal to sum of the individual demands in each group, holding cost is the minimum and penalty cost is the maximum of individual penalties.
2. Start with $|P| * |T|$ vessels in that group in total, solve the minimum cost flow problem iterating through all vessel dispatching combinations available for a group.
3. Once the optimal number of vessels required for each group are determined, repeat the procedure of dividing into groups and solving as a minimum cost flow problem for the individual groups. Demand belonging to ports in the other individual group is also added to the demand of the ports in the secondary period of the group under consideration.

Case	|V|	|N|	|T|	Other Assumptions	Method	Complexity
1	1	>1	1	- no berthing time	- sort ports based on p - solve greedily	$O(N \cdot \log(N))$
2	1	>1	1	$b_n = b, \forall n$	- case 1 where the first τ / b gets served	$O(N \cdot \log(N))$
3	1	>1	>1	- each customer has demand only once	- compute savings, $w_{nk} = p_n \cdot d_{nk} - h_n(t(n) - k)$ where $t(n)$ is the period of demand - solve greedily - move unsatisfied demand to previous period	$O(N \cdot \log(N) \cdot T)$
4	>1	>1	>1	- each customer has demand only once - each vessel must visit two ports - each demand must be satisfied by one vessel	- precompute cost of visiting pairs of ports (i, j), where $p_i \geq p_j$ and buying a vessel - solve the matching problem	$O(N^3)$
5	1	1	>1	- orders nonsubstitutable - demand nonsplittable (member of team type demand) - shortage not allowed - a vessel must satisfy consecutive demands	- Let $TT = \{t \mid b \leq \tau_t, 1 \leq t \leq T\}$ - Precompute arc costs, based on holding, overloading and vessel fixed cost - Solve the shortest path problem	$O(N^2)$
6	>1	1	>1	- vessel dispatching information available	- minimum cost flow problem	$O(N^3)$
7	>1	>1	>1	$p_n = p, h_n = h, \forall n$ - demands are divisible amongst eachother - small |T| - no overload - no berthing time	- use Detti algorithm with different |V| - Let $\Phi_{v,t}$ be the set of ports not being visited by |V| vessels in period t - go to previous period and calculate penalties carrying over the unsatisfied demand in the last period - use dynamic programming	$O((NV + N\log N)N + N^T)$
8	>1	>1	>1	$p_n = p, h_n = h, \forall n$ - demands are divisible amongst eachother - small |T| - no overload - no berthing time	$u \geq d_{n,t} = \sum_{j=0}^{\log u} \alpha_{n,t,j} \cdot 2^j, \forall n, t$ such that $\alpha_{n,t,j} \in \{0,1\}$ $b_{t,j} = \sum_{\forall n} \alpha_{n,t,j}$, $s_{t,j} = 2^j$ $v_{t,j} = penalty - holding$ Max $\sum_{\forall t} \sum_{\forall j} v_{t,j} \cdot \alpha_{n,t,j}$ $\sum_{\forall n} \sum_{\forall j} s_{t,j} \cdot \alpha_{n,t,j} \leq u \quad \forall t$ $\sum_{\forall j} \alpha_{n,t,j} \leq b_{t,j} \quad \forall t, n$ - solve with different |V| for every t	At most N^T cases

Fig. 2. Special cases deduced from the general problem under different assumptions.

4. Once the primary group has only 1 period remaining, optimal number of vessels have been determined for that group, start over.

3.2 Greedy Heuristic

1. Start with no vessels assigned to each period.
2. Add a vessel to any period and solve the problem. Remove the vessel, and add to another period, and solve again. Once the best vessel addition has been determined, move on to next vessel addition.
3. Keep determining the best vessel to add until objective function no longer improves.

Going back to the original problem with berthing constraints, we propose modified greedy heuristics and a decomposition based exact method. We first introduce the algorithms and then compare them with state of the are integer programming solver, XpressMP.

3.3 Improved Greedy Heuristic

1. Start with no vessels assigned to each period.
2. For each vessel type, compute maximal subsets of ports such that no further port can be added to a set due to berthing time constraint.
3. For each vessel type, in every period, sort the subset of ports in decreasing order based on $\sum_{n \in Maximal_v} d_{n,t} p_{n,t}$
4. Next, add any vessel to any period allowing it to only serve the top ranked subset of ports and solve the problem. Try different vessel types, for different periods in the same manner. Determine the best vessel to add to which period.
5. Once a vessel assignment has been determined, update remaining demand and sort subsets of ports accordingly.
6. Keep determining the best vessel to add until objective function no longer improves.

3.4 Bender's Type Decomposition Approach

1. Choose a feasible assignment of ports to a vessel to start with.
2. Solve the master problem to obtain a new objective function value and new port assignments to other vessels.
3. Keeping port assignments fixed, solve the dual problem.
4. STOP, if master and dual objective values are close enough, otherwise go back to the master problem.

4 Computational Results

With the minimum cost flow network formulation proposed, one question that arose was whether it was worth Table 1, the minimum cost flow problem is not submodular. *Case 1* and *Case 2* are random dispatches, where as *Case Int* refers to the scenario where minimum of number of vessels dispatched in each latter case is used, and *Case Union* refers to of *Case 1* and *Case 2* are random dispatches, where as *Case Int* refers to the scenario, where minimum of number of vessels dispatched in each latter case is used, and *Case Union* refers to the scenario, where maximum of number of vessels dispatched in each latter case is used.

Backward and Greedy Heuristic both performed well, however, as can be seen in Table 2, Backward Heuristic always takes shorter, where as Greedy Heuristic performs better by a slight margin.

However, it must be kept in mind that Table 2 reflects results for the version of the problem with no berthing time constraint. Improved Greedy Heuristic

Table 1. $|N| = 3$, $|T| = 4$, vessel type same, number of vessels in each period shown, as well as the optimal objective function value for each case, indicating that even a much simpler version of the original problem is not submodular.

Case 1	Obj. func	Case 2	Obj. func	Case int	Obj. func	Case union	Obj. func	Comp.
(3,2,2)	170	(2,1,3)	390	(2,1,2)	540	(3,2,3)	50	Lower
(1,2,1)	840	(2,1,3)	390	(1,1,1)	1140	(2,2,3)	90	Equal
(2,2,1)	580	(2,1,3)	390	(2,1,1)	880	(2,2,3)	90	Equal
(1,2,3)	350	(2,1,3)	390	(1,1,3)	650	(2,2,3)	90	Equal
(1,3,2)	390	(2,1,3)	390	(1,1,2)	800	(2,3,3)	50	Lower

Table 2. $|N| = 10$, $|T| = 10$, 3 different vessel types, number of vessels each method solves for vary for XpressMP, Backward and Greedy as 10, 100, 1 in respect. Runs are terminated after 2 h or when 0.1 % gap from the best bound is reached.

Time (s)			Objective function			Gap (%)		
Xpress	Backward	Greedy	Xpress	Backward	Greedy	Xpress	Backward	Greedy
7200	412.50	435.8	7503	7799	7733	1.7384	5.47	4.66
7200	420.40	435.20	8141	8314	8298	2.1557	4.20	4.01
7200	393.30	420.60	8270	8483	8450	0.9988	3.48	3.11
7200	332.30	390.70	7759	7806	7800	0.3519	0.95	0.88
7200	289.90	316.80	7316	7395	7375	0.1414	1.20	0.94
7200	310.30	336.60	8412	8494	8487	3.2309	4.16	4.09
7200	345.2	347.6	8270	8356	8338	3.8278	4.82	4.62
7200	429.50	443.30	7918	8159	8112	1.1641	4.08	3.53
7200	421.8	425.8	7475	7825	7768	1.4574	5.87	5.17
7200	306.70	321.10	7448	7661	7600	2.4489	5.16	4.40

designed to deal with this issue performs a bit slower than the previously mentioned heuristics, but gives good bounds for the solution of the original problem as can be seen in Table 3.

The formulation proposed for Bender's type approach is computationally efficient, as can be on Tables 4 and 5. The running time is a bit longer, but we're able to get exact solutions.

5 Delivery with Time-Windows

The problem we are interested in is related to the movement of full and empty containers in between ports. In this study we take the viewpoint of a single shipping company who delivers full containers from a single source port to a number of ports.

Table 3. $|N| = 10$, $|T| = 10$, 3 different vessel types, number of vessels each method solves for vary for XpressMP and Improved Greedy as 10 and 1 in respect. All runs are terminated after 2 h or when 0.1 % gap from the best bound is reached.

Time (s)		Objective function		Gap (%)	
Xpress	I. Greedy	Xpress	I. Greedy	Xpress	I. Greedy
7200	593.9	7503	7533	1.74	2.13
7200	582.2	8141	8226	2.16	3.17
7200	635.3	8270	8381	1.00	2.31
7200	678.2	7759	7921	0.35	2.39
7200	668.6	7316	7436	0.14	1.75
7200	594.4	8412	8548	3.23	4.77
7200	649.3	8270	8353	3.83	4.78
7200	667.3	7918	8092	1.16	3.29
7200	641.1	7475	7513	1.46	1.96
7200	662.1	7448	7565	2.45	3.96

Table 4. $|N| = 10$, $|T| = 10$, 3 different vessel types, number of vessels each method solves for vary for XpressMP and Improved Greedy as 10 and 1 in respect. All runs are terminated after 2 h or when 0.1 % percent gap from the best bound is reached.

Time (s)		Objective function		Gap (%)	
Xpress	Decomposition	Xpress	Decomposition	Xpress	Decomposition
7200	767.1	7503	7380	1.74	0.10
7200	757.5	8141	7973	2.16	0.10
7200	905.8	8270	8196	1.00	0.10
7200	1291.4	7759	7739	0.35	0.10
7200	1420.4	7316	7313	0.14	0.10
7200	959	8412	8148	3.23	0.10
7200	1183.5	8270	7961	3.83	0.10
7200	907.4	7918	7834	1.16	0.10
7200	1032.8	7475	7373	1.46	0.10
7200	1349.4	7448	7273	2.45	0.10

We assume that these ports can naturally be visited in a given cyclic order, though not all ships have to stop at each of the ports in this cycle. We also assume that at each of the ports there is a known demand or surplus of empty containers, and the ships can use their spare capacity to load these empty containers and deliver them to the other ports on their route.

We consider a single planning horizon with certain length in which we would like to optimize the schedule and loading plans for the vessels. In this planning

Table 5. $|N| = 15$, $|T| = 10$, 3 different vessel types, number of vessels each method solves for vary for XpressMP and Improved Greedy as 10 and 1 in respect. All runs are terminated after 2 h or when 0.1 % percent gap from the best bound is reached.

Time (s)		Objective function		Gap (%)	
Xpress	Decomposition	Xpress	Decomposition	Xpress	Decomposition
7200	805	7764	7619	1.98	0.10
7200	777.9	8652	8448	2.46	0.10
7200	990.5	8326	8245	1.07	0.10
7200	1401.1	8001	7978	0.39	0.10
7200	1544.6	7464	7461	0.15	0.10
7200	1015.4	8789	8496	3.43	0.10
7200	1216.9	9423	9026	4.30	0.10
7200	984.8	8128	8035	1.24	0.10
7200	1114.2	7849	7727	1.66	0.10
7200	1413.5	7652	7455	2.67	0.10

cycle, vessels are dispatched from the source port, carrying full and empty containers, within their capacity, to the other customer ports to satisfy their needs. Travel time between ports are deterministic, and the deliveries of the full containers must be made within given fixed time windows, defined uniquely for every port. If the vessels arrive prior to delivery time window, they must wait till the window opens, and incurring a waiting cost, where as if the delivery is made after the time window, a late fee is charged. Once a vessel is at a port and able to deliver, it remains there for the duration of berthing time associated with the particular port, and then the vessels can move to their next destination. Any unsatisfied handling of full containers or empty containers at a customer port incurs a penalty as well. There is no limitation on how many vessels could visit a single customer port. In addition to the capacity restrictions of vessels, vessels only have a fixed amount of time to visit ports and get back to the source port eventually. Also, the fleet size is not fixed, and we must decide how many vessels should be dispatched in the considered planning horizon.

5.1 Formulation

Let us denote by $\mathbf{V} = \{1, 2, ..., V\}$ the set of available vessels, by $\mathbf{P} = \{0, 1, ..., P\}$ the set of ports, 0 being the source, and $1, 2, ..., P$ being their natural order to get visited. Let us denote by $\mathbf{T} = \{1, 2, ..., T\}$ the set of time period, where time is measured in some natural unit, say in days. In other words, our planning horizon is T days, and all time windows are assumed to be specified as subintervals of consecutive days within $\mathbf{T}$. In the sequel we shall refer to vessels by $v, w, ...$, to ports by $p, q, ...$ and to time units by t.

We assume that each vessel $v \in \mathbf{V}$ has a capacity U_v, and a fixed cost C_v incurred if the vessel is dispatched. Travel times between ports $p, q \in \mathbf{P}$, $p < q$ are deterministic, denoted by $T_{p,q}$. Each customer port $p \in \mathbf{P}$ have a known demand for full containers, denoted by F_p, as well as demand for empty containers, denoted by E_p. We assume that $F_p \geq 0$ for all ports $p \in \mathbf{P}$, while E_p may take both negative and positive values, where $E_p < 0$ indicates the presence of a surplus of empty containers to be taken away. All deliveries made to a customer port $p \in \mathbf{P}$ expected to be within a fixed time window, $[A_p, B_p] \subset \mathbf{T}$. However vessels may arrive early or late with respect to the target time window. We assume that vessels berth only when they load/unload. To simplify our notations, we consider a particular day the "arrival" day for a vessel at a port if by that day it unloads/loads its cargo. We assume that the travel time $T_{p,q}$ includes the berthing time at port q. Accordingly, if a vessel arrives early to a port $p \in \mathbf{P}$, a holding cost of C_p^H/day/container is charged for the unloaded cargo, and similarly if a delivery is too late, a penalty cost of C_p^L/day/container is charged. Customer ports may be visited by more than one vessel, but in case full container demand, or loading/unloading of empty containers are not totally fulfilled, then costs C_p^F/container and C_p^E/container are charged, respectively. Finally, a revenue of R_p is received for satisfied full container demand at port $p \in \mathbf{P}$.

Let us note that since the total demand for full containers is given input for the considered problem, we get a mathematically equivalent formulation by simply assuming that $R_p = 0$ for all ports $p \in \mathbf{P}$.

To be able to formulate our model, we need to introduce the following decision variables:

Variable group	*Type*	Explanation
$x_{v,p}$ $v \in \mathbf{V}$ $p \in \mathbf{P}$	$\{0,1\}$	$x_{v,p} = 1$ if vessel v visits port p
y_v $v \in \mathbf{V}$	$\{0,1\}$	$y_v = 1$ if vessel v is dispatched
$a_{v,p}$ $v \in \mathbf{V}$ $p \in \mathbf{P}$	$\mathbb{Z}_+$	Number of full containers unloaded from vessel v at port p
$b_{v,p}$ $v \in \mathbf{V}$ $p \in \mathbf{P}$	$\mathbb{Z}_+$	Number of empty containers unloaded from vessel v at port p
$c_{v,p}$ $v \in \mathbf{V}$ $p \in \mathbf{P}$	$\mathbb{Z}_+$	Number of empty containers loaded by vessel v at port p
$f_{v,p}$ $v \in \mathbf{V}$ $p \in \mathbf{P}$	$\mathbb{Z}_+$	Number of full containers on vessel v when leaving port p
$e_{v,p}$ $v \in \mathbf{V}$ $p \in \mathbf{P}$	$\mathbb{Z}_+$	Number of empty containers on vessel v when leaving port p
$h_{v,p}$ $v \in \mathbf{V}$ $p \in \mathbf{P}$	$\mathbb{Z}_+$	Number of holding days of vessel v at port p due to early arrival
$\ell_{v,p}$ $v \in \mathbf{V}$ $p \in \mathbf{P}$	$\mathbb{Z}_+$	Number of late days of vessel v at port p
$t_{v,p}$ $v \in \mathbf{V}$ $p \in \mathbf{P}$	$\mathbb{Z}_+$	Arrival day (within $\mathbf{T}$) of vessel v at port p
z_p^F $p \in \mathbf{P}$	$\mathbb{Z}_+$	Unsatisfied full conatiner demand at port p
z_p^E $p \in \mathbf{P}$	$\mathbb{Z}_+$	unsatisfied empty container demand/pickup at port p

Let us note that the variables describing vessel content, $f_{v,p}$ and $e_{v,p}$ are defined for all vessels and ports, even if vessel v is not dispatched, or even if it does not visit port p. We shall make sure that these variables take value 0 for un-dispatched vessels, and we can view them, for dispatched vessel and unvisited ports, as the content of a passing by ship.

We can now start describing our model in terms of these decision variables and parameters. Let us start with a few necessary relations, stemming form the logical relations these quantities must satisfy.

Un-dispatched vessels cannot visit ports, and no loading or unloading at unvisited ports:

$$x_{v,p} \leq y_v \qquad \forall\, v \in \mathbf{V},\ p \in \mathbf{P} \tag{7a}$$
$$a_{v,p} \leq x_{v,p} \qquad \forall\, v \in \mathbf{V},\ p \in \mathbf{P} \tag{7b}$$
$$b_{v,p} \leq x_{v,p} \qquad \forall\, v \in \mathbf{V},\ p \in \mathbf{P} \tag{7c}$$
$$c_{v,p} \leq x_{v,p} \qquad \forall\, v \in \mathbf{V},\ p \in \mathbf{P} \tag{7d}$$

Number of containers on board must obey the law of conservation, and cannot exceed the ship's capacity:

$$f_{v,p} = f_{v,p-1} - a_{v,p} \qquad \forall\, v \in \mathbf{V},\ p \in \mathbf{P} \setminus \{0\} \tag{7e}$$
$$e_{v,p} = e_{v,p-1} - b_{v,p} + c_{v,p} \qquad \forall\, v \in \mathbf{V},\ p \in \mathbf{P} \setminus \{0\} \tag{7f}$$
$$e_{v,p} + f_{v,p} \leq U_v \qquad \forall\, v \in \mathbf{V},\ p \in \mathbf{P} \tag{7g}$$

Let us note that we defined all decision variables to be nonnegative. Thus, the above relations imply also that we cannot unload more than what we have on a ship, and we cannot load more than the ship's free capacity.

The next group of constraints make sure that we load/unload empty containers only where there is a demand/surplus:

$$b_{v,p} = 0 \qquad \forall\, p \in \mathbf{P} \text{ with } E_p \leq 0, \text{ and } \forall\, v \in \mathbf{V} \tag{7h}$$
$$c_{v,p} = 0 \qquad \forall\, p \in \mathbf{P} \text{ with } E_p \geq 0, \text{ and } \forall\, v \in \mathbf{V} \tag{7i}$$

Finally, we prescribe that container demands are met:

$$F_p = z_p^F + \sum_{v \in \mathbf{V}} a_{v,p} \qquad \forall\, p \in \mathbf{P} \tag{7j}$$

$$E_p = \begin{cases} z_p^E + \sum\limits_{v \in \mathbf{V}} b_{v,p} & \text{if } E_p \geq 0 \\ -z_p^E - \sum\limits_{v \in \mathbf{V}} c_{v,p} & \text{if } E_p \leq 0 \end{cases} \qquad \forall\, p \in \mathbf{P} \tag{7k}$$

Note that we defined E_p to take both positive (demand) and negative (surplus) values, and hence we had to formulate the above constraints accordingly.

The next group of constraints will help to get the right values assigned to the time related decision variables.

If the same vessel visits two ports, then the arrival times must differ by at least the travel time between these ports:

$$t_{v,p} \geq t_{v,q} + T_{q,p} + T(x_{v,p} + x_{v,q} - 2) \qquad \forall\, v \in \mathbf{V},\ p, q \in \mathbf{P},\ q < p \tag{8a}$$

Note that since T is a maximum time difference in our problem, the last term makes the above inequality irrelevant, unless vessel v visits both ports p and q.

$$A_p \leq h_{v,p} + t_{v,p} + T(1 - x_{v,p}) \qquad \forall\, v \in \mathbf{V},\ p \in \mathbf{P} \tag{8b}$$

$$B_p \geq t_{v,p} - \ell_{v,p} - T(1 - x_{v,p}) \qquad \forall\, v \in \mathbf{V},\ p \in \mathbf{P} \tag{8c}$$

Note that the last term makes the above constraints trivial, unless vessel v visits port p. Note also that both $h_{v,p}$ and $\ell_{v,p}$ are limited form below by these constraints. This, together with the fact that they will have nonnegative coefficients in our minimization objective, assures that our objective computes penalties for the true earliness/lateness.

The last group of constraints is to help to formulate our objective, which is a compositions of all expenses (penalties, etc.).

Vessel dispatching costs:

$$c^D = \sum_{v \in \mathbf{V}} C_v y_v \tag{9a}$$

Early and late arrival penalties:

$$c^H = \sum_{p \in \mathbf{P}} C_p^H \sum_{v \in \mathbf{V}} h_{v,p}(a_{v,p} + b_{v,p}) \tag{9b}$$

$$c^L = \sum_{p \in \mathbf{P}} C_p^L \sum_{v \in \mathbf{V}} \ell_{v,p}(a_{v,p} + c_{v,p}) \tag{9c}$$

Unsatisfied demands' penalties:

$$c^U = \sum_{p \in \mathbf{P}} C_p^F z_p^F + C_p^E z_p^E \tag{9d}$$

Then our problem is to minimize $c^D + c^H + c^L + c^U$, subject to the constraints (7)–(9) and the sign and type restrictions in the definitions of the decision variables.

Note that constraints (9b) and (9c) are quadratic in terms of our decision variables. All other constraints are linear.

Note also, that if variables x and y are binary, then the integrality of the other decision variables follow from the structure of the constraints and the integrality of the input parameters.

6 Conclusions

In this study, we studied a difficult real life supply chain scheduling problem encountered in oil and petrochemical industry, which involves production, inventory, and distribution operations, and requires an integrated scheduling to minimize the total operation cost. We showed the hardness of this problem, and showed that some of its special cases are polynomial time solvable. A minimum cost flow based heuristic, motivated by the observations from one of the special cases, was proposed and demonstrated to have a promising performance under the set of test cases considered in this study. Also, a new formulation of the model was developed, which made Bender's type decomposition method computationally efficient. Therefore, we're now able to get really good(exact) results for big problems at a much faster fashion then solver XpressMP.

In addition, we defined and formulated another version of the problem, for which time-windows for delivery are also set. Even though the methodology and results are not shown in this study, we have managed to apply a similar minimum cost flow approach as well as Bender's decomposition approach, both yielding promising results.

There are several interesting extensions of the work presented here. These include integrating the inland production with single or multiple refineries at different locations on the network, and multiple products needed by the same customer port. This integration would cause the supply chain to become bigger, and therefore more complex, however closer to reality, as inland production and demand satisfaction are activities that need synchronization. Furthermore, the involvement of multiple refineries and multiple products introduces the new optimization issues due to assigning refineries to customer ports and allocating vessel capacity for different products. This will make the modeling and the design of search procedures more interesting and challenging.

Also, for the simplicity of modeling, in this study, we assumed a linear penalty function for vessel overloading. However, this penalty cost is in reality very complex and is affected by many factors such as the level of overloading and navigation conditions. A nonlinear cost function would be more meaningful in this case.

Acknowledgements. This report was made possible by a National Priorities Research Program grant from the Qatar National Research Fund (a member of The Qatar Foundation). The statements made herein are solely the responsibility of the authors.

References

1. Andersson, H., Hoff, A., Christiansen, M., Hasle, G., Lkketangen, A.: Industrial aspects and literature survey: combined inventory management and routing. Comput. Oper. Res. **37**(9), 1515–1536 (2010)
2. Bendall, H., Stent, A.: A scheduling model for a high speed containership service: a hub and spoke short-sea application. Int. J. Marit. Econ. **3**(3), 262–277 (2001)
3. Bookbinder, J., Reece, K.: Vehicle routing considerations in distribution system design. Eur. J. Oper. Res. **37**(2), 204–213 (1988)
4. Cho, S., Perakis, A.: Optimal liner fleet routeing strategies. Marit. Policy Manag. **23**(3), 249–259 (1996)
5. Cho, S., Perakis, A.: An improved formulation for bulk cargo ship scheduling with a single loading port. Marit. Policy Manag. **28**(4), 339–345 (2001)
6. Choi, B., Lee, K., Leung, J., Pinedo, M., Briskorn, D.: Container scheduling: complexity and algorithms. Prod. Oper. Manag. **21**(1), 115–128 (2012)

7. Christiansen, M., Fagerholt, K., Nygreen, B., Ronen, D.: Ship routing and scheduling in the new millennium. Eur. J. Oper. Res. 228(3), 467–483 (2012)
8. Christiansen, M., Fagerholt, K., Ronen, D.: Ship routing and scheduling: status and perspectives. Transp. Sci. **38**(1), 1–18 (2004)
9. Detti, P.: A polynomial algorithm for the multiple knapsack problem with divisible item sizes. Inf. Process. Lett. **109**(11), 582–584 (2009)
10. Gheysens, F., Golden, B., Assad, A.: A comparison of techniques for solving the fleet size and mix vehicle routing problem. OR Spectr. **6**(4), 207–216 (1984)
11. Hoff, A., Andersson, H., Christiansen, M., Hasle, G., Lkketangen, A.: Industrial aspects and literature survey: fleet composition and routing. Comput. Oper. Res. **37**(12), 2041–2061 (2010)
12. Lei, L., Zhong, H., Chaovalitwongse, W.: On the integrated production and distribution problem with bidirectional flows. INFORMS J. Comput. **21**(4), 585–598 (2009)
13. Nicholson, T., Pullen, R., Dynamic programming applied to ship fleet management. Oper. Res. Quart. 22(3), 211–220 (1971)
14. Ronen, D.: Short-term scheduling of vessels for shipping bulk or semi-bulk commodities originating in a single area. Oper. Res. **34**(1), 164–173 (1986)
15. Wu, P., Hartman, J., Wilson, G.: An integrated model and solution approach for fleet sizing with heterogeneous assets. Transp. Sci. **39**(1), 87–103 (2005)
16. Xinlian, X., Tengfei, W., Daisong, C.: A dynamic model and algorithm for fleet planning. Marit. Policy Manag. **27**(1), 53–63 (2000)

Archimedean Copulas in Joint Chance-Constrained Programming

Michal Houda[(✉)] and Abdel Lisser

Laboratoire de Recherche En Informatique, Université Paris Sud – XI, Bât. 650, 91405 Orsay Cedex, France
houda@ef.jcu.cz, lisser@lri.fr

Abstract. We investigate the problem of linear joint probabilistic constraints with normally distributed constraints in this paper. We assume that the rows of the constraint matrix are dependent, the dependence is driven by a convenient Archimedean copula. We describe main properties of the problem, show how dependence modeled through copulas translates to the model formulation, and prove that the resulting problem is convex for a sufficiently high probability level. We further develop an approximation scheme for this class of stochastic programming problems based on second-order cone programming.

Keywords: Chance-constrained programming · Dependence · Archimedean copulas · Second-order cone programming

1 Introduction

Consider an *uncertain linear optimization problem*

$$\min c^T x \text{ subject to } \Xi x \leq h,\ x \in X \tag{1}$$

where $x \in X \subset \mathbb{R}^n$ is a decision vector of the problem, $\Xi \in \mathbb{R}^K \times \mathbb{R}^n$ is an uncertain (unknown) data matrix, $c \in \mathbb{R}^n$, $h = (h_1, \ldots, h_K)^T \in \mathbb{R}^K$ are fixed deterministic vectors, dimensions n, K are structural elements of the optimization problem (1). If a realization of the data element Ξ is known and fixed in advance (before a decision is taken), we can solve the problem (1) as classical linear optimization problem. This situation is rarely the case. More often, we have to consider *uncertainty of the data* as natural element of the modeling phase.

During the history of mathematical optimization, various methods were developed to deal with the uncertainty: ex-post sensitivity analysis, parametric programming, or robust optimization. In our paper, we concentrate on the *stochastic programming* approach assuming that the data matrix Ξ is a random matrix whose probabilistic characteristics are known in advance. For example, if the constraints of (1) are required to be satisfied with a prescribed sufficiently high probability $p \in [0; 1]$, then the problem (1) can be reformulated as

$$\min c^T x \text{ subject to } \mathbb{P}\{\Xi x \leq h\} \geq p,\ x \in X \tag{2}$$

E. Pinson et al. (Eds.): ICORES 2014, CCIS 509, pp. 126–139, 2015.
DOI: 10.1007/978-3-319-17509-6_9

where $p \in [0; 1]$ is a prescribed probability level. The problem (2) is known as *probabilistically* (or *chance*) *constrained linear optimization problem*. The problem was treated many times in literature; for a thorough review of methods and bibliography we refer to the classical book [1] and recent chapters [2,3].

The chance constrained optimization problems are very challenging in their general (linear or nonlinear) form. Two main issues of the stochastic optimization theory concerning these problems are the *convexity* of the set of feasible solutions, and a very high *computational effort* to be accomplished. In detail: even for the "nice" linear program (2) the feasible set may be nonconvex, and the probability $\mathbb{P}$ can result in an intractable computation of multivariate integrals.

In our paper, we restrict our consideration to a *problem with linear normally distributed constraint rows*, namely, the rows Ξ_k^T of Ξ follow n-dimensional normal distributions with means μ_k and positive definite covariance matrices Σ_k. To further simplify the situation we assume that $X = \mathbb{R}^n$ (only the probabilistic constraints are in question). Denote

$$X(p) := \{x \in \mathbb{R}^n \mid \mathbb{P}\{\Xi x \le h\} \ge p\}. \tag{3}$$

We are interested in an equivalent formulation of the set $X(p)$ convenient for numerical purposes. To this end, we first present a result for the set

$$M(p) := \{x \in \mathbb{R}^n \mid \mathbb{P}\{g_k(x) \ge \xi_k,\ k = 1, \dots, K\} \ge p\}, \tag{4}$$

where $\xi := (\xi_1, \dots, \xi_K)$ is an absolutely continuous random vector and $g_k(x)$ are continuous functions. $M(p)$ is usually referred to as the set of feasible solutions for a continuous *chance-constrained problem with random right-hand side.*

The convexity of the sets $X(p)$ and $M(p)$ is treated several times in the literature; we mention [4–6] as the first classical results, and [7–9] as recently published papers. These results are simplified either by restricting consideration to one-row problem only, or by assuming independence of matrix rows. In our paper we demonstrate the use of *copula theory* to deal with dependence of rows in (2). This was done first by [10] for the set $M(p)$ using a class of so-called logexp-concave copulas. We extend their results to another large, more usual class of copulas and formulate an equivalent description of the problem (2) convenient to be solved by methods of second-order cone programming.

2 Dependence

2.1 Basic Facts About Copulas

Theory of copulas is well known for the people of probability theory and mathematical statistics but, to our knowledge, was not used up to these days in stochastic programming to describe the structure of the problem. In this section, we mention only some basic facts about copulas necessary for our following investigation. Most of the notions here (up to Proposition 4) were taken from the book [11].

Definition 1. *A* copula *is the distribution function* $C : [0;1]^K \to [0;1]$ *of some* K*-dimensional random vector whose marginals are uniformly distributed on [0;1].*

Proposition 1. (Sklar's Theorem). *For any* K*-dimensional distribution function* $F : \mathbb{R}^K \to [0;1]$ *with marginals* $F_1, \dots, F_K$*, there exists a copula* C *such that*

$$\forall z \in \mathbb{R}^K \quad F(z) = C(F_1(z_1), \dots, F_K(z_K)). \tag{5}$$

If, moreover, F_k *are continuous, then* C *is uniquely given by*

$$C(u) = F(F_1^{-1}(u_1), \dots, F_K^{-1}(u_K)). \tag{6}$$

Otherwise, C *is uniquely determined on* $\operatorname{range} F_1 \times \cdots \times \operatorname{range} F_K$.

Through Sklar's Theorem, we have in hand an efficient general tool for handling an arbitrary dependence structure. First, if we know the marginal distributions F_k together with the copula representing the dependence we can unambiguously determine the joint distribution. On the other hand, the copula can be uniquely derived from the knowledge of the joint and all marginal distributions. Our first example is the *independent (product) copula* which is nothing else than the independence formula for distribution functions:

$$C_\Pi(u) = \prod_k u_k. \tag{7}$$

The second important example is the *Gaussian copula* which is given by Sklar's Theorem applied to a joint normal distribution and its normally distributed marginals:

$$C_\Sigma(u) = \Phi^\Sigma(\Phi^{-1}(u_1), \dots, \Phi^{-1}(u_K)) \tag{8}$$

where Φ^Σ is the distribution function of the multivariate normal distribution with zero mean, unit variance and covariance matrix Σ, and $\Phi^{-1}(u_k)$ are standard one-dimensional normal quantiles. For illustration purposes, we provide a set of figures (Figs. 1, 2, 3, 4 and 5) of some popular copulas. From the left-hand side, the reader can always find the distribution function of the copula (i. e., the copula itself), its density, and the density of the distribution given by the copula applied to the standard normal marginals. Figure 1 represents the independent copula; compare it to the Gaussian copula in Fig. 2. Note that the Gaussian copula is the *only* copula that can represent the joint normal distribution.

The following proposition provides the limits in which the copulas can be located.

Proposition 2. (The Fréchet-Hoeffding Bounds). *Every copula* C *satisfies the inequalities*

$$W(u) \le C(u) \le C_M(u) \tag{9}$$

where

$$W(u) := \max\left\{\sum u_k - K + 1, 0\right\},$$
$$C_M(u) := \min_k\{u_k\}.$$

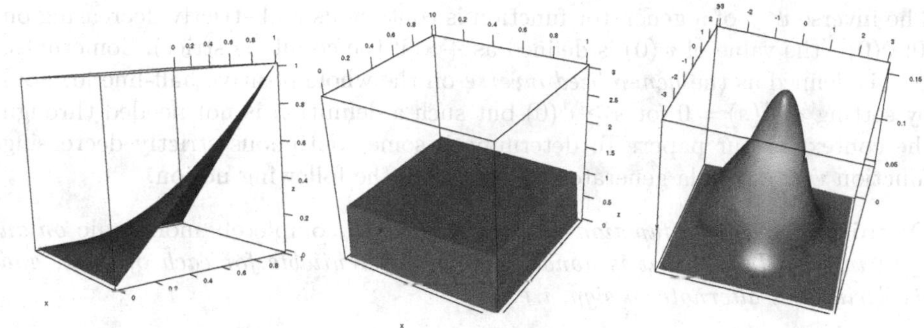

Fig. 1. Independent copula: distribution, density, and density with standard normal marginals.

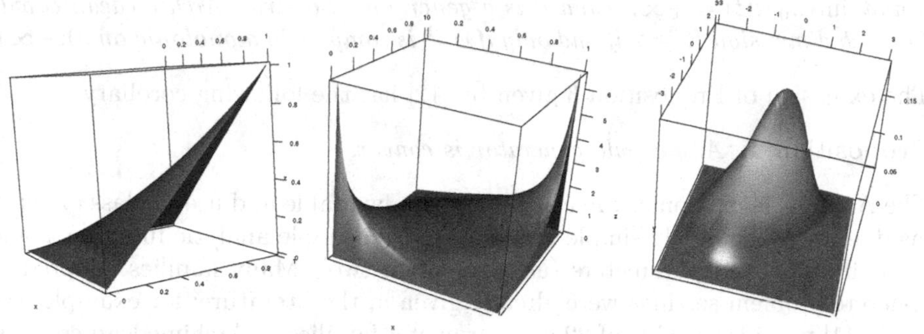

Fig. 2. Gaussian copula ($\rho = 0.55$): distribution, density, and density with standard normal marginals.

The function W represents the completely negative dependence between marginal distributions, but it is known *not* to be a copula if $K > 2$. C_M represents the completely positive dependence and it is known under the name of the *comonotone (maximum) copula.* These functions together with the independent copula are often found to be limiting cases of some other classes of copulas.

The Gaussian copula has a rather complicated structure (even it is not analytic) to be treated directly in our optimization problems. Instead, we need a different, simpler class of copulas, which we found in so-called Archimedean copulas.

Definition 2. *A copula C is called* Archimedean *if there exists a continuous strictly decreasing function $\psi : [0;1] \to [0;+\infty]$, called* generator *of C, such that $\psi(1) = 0$ and*

$$C(u) = \psi^{-1}\left(\sum_{i=1}^{n} \psi(u_i)\right). \tag{10}$$

If $\lim_{t\to 0} \psi(t) = +\infty$ then C is called a strict Archimedean copula *and ψ is called a* strict generator.

The inverse ψ^{-1} of a generator function is continuous and strictly decreasing on $[0; \psi(0)]$ (the value of $\psi(0)$ is defined as $+\infty$ if the copula is strict). Sometimes, ψ^{-1} is defined as the *generalized inverse* on the whole positive half-line $[0; +\infty)$ by setting $\psi^{-1}(s) = 0$ for $s \geq \psi(0)$ but such a definition is not needed through the context of our paper. To determine if some continuous strictly decreasing function ψ is a copula generator we introduce the following notion.

Definition 3. *A real function $f : \mathbb{R} \to \mathbb{R}$ is called* completely monotonic *on an open interval $I \subseteq \mathbb{R}$ if it is nondecreasing, differentiable for each order k, and its derivatives alternate in sign, i. e.,*

$$(-1)^k \frac{\mathrm{d}^k}{\mathrm{d}t^k} f(t) \geq 0 \ \forall k = 0, 1, \ldots, \ \textit{and} \ \forall t \in I. \tag{11}$$

Proposition 3. *Let $\psi : [0; 1] \to \mathbb{R}_+$ be a strictly decreasing function with $\psi(1) = 0$ and $\lim_{t \to 0} \psi(t) = +\infty$. Then it is a generator of a strict Archimedean copula for each dimension $K \geq 2$ if and only if ψ^{-1} is completely monotonic on* $(0; +\infty)$.

The extension of Proposition 3 given by [12] has the following corollary:

Proposition 4. *Any copula generator is convex.*

The Archimedean copulas are considered as a favorable and useful class of copulas due to their possibly simple formulation by a simple analytic function ψ and a small number of parameters (usually one or two). Many families adapted to concrete problem settings were already given in the literature; for example, the book [11] provides a table of 22 one-parameter families of Archimedean copulas. We give some examples in Table 1 and Figs. 3, 4 and 5. The Gumbel-Hougaard and Joe copulas are asymmetric (in the sense of density contours for normal marginals) stressing the dependence of positive random variables; the Clayton copula is in a similar view useful to model the positive dependence of negative random variables. The Frank copulas have symmetric density contours for normal marginals.

The Archimedean copulas provide an easy equivalent formulation for feasible sets (3) and (4). We start with the set $M(p)$; assume (for each $k = 1, \ldots, K$) that the elements ξ_k of ξ have continuous distribution functions F_k, and the

Table 1. Selected Archimedean copulas with completely monotonic inverse generators.

Copula family	Param. θ	Gen. $\psi_\theta(t)$
Independent (product)	–	$-\ln t$
Gumbel-Hougaard	$\theta \geq 1$	$(-\ln t)^\theta$
Clayton	$\theta > 0$	$\frac{1}{\theta}(t^{-\theta} - 1)$
Joe	$\theta \geq 1$	$-\ln[1 - (1-t)^\theta]$
Frank	$\theta > 0$	$-\ln\left(\frac{e^{-\theta t} - 1}{e^{-\theta} - 1}\right)$

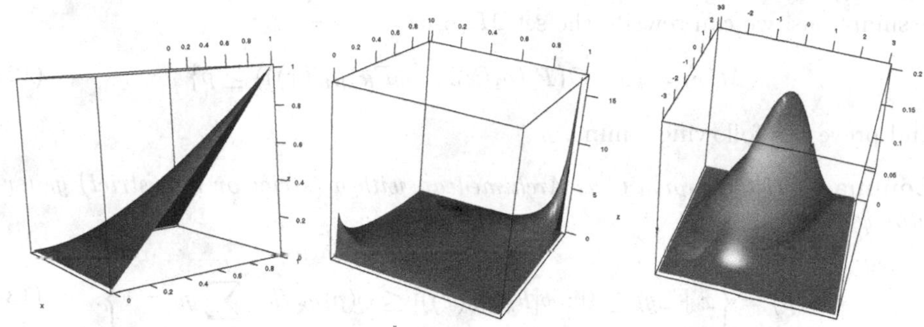

Fig. 3. Gumbel-Hougaard copula ($\theta = 1.6$): distribution, density, and density with standard normal marginals.

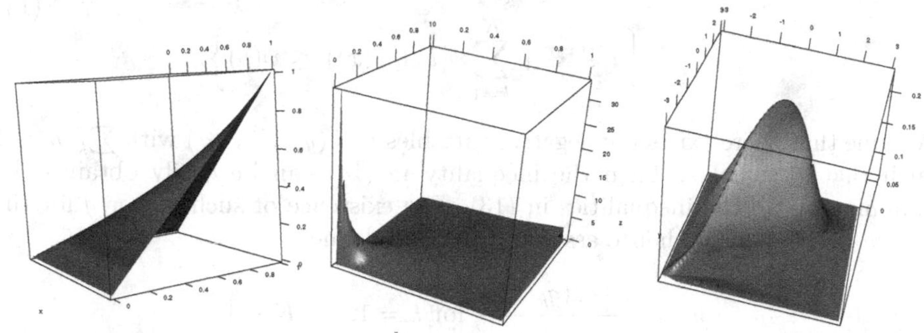

Fig. 4. Clayton copula ($\theta = 1.8$): distribution, density, and density with standard normal marginals.

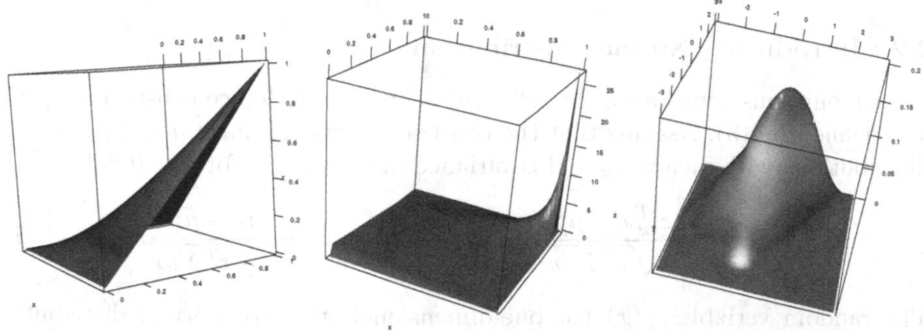

Fig. 5. Joe copula ($\theta = 2.1$): distribution, density, and density with standard normal marginals.

whole vector ξ has the joint distribution induced by a copula C. With these assumptions, we can rewrite the set $M(p)$ as

$$M(p) = \{x \mid C\big(F_1(g_1(x)), \dots, F_K(g_K(x))\big) \geq p\} \tag{12}$$

and prove the following lemma.

Lemma 1. *If the copula C is Archimedean with a (strict or non-strict) generator ψ then*

$$M(p) = \Big\{x \mid \exists y_k \geq 0 : \psi[F_k(g_k(x))] \leq \psi(p) y_k \forall k,\ \sum_{k=1}^{K} y_k = 1\Big\}. \tag{13}$$

Proof. From basic properties of ψ it is easily seen that

$$\begin{aligned} M(p) &= \Big\{x \mid \psi^{-1}\Big(\sum_{k=1}^{K} \psi[F_k(g_k(x))]\Big) \geq p\Big\} \\ &= \Big\{x \in \mathbb{R}^n \mid \sum_{k=1}^{K} \psi[F_k(g_k(x))] \leq \psi(p)\Big\}. \end{aligned} \tag{14}$$

Assume that there exists nonnegative variables $y = (y_1, \dots, y_K)$ with $\sum_k y_k = 1$ such that (13) holds. Then the inequality in (14) can be easily obtained by summing up all the inequalities in (13). The existence of such vector y for the case $p = 1$ is obvious; hence assume $p < 1$ and define

$$\begin{aligned} y_k &:= \frac{\psi\big[F_k(g_k(x))\big]}{\psi(p)} \text{ for } k = 1, \dots, K-1, \\ y_K &:= 1 - \sum_{k=1}^{K-1} y_k. \end{aligned}$$

It is now easy to verify that such definition of y_k satisfies (13). □

2.2 Introducing Normal Distribution

Return our consideration to the set $X(p)$ of the linear chance constrained problem defined by (3). Assume that the constraint rows Ξ_k^T have n-variate normal distributions with means μ_k and covariance matrices Σ_k. For $x \neq 0$ define

$$\xi_k(x) := \frac{\Xi_k^T x - \mu_k^T x}{\sqrt{x^T \Sigma_k x}}, \qquad g_k(x) := \frac{h_k - \mu_k^T x}{\sqrt{x^T \Sigma_k x}}. \tag{15}$$

The random variable $\xi_k(x)$ has one-dimensional standard normal distribution (in particular, this distribution is independent of x). Therefore the feasible set can be written as

$$X(p) = \Big\{x \mid \mathbb{P}\big[\xi_k(x) \leq g_k(x)\ \forall k\big] \geq p\Big\}. \tag{16}$$

If $K = 1$ (i. e., there is only one row constraint), the feasible set can be simply rewritten as

$$X(p) = \left\{x \mid \mu_1^T x + \Phi^{-1}(p)\sqrt{x^T \Sigma_1 x} \leq h_1\right\}. \tag{17}$$

where, again, Φ^{-1} is the one-dimensional standard normal quantile function. Introducing auxiliary variables y_k and applying Lemma 1, we derive the following lemma which gives us an equivalent description of the set $X(p)$ using the copula notion.

Lemma 2. *Suppose, in* (3), *that* $\Xi_k^T \sim N(\mu_k, \Sigma_k)$ *(with appropriate dimensions) where* $\Sigma_k \succ 0$. *Then the feasible set of the problem* (2) *can be equivalently written as*

$$X(p) = \Big\{x \mid \exists y_k \geq 0 : \sum_k y_k = 1,$$
$$\mu_k{}^T x + \Phi^{-1}\left(\psi^{-1}(y_k \psi(p))\right) \sqrt{x^T \Sigma_k x} \leq h_k \ \forall k\Big\} \tag{18}$$

where Φ *is the distribution function of a standard normal distribution and* ψ *is the generator of an Archimedean copula describing the dependence properties of the rows of the matrix* Ξ.

Proof. Straightforward using the arguments given above. The remaining case $x = 0$ is obvious. □

2.3 Convexity

It is not easy to show the convexity of the sets $M(p)$ and $X(p)$. The technique is based on the theory presented in [8] for the case of independence, and [10] for the case of dependence modeled via logexp-concave copulas. Our approach is different and makes direct use of the convexity property of Archimedean generators. Before do that, we recall some necessary definitions needed to formulate the convexity result.

Definition 4 ([1], Chap. 4 of [13]). *A function* $f : \mathbb{R}^s \to (0; +\infty)$ *is called* r-concave *for some* $r \in [-\infty; +\infty]$ *if*

$$f(\lambda x + (1-\lambda)y) \geq [\lambda f^r(x) + (1-\lambda) f^r(y)]^{1/r} \tag{19}$$

is fulfilled for all $x, y \in \mathbb{R}^s$ *and all* $\lambda \in [0; 1]$. *The cases* $r = -\infty, 0, +\infty$ *are to be interpreted by continuity.*

The case $r = 1$ is concavity in the usual sense. The case $r = 0$ correspond to the so-called log-concavity, i.e., to the case in which the function $\ln f$ is concave. The case $r = -\infty$ is known as quasi-concavity and corresponding right-hand side of (19) takes the form of $\min\{f(x), f(y)\}$. If f is r-concave for some r, then it is r'-concave for all $r' \leq r$; in particular, all r-concave functions are quasi-concave.

Definition 5 ([8]). *A function* $f\colon \mathbb{R} \to \mathbb{R}$ *is called* r-decreasing *for some* $r \in \mathbb{R}$ *with the threshold* $t^* > 0$ *if it is continuous on* $(0; +\infty)$ *and the function* $t \mapsto t^r f(t)$ *is strictly decreasing for all* $t > t^*$.

The threshold t^* depends on the value of r, hence, in this view, it can be considered as a function of r. For simplicity, we have dropped this implicit dependence from the notation. If the function $f(t)$ is non-negative and r-decreasing for some r, then it is r'-decreasing for all $r' \leq r$. In particular, if $r > 0$ then $f(t)$ is 0-decreasing, hence strictly decreasing for $t > t^*$. The table of prominent one-dimensional r-decreasing densities together with their thresholds has been given in [8].

Now we are ready to formulate a convexity result for $M(p)$, the feasible set of a chance-constrained programming problem with random right-hand side defined by (4).

Theorem 1. *Consider the set* $M(p)$ *and the following assumptions for* $k = 1, \ldots, K$:

1. *there exist* $r_k > 0$ *such that* g_k *are* $(-r_k)$*-concave,*
2. *the marginal distribution functions* F_k *have* (r_k+1)*-decreasing densities with the thresholds* t_k^*, *and*
3. *the copula* C *is Archimedean with a strict generator* ψ, *and* ψ^{-1} *is completely monotonic function.*

Then $M(p)$ *is convex for all* $p > p^* := \max_k F_k(t_k^*)$.

Proof. Let $p > p^*$, $\lambda \in [0;1]$, and $x, y \in M(p)$. We have to show that $\lambda x + (1-\lambda)y \in M(p)$, that is

$$\begin{aligned} C\big(F_1[g_1(\lambda x + (1-\lambda)y)], \ldots, F_K[g_K(\lambda x + (1-\lambda)y)]\big) \\ = \psi^{-1}\left\{\sum_{k=1}^{K} \psi\big(F_k[g_k(\lambda x + (1-\lambda)y)]\big)\right\} \geq p \end{aligned}$$

or, equivalently,

$$\sum_{k=1}^{K} \psi\big(F_k[g_k(\lambda x + (1-\lambda)y)]\big) \leq \psi(p)$$

(c.f. the proof of Lemma 1). Denote, for $k = 1, \ldots, K$,

$$q_k^x := F_k[g_k(x)], \qquad q_k^y := F_k[g_k(y)].$$

In the first part of the proof of Theorem 1 in [10] it has been shown, based on assumptions 1 and 2, that

$$F_k[g_k(\lambda x + (1-\lambda)y)] \geq [q_k^x]^\lambda [q_k^y]^{1-\lambda},$$

hence

$$\begin{aligned} \psi\{F_k[g_k(\lambda x + (1-\lambda)y)]\} &\leq \psi\{[q_k^x]^\lambda [q_k^y]^{1-\lambda}\} \\ &= \psi\{\exp[\lambda \ln q_k^x + (1-\lambda)\ln q_k^y]\}. \end{aligned}$$

Assumption 3 implies that the function $u \mapsto \psi(e^u)$ is convex function on $(-\infty; 0]$. This allows us to continue

$$\begin{aligned}\psi\{F_k[g_k(\lambda x + (1-\lambda)y)]\}& \\ &\leq \lambda\psi(e^{\ln q_k^x}) + (1-\lambda)\psi(e^{\ln q_k^y}) \\ &= \lambda\psi(q_k^x) + (1-\lambda)\psi(q_k^y).\end{aligned}$$

Introducing auxiliary variables y_k with $\sum y_k = 1$, and applying Lemma 1, we conclude on

$$\begin{aligned}\sum_{k=1}^{K}\psi(F_k[g_k(\lambda x + (1-\lambda)y)])& \\ &\leq \sum_{k=1}^{K}(\lambda\psi(q_k^x) + (1-\lambda)\psi(q_k^y)) \\ &\leq \sum_{k=1}^{K}(\lambda\psi(p)y_k + (1-\lambda)\psi(p)y_k) = \psi(p).\end{aligned}$$

□

3 Main Result

3.1 Convex Reformulation

In Lemma 2 we have already stated an equivalent formulation of the feasible set $X(p)$. Together with Theorem 1 we can formulate the following theorem.

Theorem 2. *Consider the problem* (2) *where*

1. *the matrix Ξ has normally distributed rows Ξ_k^T with means μ_k and positive definite covariance matrices Σ_k;*
2. *the joint distribution function of $\xi_k(x)$ given by* (15) *is driven by an Archimedean copula with the generator ψ.*

Then the problem (2) *can be equivalently written as*

$$\begin{aligned}&\min c^T x \text{ subject to} \\ &\quad {\mu_k}^T x + \Phi^{-1}\left(\psi^{-1}(y_k\psi(p))\right)\sqrt{x^T\Sigma_k x} \leq h_k, \\ &\quad \sum_k y_k = 1 \\ &\quad x \in X,\ y_k \geq 0 \text{ with } k = 1,\ldots,K.\end{aligned} \tag{20}$$

Moreover, if

3. *the function ψ^{-1} is completely monotonic;*

4. $p > p^* := \Phi\left(\max\{\sqrt{3}, 4\lambda_{max}^{(k)}[\lambda_{min}^{(k)}]^{-3/2}\|\mu_k\|\}\right)$, *where* $\lambda_{max}^{(k)}, \lambda_{min}^{(k)}$ *are the largest and lowest eigenvalues of the matrices* Σ_k, *and* Φ *is the one-dimensional standard normal distribution function, then the problem is convex.*

Proof. The first part of the theorem was proven as Lemma 2. Concerning convexity assertion, transformation (15) converts the problem to that of Theorem 1. Notice that the function g_k of (15) is -2-concave and normal density is 3-decreasing with the threshold $\sqrt{3}$. The remaining part of the proof (together with the exact form of the required probability level p^*) repeats arguments of the proof of Theorem 5.1 of [8] without considerable changes. □

3.2 SOCP Approximation

Second-order cone programming (SOCP) is a subclass of convex optimization in which the problem constraint set is the intersection of an affine linear manifold and the Cartesian product of second-order (Lorentz) cones [14]. Formally, a constraint of the form

$$\|Ax + b\|_2 \leq e^T x + f$$

is a second-order cone constraint as the affine function $(Ax + b, e^T x + f)$ is required to lie in the second-order cone $\{(y, t) \mid \|y\|_2 \leq t\}$. The linear and convex quadratic constraints are nominal examples of second-order cone constraints. It is easy to see that the constraint (17) is SOCP constraint with $A := \Sigma_1$, $b := 0$, $e := -\frac{1}{\Phi^{-1}(p)}\mu_1$, and $f := \frac{1}{\Phi^{-1}(p)}h_1$ provided $p \geq \frac{1}{2}$. For a details about SOCP methodology we refer the reader to [14], and to the monograph [15].

Theorem 2 provides us an equivalent nonlinear convex reformulation of the linear chance-constrained problem (2). Due to the decision variables y_k appearing as arguments to the (nonlinear) quantile function Φ^{-1}, it is not still a second-order cone formulation. To resolve this computational issue, we formulate a lower and upper approximation to the problem (20) using the favorable convexity property of the Archimedean generator. We first formulate an auxiliary convexity lemma which gives us a possibility to find these approximations.

Lemma 3. *If* $p > p^*$ *(given in Theorem 2), and* ψ *is a generator of an Archimedean copula, then the function*

$$y \mapsto H(y) := \Phi^{-1}\left(\psi^{-1}(y\psi(p))\right) \tag{21}$$

is convex on $[0; 1]$.

Proof. The function $\psi^{-1}(\cdot)$ is a strictly decreasing convex function on $[0; \psi(0)]$ with values in $[p; 1]$; the function $\Phi^{-1}(\cdot)$ is non-decreasing convex on $(p^*; 1]$. Hence, the function $H(y)$ is convex. □

The proposed approximation technique follows the outline appearing in [16,17]. For both the approximations that follow, we consider a partition of the interval $(0; 1]$ in the form $0 < y_{k1} < \ldots < y_{kJ} \leq 1$ (for each variable y_k).

Remark 1. The number J of partition points can differ for each row index k but, to simplify the notation and without loss of generality, we consider this number to be the same for each index k.

Lower Bound: Piecewise Tangent Approximation. We approximate the function $H(y_k)$ using the first order Taylor approximation at each of the partition points; the calculated Taylor coefficients a_{kj}, b_{kj} translate into the formulation of the problem (20) as the linear and SOCP constraints with additional auxiliary nonnegative variables z^k and w^k. The convexity of $H(\cdot)$ ensures that the resulting optimal solution is a lower bound for the original problem.

Theorem 3. *Given the partition points y_{kj}, consider the problem*

$$
\begin{aligned}
&\min c^T x \text{ subject to} \\
&\quad {\mu_k}^T x + \sqrt{z^{kT} \Sigma_k z^k} \leq h_k, \\
&\quad z^k \geq a_{kj} x + b_{kj} w^k \ (\forall k, \forall j) \\
&\quad \sum_k w^k = x, \\
&\quad w^k \geq 0, \ z^k \geq 0 \ (\forall k),
\end{aligned}
\tag{22}
$$

where

$$
\begin{aligned}
a_{kj} &:= H(y_{kj}) - b_{kj} y_{kj}, \\
b_{kj} &:= \frac{\psi(p)}{\phi(H(y_{kj}))\psi'\left(\psi^{-1}(y_{kj}\psi(p))\right)},
\end{aligned}
$$

and ϕ be the standard normal density. Then the optimal value of the problem (22) *is a lower bound for the optimal value of the problem* (2).

Remark 2. The linear functions $a_{kj} + b_{kj} y$ are tangent to the (quantile) function H_k at the partition points; hence the origin of the name *tangent approximation.* This approximation leads to an outer bound for feasible solution set $X(p)$.

Upper Bound: Piecewise Linear Approximation. The line passing through the two successive partition points with their corresponding values $H(y_{kj})$ is an upper linear approximation of $H(y_k)$ between these two successive points. Taking pointwise maximum of these linear functions we arrive to an upper approximation of the function H, hence to an upper bound for the optimal value of the original problem.

Theorem 4. *Given partition points y_{kj}, consider the problem*

$$
\begin{aligned}
&\min c^T x \text{ subject to} \\
&\quad {\mu_k}^T x + \sqrt{z^{kT} \Sigma_k z^k} \leq h_k, \\
&\quad z^k \geq a_{kj} x + b_{kj} w^k \ (\forall k, j < J) \\
&\quad \sum_k w^k = x, \\
&\quad w^k \geq 0, \ z^k \geq 0 \ (\forall k)
\end{aligned}
\tag{23}
$$

where

$$a_{kj} := H(y_{kj}) - b_{kj} y_{kj},$$
$$b_{kj} := \frac{H(y_{k,j+1}) - H(y_{kj})}{y_{k,j+1} - y_{kj}}.$$

If an optimal solution of the problem (23) *is feasible in* (2) *then the optimal value of the problem* (23) *is an upper bound for the optimal value of the problem* (2).

The last two problems are second-order cone programming problems and they are solvable by standard algorithms of SOCP. We do not provide further details in our paper; some promising numerical experiments were done by [16] for the problem with independent rows. If the dependence level is not too high (for example, if the parameter θ of the Gumbel-Hougaard copula approaches to one) the resulting approximation bounds are comparable to this independent case.

The second-order cone programming approach to solve chance-constrained programming problems opens a great variety of ways how to solve real-life problems. Many applications are modeled through chance-constrained programming: among them we can choose for example

- applications from finance: asset liability management, portfolio selection (covering necessary payments through an investment period with high probability),
- engineering applications in energy and other industrial areas (dealing with uncertainties in energy markets and/or weather conditions),
- water management (designing reservoir systems with uncertain stream inflows),
- applications in supply chain management, production planning, etc.

We refer the reader to the book [18] for a diversified set of applications from these (and other) areas and for ideas how uncertainty is incorporated into the models by the stochastic programming approach. The method proposed in this paper shifts the research and open new possibilities as the constraint dependence is in fact a natural property of constraints involved in all mentioned domains.

4 Conclusions

In our paper, we have presented an innovative use of the copula theory that is used to translate a known result developed for chance-constrained optimization problems with independent constraint rows to the case where the constraint rows exhibit some dependence. In particular, we assume that the dependence can be represented by a strict Archimedean copula with the generator which inverse is completely monotonic. Then the convexity of the feasible set is proven for sufficiently high values of p, and an equivalent deterministic formulation can be given. Furthermore, a lower and an upper bound for the optimal value of the problem can be calculated by introducing the piecewise tangent and piecewise linear approximations of the quantile function, and by solving the associated second-order cone programming problems.

Acknowledgements. This work was supported by the Fondation Mathématiques Jacques Hadamard, PGMO/IROE grant No. 2012-042H.

References

1. Prékopa, A.: Stochastic Programming. Akadémiai Kiadó, Budapest (1995)
2. Prékopa, A.: Probabilistic programming. In: Ruszczyński, A., Shapiro, A. (eds.) Stochastic Programming. Handbooks in Operations Research and Management Science, pp. 267–352. Elsevier, Amsterdam (2003)
3. Dentcheva, D.: Optimization models with probabilistic constraints. In: Shapiro, A., Dentcheva, D., Ruszczyński, A. (eds.) Lectures on Stochastic Programming: Modeling and Theory. MOS-SIAM Series on Optimization, pp. 87–153. SIAM, Philadelphia (2009)
4. Miller, B.L., Wagner, H.M.: Chance constrained programming with joint constraints. Oper. Res. **13**, 930–945 (1965)
5. Prékopa, A.: Logarithmic concave measures with applications to stochastic programming. Acta Scientiarium Mathematicarum (Szeged) **32**, 301–316 (1971)
6. Jagannathan, R.: Chance-constrained programming with joint constraints. Oper. Res. **22**, 358–372 (1974)
7. Henrion, R.: Structural properties of linear probabilistic constraints. Optimization **56**, 425–440 (2007)
8. Henrion, R., Strugarek, C.: Convexity of chance constraints with independent random variables. Comput. Optim. Appl. **41**, 263–276 (2008)
9. Prékopa, A., Yoda, K., Subasi, M.M.: Uniform quasi-concavity in probabilistic constrained stochastic programming. Oper. Res. Lett. **39**, 188–192 (2011)
10. Henrion, R., Strugarek, C.: Convexity of chance constraints with dependent random variables: the use of copulae. In: Bertocchi, M., Consigli, G., Dempster, M.A.H. (eds.) Stochastic Optimization Methods in Finance and Energy. International Series in Operations Research & Management Science, pp. 427–439. Springer, New York (2011)
11. Nelsen, R.B.: An Introduction to Copulas, 2nd edn. Springer, New York (2006)
12. McNeil, A.J., Nešlehová, J.: Multivariate Archimedean copulas, d-monotone functions and ℓ_1-norm symmetric distributions. Ann. Stat. **37**, 3059–3097 (2009)
13. Shapiro, A., Dentcheva, D., Ruszczyński, A.: Lectures on Stochastic Programming: Modeling and Theory. MOS-SIAM Series on Optimization. SIAM, Philadelphia (2009)
14. Alizadeh, F., Goldfarb, D.G.: Second-order cone programming. Math. Program. **95**, 3–51 (2003)
15. Boyd, S.P., Vandenberghe, L.: Convex Optimization. Cambridge University Press, Cambridge (2004). Seventh printing with corrections (2009)
16. Cheng, J., Lisser, A.: A second-order cone programming approach for linear programs with joint probabilistic constraints. Oper. Res. Lett. **40**, 325–328 (2012)
17. Cheng, J., Gicquel, C., Lisser, A.: A second-order cone programming approximation to joint chance-constrained linear programs. In: Mahjoub, A.R., Markakis, V., Milis, I., Paschos, V.T. (eds.) ISCO 2012. LNCS, vol. 7422, pp. 71–80. Springer, Heidelberg (2012)
18. Wallace, S.W., Ziemba, W.T. (eds.): Applications of Stochastic Programming. MPS-SIAM Series on Optimization. SIAM, Philadelphia (2005)

Comparison of Stochastic Programming Approaches for Staffing and Scheduling Call Centers with Uncertain Demand Forecasts

Mathilde Excoffier(✉), Céline Gicquel, Oualid Jouini, and Abdel Lisser

Laboratoire de Recherche en Informatique lri, 91405 Orsay Cedex, France
mathilde.excoffier@lri.fr

Abstract. We consider the staffing and shift-scheduling problems in call centers and propose a solution in one step. It consists in determining the minimum-cost number of agents to be assigned to each shift of the scheduling horizon so as to reach the required customer quality of service. We assume that the mean call arrival rate in each period of the horizon is a random variable following a continuous distribution. We model the resulting optimization problem as a stochastic program involving joint probabilistic constraints. We propose a solution approach based on linear approximations to provide approximate solutions of the problem. We finally compare them with other approaches and give numerical results carried out on a real-life instance. These results show that the proposed approach compares well with previously published approaches both in terms of risk management and cost minimization.

Keywords: Queuing systems · Stochastic optimization · Joint chance contraints · Staffing · Shift-scheduling · Call centers

1 Introduction

Staffing and shift-scheduling in call centers is a very challenging problem in Operations Research. Call centers are expensive infrastructures for companies, in which the staff agents represent 60 %–80 % of the total operating budget [1]. Thus an efficient workforce management is of primary importance to achieve profitability of a call center. One of the most important problem is the short-term staffing and scheduling problem: it consists in deciding how many staff members handling the phone calls, i.e. "agents", should work during the forthcoming days or weeks in order to minimize manpower costs while ensuring that the required customer quality of service is reached. The Quality of Service (QoS) can be for instance a maximum expected abandonment rate, i.e. number of clients hanging up without being served, or a maximum expected waiting time before entering service in the queue.

The problem here is to decide how many people answering the phone, that is to say agents, we need to assign each day. This problem comprises two steps.

The first step is the staffing problem, which involves computing the number of required agents. These values come from a calculation based on an objective

E. Pinson et al. (Eds.): ICORES 2014, CCIS 509, pp. 140–156, 2015.
DOI: 10.1007/978-3-319-17509-6_10

service level and estimations of arrival rates. The objective service level considered here is the maximum expected time of waiting before being served. The estimations of arrival rates come from forecasts using historical data, in which usually the main (and often only) information available is the number of calls per period. As arrival rates strongly vary in time, estimations are given for short periods of time (usually 30-minute periods).

In order to use all this information and compute the values of required agents at each period, we model the call center at each period as an Erlang C queuing system in stationary state and we use the Erlang C results, as commonly done practice.

The second step is the scheduling problem. This optimization problem involves scheduling enough agents with respect to a given Quality of Service (QoS) while respecting the inherent constraints of manpower work, like hiring a whole number of agents, or following some working hours. The goal here is to assign established shifts to the working agents through a given period.

There are several criteria in the establishment of the problem:

- Uncertainty management: how uncertainty is dealt with in the model?
- Risk management: how to modelize the penalty of not reaching the expected QoS?
- Recourse: what possibility do we have to correct the solution in a second-stage after observation?

Several approaches for staffing call centers considering uncertainty of arrival rates forecasts exist in the literature. Jongbloed and Koole [9] focus on giving a prediction interval for possible arrival rates, and then give an interval for the associate required numbers of agents. Gurvich et al. [8] present and compare two different approaches for dealing with uncertainty: the average-performance constraints problem considers the average of the uncertain variables and the chance constraints problem deals with uncertainty and risk both together.

Robbins and Harrison [15] choose to model the mixed integer linear program with several scenarios each defining a probability of reaching the QoS. Moreover, they use piecewise linear approximations of their QoS function. The idea of using scenarios through discretization of the probability distributions is used in several papers, such as [6,11] for example, or [12] who consider a finite distribution from the beginning.

The risk management can be modeled by a penalty cost in the objective function, as in [15], or with joint chance constraints programs, as in [8].

Liao et al. [10] introduce a distributionally robust approach for the scheduling problem and use discrete distributions for uncertainty.
While [8,11,15] use a one-stage approach, [4,6,13] allow a recourse on the solution, with a two-stage approach.

The contributions of the present paper are thus threefold. First, we model the call arrival rate in each period as a random variable following a continuous normal distribution. This is in contrast with most previously published approaches which rely on a discrete representation of the uncertainty through a finite set of scenarios. Since we consider every possible variation of the arrival rate instead of a limited number, our approach leads to a better accuracy of the final solution. Moreover, we keep this idea of continuity during all the process until the final solving of the linear programs. Second, we propose a solution in order to solve the staffing problem and the shift-scheduling problem as a one-stage stochastic program involving joint probabilistic constraints. It allows to manage the risk of not reaching the required quality of service at the scheduling horizon level rather than on a period by period basis. Moreover, our approach relies on a dynamic sharing out of the risk among all the periods and thus provides flexibility in the risk management. Third we present a linear-approximation based solution approach leading to approximate solutions for the problem.

The rest of this paper is organized as follows. In Sect. 2 we describe the call center queuing model. In Sects. 3 and 4 we present our approach to model and solve the stochastic staffing and shift-scheduling problem. First we define the joint chance constraints program and then we linearize the inverse of the cumulative distribution function in the constraints. Then we give in Sect. 5 computational results on several instances and we compare them to results given by simpler models. We propose to compare our results with another approach based on two steps in Sect. 6. Finally we conclude and highlight future research in Sect. 7.

2 Staffing Problem Modeling

The problem here is to decide how many people answering the phone, that is to say agents, we need to assign each day. In order to do that, we are given a number of required agents. These values of requirements come in fact from a calculation based on an objective service level and estimations of arrival rates. The objective service level is the customer Quality of Service. The estimations of arrival rates come from forecasts using historical data. As arrival rates vary in time, estimations are given for short periods of time, e.g. 30-minute periods.

In order to compute the values of required agents at each period, we model the call center as a queueing system in stationary state and we use the Erlang C model.

Call centers are typically modeled as queuing systems as we can see for example in [7]. The day is divided into T periods. During each period, we assume that the stationary regime is reached. Customer arrival process during each period is Poisson and service times are assumed to be independent and exponentially distributed with rate μ. This is a non-stationary queue $M_t/M/N_t$ where N_t represents the number of servers, i.e. the number of agents in our problem.

Customers are served in the order of their arrivals, i.e. under the First Come-First Served (FCFS) discipline of service. The queue capacity is assumed to be infinite. Finally, customers abandonment and retrials are ignored.

Since we consider uncertainty on arrival rates, we have to deal with stochastic programs. We assume that the arrival rates are random variables, denoted by Λ_t for the period t, following normal distributions where the expected values are the forecast values.

3 Problem Formulation

We propose here to describe and solve a mixed integer linear stochastic program able to solve a joint staffing and scheduling problem.

As explained, we consider that the values we use are forecasts obtained from historical data and may differ from the reality. We still want to guarantee a Quality of Service, expressed with a risk of how much can be the forecasts and reality different. In order to deal with a global problem, this risk will be set for the entire horizon (for example one week or one month).

In Sect. 2 we explained that we computed the number of required agents for each 30-minute-period (or 1-hour). We create several possible shifts, according to real work days, which cover the schedule of the call center. As it is inconvenient to ask an agent to come for only short periods of time, they have to follow typical shifts (like full-time or part-time). This may lead to over-staffing on some periods. In this model we consider shifts with breaks, like lunch breaks.

We want to define a risk level for the whole horizon. In order to deal with this condition, we model our problem with joint chance constraints [14]:

$$\begin{aligned} &\min c^t x \\ &\text{s.t. } P\{Ax \geqslant B\} \geqslant 1-\epsilon \\ &x_i \in \mathbb{Z}^+, \epsilon \in]0;1] \end{aligned} \tag{1}$$

where x is the agents vector, and x_i is the number of agents assigned to the shift i ; c is the cost vector and $A = (a_{i,j})$ is the matrix of shifts for $i \in [\![1;T]\!]$ and $j \in [\![1;S]\!]$. The vector B is the vector of the staffed agents random variables B_t.

This program optimizes the cost of hired agents under the constraint that the probability of reaching the requirement for the whole schedule is higher than the quality interval $1-\epsilon$.

We denote by A the matrix of possible shifts. The term $a_{i,j}$ is equal to 1 if agents working during period i according to shift j and 0 if not. Thus Ax is the vector defining the number of agents working at each period (Fig. 1).

Shift	1	2	3	4	5	6	7	8
08:00 - 09:00	1	0	0	1	0	0	1	0
09:00 - 10:00	1	1	0	1	1	0	1	1
10:00 - 11:00	1	1	1	1	1	1	1	1
11:00 - 12:00	1	1	1	0	1	1	0	1
12:00 - 13:00	0	1	1	1	0	1	0	0
13:00 - 14:00	1	0	1	1	1	0	1	0
14:00 - 15:00	1	1	0	1	1	1	1	1
15:00 - 16:00	1	1	1	1	1	1	1	1
16:00 - 17:00	0	1	1	0	1	1	1	1
17:00 - 18:00	0	0	1	0	0	1	0	0

Fig. 1. Example of a simple shifts matrix

The variables B_t are computed with an Erlang C model. The arrival rates values are independent random variables following continuous normal distributions for which the means are the forecast values. Since the B_t are function of Λ_t, they are random variables.

Thus we now consider B a vector of random variables; for each period t, B_t is function of the arrival rate Λ_t, and so we have to deal with the unknown continuous distribution of B_t.

Since we consider independent random variables, we split the product of probabilities and obtain the following Mixed Integer Program:

$$\begin{aligned} &\min c^t x \\ &\text{s.t. } \prod_{t=1}^{T} F_{B_t}(A_t * x) \geqslant (1-\epsilon) \\ &x_i \in \mathbb{Z}^+, \epsilon \in]0;1] \end{aligned} \tag{2}$$

where A_t is the t-th row of A matrix and $F_{B_t}(A_t * x) = P(B_t \leqslant A_t * x)$.

In order to solve this program, we need to separate the chance constraint into several constraints for each period. This means dividing up the risk along the horizon.

The simplest way is equally dividing the risk through the T periods, according to Bonferroni method:

$$\begin{aligned} &\min c^t x \\ &\text{s.t. } \forall t \in [\![1;T]\!],\ (A_t x) \geqslant F_{B_t}^{-1}\left((1-\epsilon)^{\frac{1}{T}}\right) \\ &x_i \in \mathbb{Z}^+, \epsilon \in]0;1], \forall t \in [\![1;T]\!],\ y_t \in]0;1] \end{aligned} \tag{3}$$

We divide the quality interval and then apply the inverse of the normal cumulative distribution function. The drawback of this idea is that we have to decide how to distribute the risk in advance, before the optimization process.

In order to be able to optimize the risk through the periods, we decide to include the sharing out of the risk in the optimization and put the risk levels as problem variables. Instead of considering $\frac{1}{T}$ as the proportion of the risk for one period, we introduce proportion variables denoted as y_t. They are now variables and the sum of y_t still should be 1 in order to reach the global risk level.

The new problem, with a flexible sharing out of the risk is now:

$$\begin{aligned} &\min c^t x \\ &\text{s.t. } \forall t \in [\![1;T]\!],\ (A_t x) \geqslant F_{B_t}^{-1}\left((1-\epsilon)^{y_t}\right) \\ &\sum_{t=1}^{T} y_t = 1 \\ &x_i \in \mathbb{Z}^+, \epsilon \in]0;1], \forall t \in [\![1;T]\!],\ y_t \in]0;1] \end{aligned} \tag{4}$$

In order to solve this problem, we propose two linearizations which give an upper bound and a lower bound. These linearizations are based on piecewise approximations of $y \mapsto F_{B_t}^{-1}(p^y)$. We cannot compute exact values of this function, thus we focus on linear approximations. This function is continuous and differentiable (except on a countable number of points). We need to deal with a convex function in order to apply the approximations.

4 Solution Approximations

4.1 Definition of ψ Function

We first introduce the function ψ which gives a relation between b and λ. We consider the following continuous function ψ:

$$\begin{aligned} \psi : \mathbb{R} &\rightarrow \mathbb{R}^+ \\ \lambda &\mapsto \psi(\lambda) = b(\lambda, ASA^*, \mu) \end{aligned} \tag{5}$$

The function ψ gives the minimum number of agents b required to ensure that the targeted QoS ASA^* is reached when the call arrival rate is λ and the expected service rate is μ. The chosen QoS is the Average Speed of Answer (ASA). The computed value of b is a real number and not an integer, which is necessary to allow the linear approximations in the next parts: we need the inverse of the cumulative distribution function to be continuous.

To the best of our knowledge, an analytical expression computing ψ is not known. However, for a given value of λ, we propose to compute $\psi(\lambda)$ with the following algorithm.

First we consider this well-known Erlang C model's function:

$$ASA(N, \lambda, \mu) = \mathbb{E}[Wait] = \frac{1}{N * \mu * (1 - \frac{\lambda}{N*\mu}) \left(1 + (1 - \frac{\lambda}{N*\mu}) \sum_{m=0}^{N-1} \frac{N!}{m!} (\frac{\mu}{\lambda})^{N-m}\right)}$$

This formula gives the expectation of waiting time (ASA: Average Speed of Answer) given the arrival rate λ, the service rate μ and the number of servers N which is an integer. In order to consider ψ as function of a positive real value of b, we use the algorithm below:

- We compute $ASA(N, \lambda, \mu)$ and $ASA(N+1, \lambda, \mu)$ such that

$$ASA(N, \lambda, \mu) \geqslant ASA^* \text{ and } ASA(N+1, \lambda, \mu) < ASA^*$$

  We denote $ASA(N, \lambda, \mu)$ as $ASA_{N,\lambda}$.
- The real value of N is computed by a linearization in the $[ASA_{N,\lambda}; ASA_{N+1,\lambda}]$ segment. The affine function is:

$$ASA^* = (ASA_{N+1,\lambda} - ASA_{N,\lambda}) * b + (N+1) * ASA_{N,\lambda} - N * ASA_{N+1,\lambda}$$

  and b is the real value $\psi(\lambda)$ we are looking for. □

Using this algorithm for the value of λ we are considering in the ψ function, finally we obtain b.

$$\psi(\lambda) = b = \frac{ASA^* + N * ASA_{N+1,\lambda} - (N+1) * ASA_{N,\lambda}}{ASA_{N+1,\lambda} - ASA_{N,\lambda}}$$

The ψ function allows us to determine the values of b as a function of λ, μ and an objective QoS ASA^*. In a nutshell, we determine the number of agents required to deal with the arrival rates of clients λ with respecting a Quality of Service previously defined. This function is strictly increasing.

We can denote then

$$F_B(b) = F_A(\psi^{-1}(b))$$

4.2 Convexity of $y \mapsto F_{B_t}^{-1}(p^y)$

We have previously defined $F_B(b) = F_\Lambda(\psi^{-1}(b))$.

Thus we have $F_B(b) = F_\Lambda(\psi^{-1}(b)) = 1-\epsilon$ and so $F_B^{-1}(1-\epsilon) = \psi(F_\Lambda^{-1}(1-\epsilon))$.

In our problem, we split the risk $1-\epsilon$. Since $1-\epsilon$ represents a probability, let's call it p in this part. In our problem we want a high quality interval and thus a small value of ϵ. We can consider from here that $p > 0.5$, which is necessary for the following proof of convexity.

We need to consider $p^y, y \in]0;1]$ in our optimization problem. So we consider the equality, with $y \in]0;1]$:

$$F_B^{-1}(p^y) = \psi(F_\Lambda^{-1}(p^y))$$

Lemma. $y \mapsto F_\Lambda^{-1}(p^y)$ is convex.

Proof. Since $f : y \mapsto p^y$ is convex and $g : p \mapsto F_\Lambda^{-1}(p)$ is convex for $p > 0.5$ and increasing, thus $y \mapsto F_\Lambda^{-1}(p^y)$ is convex. □

This previous result and the strictly increasing function $\lambda \mapsto \psi(\lambda)$ helped us to note that $y \mapsto F_B^{-1}(p^y)$ is a generally convex function. We then consider an approximated function of $y \mapsto F_B^{-1}(p^y)$ which is convex.

With this result we are able to linearize the approximated convex function as in [3].

4.3 Piecewise Linear Approximation

Here we give an upper approximation of $y \mapsto F_{B_t}^{-1}(p^y)$.

Let $y_j \in]0;1]$, $j \in [\![1;n]\!]$ be n points such that $y_1 < y_2 < \ldots < y_n$.

Let's denote $\hat{F}_{Bj}^{-1}(p^y)$ the linearized approximation of $F_B^{-1}(p^y)$ between y_j and y_{j+1}.

$$\begin{aligned} \forall j \in [\![1;n-1]\!], \hat{F}_{Bj}^{-1}(p^y) &= F_B^{-1}(p^{y_j}) + \frac{y-y_j}{y_{j+1}-y_j}(F_B^{-1}(p^{y_{j+1}}) - F_B^{-1}(p^{y_j})) \\ &= \delta_j * y + \alpha_j \end{aligned} \qquad (6)$$

Since $F_B(b) = F_\Lambda(\psi^{-1}(b))$, we have $\forall p \in]0;1[,\ F_{Bj}^{-1}(p) = \psi(F_{\Lambda j}^{-1}(p))$, thus the coefficients are:

$$\delta_j = \frac{\psi(F_{\Lambda j}^{-1}(p^{y_{j+1}}) - \psi(F_{\Lambda j}^{-1}(p^{y_j}))}{y_{j+1} - y_j} \tag{7}$$

$$\alpha_j = \psi(F_\Lambda^{-1}(p^{y_j})) - y_j * \delta_j \tag{8}$$

Because of the convexity of the approximation, the condition in our program would be:

$$\forall y \in]\,0;1], \hat{F_B}^{-1}(p^y) = \max_{j \in [[1;n-1]]} \{\hat{F}_{Bj}^{-1}(p^y)\} \tag{9}$$

So our approximated program is now:

$$\begin{aligned} &\min c^t x \\ &\text{s.t. } \forall t \in [\![1;T]\!], \forall j \in [\![1;n-1]\!],\ A_t x \geqslant \alpha_j + \delta_j * y_t \\ &\sum_{t=1}^{T} y_t = 1 \\ &\forall i \in [\![1;S]\!], x_i \in \mathbb{Z}^+, \forall t \in [\![1;T]\!], y_t \in]\alpha_1; 1] \end{aligned} \tag{10}$$

with n points for linear approximation with (α_j, δ_j) coordinates. S is the number of shifts and T the total number of periods.

4.4 Piecewise Tangent Approximation

Let's now express a lower approximation of $y \mapsto F_{B_t}^{-1}(p^y)$.

Let $y_j \in]0;1],\ j \in [\![1;n]\!]$ be n points such that $y_1 < y_2 < \ldots < y_n$.

We apply a first-order Taylor series expansion around these n tangents points. Let's denote $\hat{F}_{Bj}^{-1}(p^y)$ the linearized approximation of $F_{Bj}^{-1}(p^y)$ around y_j.Then

$$\begin{aligned} \forall j \in [\![1;n]\!], F_{Bj}^{-1}(p^y) &= F_B^{-1}(p^{y_j}) + (y - y_j)(F_B^{-1})'(p^{y_j}) * ln(p) * p^{y_j} \\ &= \delta_j * y + \alpha_j \end{aligned} \tag{11}$$

with $(F_B^{-1})'(p^{y_j}) = \frac{1}{F_B'(F_B^{-1})(p^{y_j})} = \frac{1}{f_B(F_B^{-1}(p^{y_j}))}$

And since

$$f_b(b) = \frac{f_\Lambda(\psi^{-1}(b))}{\psi'(\psi^{-1}(b))}$$

as a definition of composition of random variables:

$$\begin{aligned} f_B(F_B^{-1})(p^{y_j}) &= \frac{f_\Lambda(\psi^{-1}(F_B^{-1}(p^{y_j})))}{\psi'(\psi^{-1}(F_B^{-1}(p^{y_j})))} = \frac{f_\Lambda(\psi^{-1}(\psi(F_\Lambda^{-1}(p^{y_j}))))}{\psi'(\psi^{-1}(\psi(F_\Lambda^{-1}(p^{y_j}))))} \\ &= \frac{f_\Lambda(F_\Lambda^{-1}(p^{y_j}))}{\psi'(F_\Lambda^{-1}(p^{y_j}))} \end{aligned}$$

The coefficients are:

$$\delta_j = ln(p) * p^{y_j} * \frac{\psi'(F_\Lambda^{-1}(p_j^y))}{f_\Lambda(F_\Lambda^{-1}(p_j^y))} \tag{12}$$

$$\alpha_j = \psi(F_\Lambda^{-1}(p^{y_j})) - y_j * \delta_j \tag{13}$$

Like previously for Eq. (9), we can assure that this piecewise tangent approximations are always below the curve. Thus the condition in our approximated program is:

$$\hat{F_B}^{-1}(p^y) = \max_{j \in [\![1;n]\!]} \{F_{Bj}^{-1}(p^y)\} \tag{14}$$

Finally the piecewise tangent approximated program is:

$$\begin{aligned} &\min c^t x \\ &\text{s.t. } \forall t \in [\![1;T]\!], \forall j \in [\![1;n]\!],\ A_t x \geqslant \alpha_j + \delta_j * y_t \\ &\sum_{t=1}^{T} y_t = 1 \\ &\forall i \in [\![1;S]\!], x_i \in \mathbb{Z}^+, \forall t \in [\![1;T]\!], y_t \in]0;1] \end{aligned} \tag{15}$$

with n points for tangent approximation with (α_j, δ_j) coordinates.

5 Numerical Experiments

5.1 Instance

We apply our model to an instance from a health insurance call center. We use 19 different shifts, both full-time and part-time and consider the scheduling for one week (5.5 days, from Monday to Saturday midday).

We split the time horizon into 30-minute periods, considering 10 h a day from Monday to Friday and 3.5 h for Saturday morning, which gives 107 periods.

We consider that the agents are paid according to the number of worked hours. The cost of one agent is proportional to the number of periods worked. Thus it depends on the shift the agent works on. Here we set the cost to 1 for the fullest shifts (with the highest number of periods) and the costs of other shifts are a strict proportionality of the number of worked periods.

The data used to staff and shift-schedule are arrival rates varying between 3 calls/min and 43 calls/min, following a typical daily seasonality, as described in [5]. We denote $\forall t \in [\![1;T]\!], \lambda_t$ the mean of the T random variables, which are the given data. The variances σ_t^2 are random values generated between $[\frac{\lambda_t}{4}; \frac{\lambda_t}{2}]$.

We apply the same instance to the programs (10) and (15) and, as a comparison, to the program (3) in which we divided the risk through the periods in a pre-treatment.

5.2 Comparison with Other Programs

We also add the results from simple programs:

- In the disjoint approach we want to reach the risk level at each period, not through the whole horizon:

$$\begin{aligned} &\min c^t x \\ &\text{s.t. } \forall t \in [\![1;T]\!],\ P\{A_t x \geqslant b_t\} \geqslant 1-\epsilon \\ &\quad x_i \in \mathbb{Z}^+, \epsilon \in]0;1] \end{aligned} \tag{16}$$

Thus we got this final program:

$$\begin{aligned} &\min c^t x \\ &\text{s.t. } \forall t \in [\![1;T]\!],\ (A_t x) \geqslant F_{B_t}^{-1}((1-\epsilon)) \\ &\quad \sum_{t=1}^{T} y_t = 1 \\ &\quad x_i \in \mathbb{Z}^+, \epsilon \in]0;1] \end{aligned} \tag{17}$$

- We also compare with the results of the deterministic program:

$$\begin{aligned} &\min c^t x \\ &\text{s.t. } \forall t \in [\![1;T]\!],\ (A_t x) \geqslant b_t \\ &\quad x_i \in \mathbb{Z}^+ \end{aligned} \tag{18}$$

The values of b_t here are computed with the Erlang C formula using the mean forecasted values, considered as certain.

5.3 Results

We apply our models with the following parameters:

- $\mu = 1$
- $ASA^* = 1$
- $\epsilon = 0.05$

In Table 1 we show the solutions for staffing and shift-scheduling of this instance for 5 programs: column 1 gives the shift, column 2 (Deter) presents the x vector obtained with the deterministic model (18), column 3 (Disjoint) gives the results with the disjoint chance constraints model (17) and column 4 (Fixed) with the the fixed-risked model (3). Finally, columns 5 (LowerB) and 6 (UpperB) present the results obtained with the lower bound (15) and the upper bound (10) approximations.

We used 5 points for computing both lower and upper bounds. The gap between the two bounds is $D = 5\,\%$.

Table 1. Result for staffing and shift-scheduling.

Shift	Deter	Disjoint	Fixed	LowerB	UpperB
1	0	0	0	0	0
2	0	0	0	0	0
3	0	0	0	0	0
4	0	0	0	0	0
5	0	0	0	0	0
6	2	2	1	1	1
7	0	0	0	0	0
8	0	0	0	0	0
9	0	0	0	0	0
10	0	0	0	0	0
11	13	14	18	18	19
12	9	10	9	9	12
13	5	6	8	9	11
14	0	1	3	2	4
15	5	6	7	6	3
16	6	7	8	7	7
17	0	0	0	0	0
18	4	5	6	4	3
19	4	4	5	4	3
Total	48	55	65	60	63
Cost function	47.44	54.44	64.72	59.72	62.72

Table 2. Percentage of violated scenarios.

Model	% of violation
Deterministic model	100 %
Disjoint chance constraints model	49 %
JCC model with fixed risk level	0 %
JCC lower bound and flexible risk	3 %
JCC upper bound and flexible risk	1 %
Targeted maximal risk	5 %

In order to check the efficiency of these solutions, we randomly generated 100 scenarios according to the historical data we previously used and checked the feasibility of the 5 solutions. If the number of agents scheduled in at least one period of a scenario is insufficient, the latter is considered as violated.

In Table 2 are the results for a batch of 100 scenarios. JCC stands for "Joint Chance Constraints".

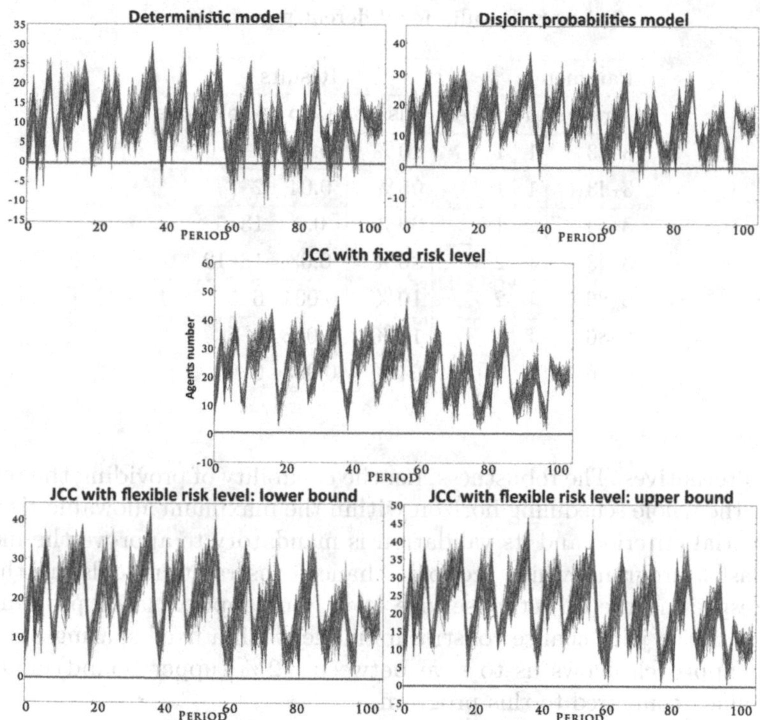

Fig. 2. Violated scenarios for each model.

In the Fig. 2 we plotted for each model and each scenario the difference between the number of agents in the previous solutions and the number of agents actually needed $b_{present} - b_{needed}$. When this value is negative for at least one period, the model is invalidated for the scenario.

First we compare the schedules obtained by using the various models discussed in the paper based on the total staffing cost.

We note that the deterministic (18) and disjoint (17) models provide less expensive schedules than the joint chance constraint models (10), (15) and (3).

However our approximated programs where the risk is dynamically divided through the periods provide less expensive schedules than the joint chance constraint model where the risk is a priori divided equally between the scheduling periods (3). This shows the interest of allowing some flexibility in the way the risk is allocated between the scheduling periods.

In small instances we could have chosen another sharing out a priori of the risk (by analysing wisely the risky periods) but it remains too complex on instances like ours.

Second, we note that all the cheaper solutions than the two bounds of our new model of joint chance contraints with a flexible sharing out of the risk does not validate the condition of the targeted QoS. Thus they cannot be considered as

Table 3. Results for different parameters.

Parameters				Results	
λ range	μ	ASA^*	Risk ϵ	Gap	Violations
3–43	1	1	10 %	0.05	3–1
3–43	1	1	05 %	0.04	2–1
3–43	2	1	30 %	0.0	13–5
3–43	1	2	30 %	0.03	18–12
6–86	1	2	10 %	0.004	6–5
6–86	1	1	10 %	0.005	6–5
6–86	1	1	01 %	0.003	0–0

possible alternatives. The robustness, i.e. the capability of providing the required QoS over the whole scheduling horizon within the maximum allowable risk level, is an essential criterion and its validation is mandatory to approve the model.

The last 3 programs which are joint chance constraints models are the only models respecting the objective service level. Our approximated programs are cheaper than the joint chance constraints model with a fixed sharing out of the risk. Our approach allows us to save between 3.2 % (upper bound) and 8.4 % (lower bound) compared to this program.

This shows the practical interest of the proposed modelling and solution approach as we provide robust schedules at a lesser cost than previously joint constraint models.

In Table 3, we present results for different values of λ, μ, ASA^* and ϵ (illustrated in the table by the risk interval).

The "Gap" column gives the gap between the lower bound and the upper bound solutions. The "violation" column gives the numbers of violated scenarios for the lower and the upper bounds. We note that for higher values of λ, the bounds are really close and give good results.

6 Staffing and Scheduling Problems in Two Steps

6.1 The Requirements B_t as Random Normal Variables

Since the consideration of the ψ function is rather advanced, we do not know the exact distribution followed by the random variables B_t. In this section we consider the initial problem as a two-step problem:

- Using the given data λ and Erlang C model' formula, we compute the number of required agents $b = \min\{N | ASA(N, \lambda, \mu) \leqslant ASA^*\}$. We denote the function

$$\begin{aligned} \phi :& \mathbb{R} && \rightarrow \mathbb{N} \\ & \lambda && \mapsto \psi(\lambda) = b(\lambda, ASA^*, \mu) = \min\{N | ASA(N, \lambda, \mu) \leqslant ASA^*\} \end{aligned} \quad (19)$$

- we consider that b is a random variable following a continuous normal distribution and we solve a problem similar to the previous one.

Note. The function ϕ is a strictly increasing piecewise constant function.

With this new approach, we would like to spot the differences and similarities between our previous random variables whose distributions are complex and the random variables following normal distributions.

For easier consideration, we decide to standardize the normal random variable. The resulting program is the following:

$$\begin{aligned} &\min c^t x \\ &\text{s.t. } \forall t \in [\![1;T]\!],\ (\frac{A_t x - \overline{b}_t}{\sigma_t}) \geqslant F^{-1}_{\beta_t \sim \mathcal{N}(0,1)}((1-\epsilon)^{y_t}) \\ &\sum_{t=1}^{T} y_t = 1 \\ &x_i \in \mathbb{Z}^+, \epsilon \in]0;1], \forall t \in [\![1;T]\!],\ y_t \in]0;1] \end{aligned} \tag{20}$$

With $\overline{b}_t = \min\{N | ASA(N, \lambda_t, \mu) \leqslant ASA^*\}$, computed for each period, and β_t a random variable following a standard normal distribution.

Since we are considering the common normal distribution, the convexity of the RHS is known for a probability $p \geqslant 0.5$. Hence, the same approximations as in Sect. 4 are possible and a bit simpler. We obtain similarly the coefficients of the approximations and the final programs have the exact same form.

The coefficients for the upper bound program are

$$\begin{aligned} \forall j \in [\![1; n-1]\!],\ \hat{F}^{-1}_{u,\beta j}(p^y) &= F^{-1}_{\beta}(p^{y_j}) + \frac{y - y_j}{y_{j+1} - y_j} * (F^{-1}_{\beta}(p^{y_{j+1}}) - F^{-1}_{\beta}(p^{y_j})) \\ &= \delta_j * y + \alpha_j \end{aligned} \tag{21}$$

Again, the coefficients for the lower bound program are

$$\forall j \in [\![1; n]\!],\ \hat{F}^{-1}_{l,\beta j}(p^y) = F^{-1}_{\beta}(p^{y_j}) + (y - y_j) * (F^{-1}_{\beta})'(p^{y_j}) * p^{y_j} * \ln(p) = \delta_j * y + \alpha_j$$

$$\text{where } (F^{-1}_{\beta})'(p^{y_j}) = \frac{1}{F'_{\beta}(F^{-1}_{\beta})(p^{y_j})} = \frac{1}{f_{\beta}(F^{-1}_{\beta}(p^{y_j}))}.$$

6.2 Results Comparison

We compared the costs for the one-step programs of Sect. 4 and two-step programs of Sect. 6, for several parameters. Results are given in Table 4. Several other computations were made, with similar results.

We notice that the gap between the bounds is really tight for the one-step approach and is less satisfactory for the two-step approach. The closer the bounds, the less confusing it is for the manager to decide for a final solution.

Table 4. Results for different parameters.

Resolution	Parameters				Cost functions		Gap
	λ range	μ	ASA^*	Risk ϵ	Lower	Upper	
JCC 1Step	3–43	1	1	10 %	60.44	61.72	2.1 %
JCC 2Step	3–43	1	1	10 %	59.72	64.44	7.9 %
JCC 1Step	3–43	1	1	05 %	61.44	63.44	3.3 %
JCC 2Step	3–43	1	1	05 %	60.44	64.72	7.1 %
JCC 1Step	3–43	1	1	20 %	59.00	60.00	1.7 %
JCC 2Step	3–43	1	1	20 %	57.44	62.44	8.7 %
JCC 1Step	6–46	1	1	05 %	112.2	113.9	1.5 %
JCC 2Step	6–46	1	1	05 %	110.7	115.6	4.4 %

We also notice that the lower bound of the two-step approach is a bit better than the lower solution of the one-step approach, while this is the contrary for the upper bounds. On condition that all these solutions are robust, it's safer to consider the set of solutions with the cheapest upper bound.

Furthermore, we estimated the robustness of the solutions, like in Sect. 5.3. The same process as in previous section was used, and we tested the solutions of both approaches. Table 5 gives some results of our simulations, for which we extract the percentage of violated scenarios. We ran between 4 and 8 sets of simulations, each consisting in 100 of generated scenarios.

Table 5. Results for different parameters.

Resolution	Parameters				Violations	
	λ range	μ	ASA^*	Risk ϵ	Lower	Upper
JCC 1Step	6–46	1	1	01 %	0.5	0.33
JCC 2Step	6–46	1	1	01 %	2.5	1.67
JCC 1Step	6–46	1	1	05 %	3.4	2
JCC 2Step	6–46	1	1	05 %	6	2.8
JCC 1Step	6–46	1	1	20 %	15	10.75
JCC 2Step	6–46	1	1	20 %	21.25	9
JCC 1Step	6–46	1	1	30 %	20.33	19
JCC 2Step	6–46	1	1	30 %	37.33	16.67

According to Table 5, we notice that the upper bound of our simplified two-step approach does not respect the accepted risk level. It is in adequation with the idea of proposing a lower bound of a problem, but it may be confusing when we need to actually build a schedule that will indeed respect the set risk level. This test however did not enable us to show a strong similarity between our unknown random distribution for B_t (introduced in Sect. 3) and the normal distribution.

We noticed an interesting feature on one simulation: the costs of the upper bounds for the two different approaches were very close (118.2 for the two-step and 118.4 for the one-step), but the number of violated scenarios were very different. The one-step solution did respect the risk level whereas the two-step solution was often jeopardised. This is actually the consequence of a different distribution of the workforce through the possible shifts. The two-step solution seemed to have chosen a worse distribution than the one-step solution.

7 Conclusion

In this paper we proposed a new procedure for solving the staffing and shift-scheduling problem in one step with uncertain arrival rates. We introduced the modelization of arrival rates as continuous normal distributions and we were able to propose linear approximations and upper and lower bounds for our staffing and shift-scheduling problem in call centers. The construction of the ψ function made possible these piecewise approximations. Computational results show that the two bounds give quite close results and both propose cheaper solutions than the other chance constraints program, while respecting our targeted Quality of Service. However we used a general convex shape for approximating the inverse of the cumulative function, even though the real shape of this function can be difficult to analyze.

We compared our one-step approach with a two-step approach in order to check similarities between the unknown distribution of B_t and the normal distribution. Although we could not bring out strong resemblance between them, we highlighted some interesting results.

As an improvement of our work in the future, we wish to theoretically analyse and give a precise model of the shape of $y \mapsto F_B^{-1}(p^y)$ in order to improve the precision of our approximated model. As we can see in our results, the two bounds we proposed give a better QoS than expected and thus, probably an over-staffing.

Moreover, we have several possibilities for improving the queuing system model for the call center, for instance:

- Skills-based call centers: we can assume the agents are specialized in specific fields and will answer the appropriate calls according to these skills. This implies the creation of multiple queues which can be connected (see [2] for example).
- Abandonments and retrials: some people may hung up before being served; if on purpose, we consider this as abandonments (when people have reach their patience limit) or if by accident, they may want to call again and we consider this as retrials.

Acknowledgements. Support for this research was provided by DIGITEO Research Foundation under Grant 2012-060D.

References

1. Aksin, Z., Armony, M., Mehrotra, V.: The modern call center: a multi-disciplinary perspective on operations management research. Prod. Oper. Manage. **16**, 665–688 (2007)
2. Cezik, M.T., L'Ecuyer, P.: Staffing multiskill call centers via linear programming and simulation. Manage. Sci. **54**, 310–323 (2008)
3. Cheng, J., Lisser, A.: A second-order cone programming approach for linear programs with joint probabilistic constraints. Oper. Res. Lett. **40**, 325–328 (2012)
4. Erdoğan, E., Iyengar, G.: On two-stage convex chance constrained problems. Math. Methods of Oper. Res. **65**, 115–140 (2007)
5. Gans, N., Koole, G., Mandelbaum, A.: Telephone call centers: tutorial, review, and research prospects. Manuf. Serv. Oper. Manage. **5**, 79–141 (2003)
6. Gans, N., Shen, H., Zhou, Y.P.: Parametric stochastic programming models for call-center workforce scheduling. working paper April 2012
7. Gross, D., Shortle, J.F., Thompson, J.M., Harris, C.M.: Fundamentals of Queueing Theory. Wiley, New York (2008)
8. Gurvich, I., Luedtke, J., Tezcan, T.: Staffing call centers with uncertain demand forecasts: a chance-constrained optimization approach. Manage. Sci. **56**, 1093–1115 (2010)
9. Jongbloed, G., Koole, G.: Managing uncertainty in call centers using poisson mixtures. Appl. Stoch. Models Bus. Indus. **17**, 307–318 (2001)
10. Liao, S., van Delft, C., Vial, J.P.: Distributionally robust workforce scheduling in call centers with uncertain arrival rates. Optim. Methods Softw. **28**, 501–522 (2013)
11. Liao, S., Koole, G., van Delft, C., Jouini, O.: Staffing a call center with uncertain non-stationary arrival rate and flexibility. OR Spectr. **34**, 691–721 (2012)
12. Luedtke, J., Ahmed, S., Nemhauser, G.L.: An integer programming approach for linear programs with probabilistic constraints. In: Fischetti, M., Williamson, D.P. (eds.) IPCO 2007. LNCS, vol. 4513, pp. 410–423. Springer, Heidelberg (2007)
13. Mehrotra, V., Ozlük, O., Saltzman, R.: Intelligent procedures for intra-day updating of call center agent schedules. Prod. Oper. Manage. **19**, 353–367 (2009)
14. Prékopa, A.: Probabilistic programming. Handbooks Oper. Res. Manage. Sci. **10**, 267–351 (2003)
15. Robbins, T.R., Harrison, T.P.: A stochastic programming model for scheduling call centers with global service level agreements. Eur. J. Oper. Res. **207**, 1608–1619 (2010)

A New Look at the Covariance Matrix Estimation in Evolution Strategies

Silja Meyer-Nieberg(✉) and E. Kropat

Universität der Bundeswehr München,
Werner-Heisenberg-Weg 39, 85577 Neubiberg, Germany
{silja.meyer-nieberg,erik.kropat}@unibw.de

Abstract. Evolution strategies belong to the best performing modern natural computing methods for continuous optimization. This paper takes a new look at the covariance matrix adaptation, a mechanism which is central to the algorithm. The adaptation focusses strongly on the sample covariance. However, as known from modern statistics, this estimate may be of poor quality if certain conditions are not fulfilled. Unfortunately, this is often the case in practice. This paper compares the established methods for the covariance correction in evolution strategies with the approaches in modern statistics. Furthermore, it introduces and evaluates new covariance correction schemes.

Keywords: Evolutionary algorithms · Continuous optimization · Evolution strategies · Covariance · Shrinkage

1 Introduction

Black-Box optimization is an important subcategory of optimization. Over the years, several methods have been developed - ranging from simple pattern search over mesh adaptive methods to natural computing, see e.g. [1,8,10]. This paper focuses on evolution strategies (ESs) which represent well-performing metaheuristics for continuous, non-linear optimization. In recent workshops on black-box optimization, see e.g. [15], variants of this particular subtype of evolutionary algorithms have emerged as one the best performing methods among a broad range of competitors stemming from natural computing. Evolution strategies rely primarily on random changes to move through the search space. These random changes, usually normally distributed random variables, must be controlled by adapting both, the extend and the direction of the movements.

Modern evolution strategies apply therefore covariance matrix and step-size adaptation – with great success. However, most methods use the common estimate of the population covariance matrix as one component to guide the search. Here, there may be room for further improvement, especially with regard to common application cases of evolution strategies which usually concern optimization in high-dimensional search spaces. For efficiency reasons, the population size λ, that is, the number of candidate solutions, is kept below the search space dimensionality N and scales usually with $\mathcal{O}(\log(N))$ or with $\mathcal{O}(N)$. In other words,

E. Pinson et al. (Eds.): ICORES 2014, CCIS 509, pp. 157–172, 2015.
DOI: 10.1007/978-3-319-17509-6_11

either $\lambda \ll N$ or $\lambda \approx N$ which may represent a problem when using the sample covariance matrix. This even more so, since the sample size used in the estimation is just a fraction of the population size. Furthermore, the result is not robust against outliers which may appear in practical optimization which has often to cope with noise. This paper introduces and explores new approaches addressing the first problem by developing a new estimate for the covariance matrix. To our knowledge, these estimators have not been applied to evolution strategies before.

The paper is structured as follows: First, evolution strategies are introduced and common ways to adapt the covariance matrix are described and explained. Afterwards, we point out a potential dangerous weakness of the traditionally used estimate of the population covariance. Candidates for better estimates are presented and described in the following section. We propose and investigate several approaches ranging from a transfer of shrinkage estimators over a maximum entropy covariance selection principle to a new combination of both approaches. The quality of the resulting algorithms is assessed in the experimental test section. Conclusions and possible further research directions constitute the last part of the paper.

1.1 Evolution Strategies

Evolutionary algorithms (EAs) [10] are population-based stochastic search and optimization algorithms including today genetic algorithms, genetic programming, (natural) evolution strategies, evolutionary programming, and differential evolution. As a rule, they require only weak preconditions on the function to be optimized. Therefore, they are applicable in cases when only point-wise function evaluations are possible.

An evolutionary algorithm starts with an initial population of candidate solutions. The individuals are either drawn randomly from the search space or are initialized according to previous information on good solutions. A subset of the parent population is chosen for the creation of the offspring. This process is termed parent selection. Creation normally consists of recombination and mutation. While recombination combines traits from two or more parents, mutation is an unary operator and is realized by random perturbations. After the offspring have been created, survivor selection is performed to determine the next parent population. Evolutionary algorithms differ in the representation of the solutions and in the realization of the selection, recombination, and mutation operators.

Evolution strategies (ESs) [20,22] are a variant of evolutionary algorithms that is predominantly applied in continuous search spaces. Evolution strategies are commonly notated as $(\mu/\rho, \lambda)$-ESs. The parameter μ stands for the size of the parent population. In the case of recombination, ρ parents are chosen randomly and are combined for the recombination result. While other forms exist, recombination usually consists of determining the weighted mean of the parents [4]. The result is then mutated by adding a normally distributed random variable with zero mean and covariance matrix $\sigma^2\mathbf{C}$. While there are ESs that operate without recombination, the mutation process is essential and can be seen as the main search operator. Afterwards, the individuals are evaluated using the function to be optimized or a derived function which allows an easy ranking of the population. Only the rank of an individual is important for the selection.

There are two main types of evolution strategies: Evolution strategies with "plus"-selection and ESs with "comma"-selection. The first select the μ-best offspring and parents as the next parent population, where ESs with "comma"-selection discard the old parent population completely and take only the best offspring. Methods for adapting the scale factor σ or the full covariance matrix have received a lot of attention (see [19]). The main approaches are described in the following section.

1.2 Covariance Matrix Adaptation

First, the update of the covariance matrix is addressed. In evolution strategies two types exist: one applied in the *covariance matrix adaptation evolution strategy* (CMA-ES) [14] which considers past information from the search and an alternative used by the *covariance matrix self-adaptation evolution strategy* (CMSA-ES) [5] which focusses more on the present population.

The covariance matrix update of the CMA-ES is explained first. The CMA-ES uses weighted intermediate recombination, in other words, it computes the weighted centroid of the μ best individuals of the population. This mean $\mathbf{m}^{(g)}$ is used for creating all offspring by adding a random vector drawn from a normal distribution with covariance matrix $(\sigma^{(g)})^2\mathbf{C}^{(g)}$, i.e., the actual covariance matrix consists of a general scaling factor (or step-size or mutation strength) and the matrix denoting the directions. Following usual notation in evolution strategies this matrix $\mathbf{C}^{(g)}$ will be referred to as *covariance matrix* in the following.

The basis for the CMA update is the common estimate of the covariance matrix using the newly created population. Instead of considering the whole population for deriving the estimates, though, it introduces a bias towards good search regions by taking only the μ best individuals into account. Furthermore, it does not estimate the mean anew but uses the weighted mean $\mathbf{m}^{(g)}$. Following [14],

$$\mathbf{y}_{m:\lambda}^{(g+1)} := \frac{1}{\sigma^{(g)}}\left(\mathbf{x}_{m:\lambda}^{(g+1)} - \mathbf{m}^{(g)}\right) \tag{1}$$

are determined with $\mathbf{x}_{m:\lambda}$ denoting the mth best of the λ particle according to the fitness ranking. The rank-μ update the obtains the covariance matrix as

$$\mathbf{C}_{\mu}^{(g+1)} := \sum_{m=1}^{\mu} w_m \mathbf{y}_{m:\lambda}^{(g+1)}(\mathbf{y}_{m:\lambda}^{(g+1)})^{\mathrm{T}} \tag{2}$$

To derive reliable estimates larger population sizes are usually necessary which is detrimental with regard to the algorithm's speed. Therefore, past information, that is, past covariance matrizes are usually also considered

$$\mathbf{C}^{(g+1)} := (1 - c_{\mu})C^{(g)} + c_{\mu}\mathbf{C}_{\mu}^{(g+1)} \tag{3}$$

with parameter $0 \le c_{\mu} \le 1$ determining the effective time-horizon. In CMA-ESs, it has been found that an enhance of the general search direction in the covariance matrix is usual beneficial. For this, the concepts of the *evolutionary path* and the *rank-one-update* are introduced. As its name already suggests, an evolutionary path considers the path in the search space the population has taken so far. The weighted means serve as representatives. Defining

$$\mathbf{v}^{(g+1)} := \frac{\mathbf{m}^{(g+1)} - \mathbf{m}^{(g)}}{\sigma^{(g)}}$$

the evolutionary path reads

$$\mathbf{p}_c^{(g+1)} := (1 - c_c)\mathbf{p}_c^{(g)} + \sqrt{c_c(2 - c_c)\mu_{\text{eff}}}\Big(\frac{\mathbf{m}^{(g+1)} - \mathbf{m}^{(g)}}{\sigma^{(g)}}\Big). \tag{4}$$

For details on the parameters, see e.g. [12]. The evolutionary path gives a general search direction that the ES has taken in the recent past. In order to bias the covariance matrix accordingly, the rank-one-update

$$\mathbf{C}_1^{(g+1)} := \mathbf{p}_c^{(g+1)}(\mathbf{p}_c^{(g+1)})^{\mathrm{T}} \tag{5}$$

is performed and used as a further component of the covariance matrix. A normal distribution with covariance $\mathbf{C}_1^{(g+1)}$ leads towards a one-dimensional distribution on the line defined by $\mathbf{p}_c^{(g+1)}$. With (5) and (3), the final covariance update of the CMA-ES reads

$$\mathbf{C}^{(g+1)} := (1 - c_1 - c_\mu)\mathbf{C}^{(g)} + c_1\mathbf{C}_1^{(g+1)} + c_\mu\mathbf{C}_\mu^{(g+1)}. \tag{6}$$

The CMA-ES is one of the most powerful evolution strategies. However, as pointed out in [5], its scaling behavior with the population size is not good. The alternative approach of the CMSA-ES [5] updates the covariance matrix differently. Considering again the definition (1), the covariance update is a convex combination of the old covariance and the population covariance, i.e., the rank-μ update

$$\mathbf{C}^{(g+1)} := (1 - \frac{1}{c_\tau})\mathbf{C}^{(g)} + \frac{1}{c_\tau}\sum_{m=1}^{\mu} w_m \mathbf{y}_{m:\lambda}^{(g+1)}(\mathbf{y}_{m:\lambda}^{(g+1)})^{\mathrm{T}} \tag{7}$$

with the weights usually set to $w_m = 1/\mu$. See [5] for information on the free parameter c_τ.

1.3 Step-Size Adaptation

The CMA-ES uses the so-called *cumulative step-size adaptation* (CSA) to control the scaling parameter (also called *step-size*, *mutation strength* or *step-length*) [12]. To this end, the CSA determines again an evolutionary path by summating the movement of the population centers

$$\mathbf{p}_\sigma^{(g+1)} = (1 - c_\sigma)\mathbf{p}_\sigma^{(g)} + \sqrt{c_\sigma(2 - c_\sigma)\mu_{\text{eff}}}(\mathbf{C}^{(g)})^{-\frac{1}{2}}\frac{\mathbf{m}^{(g+1)} - \mathbf{m}^{(g)}}{\sigma^{(g)}} \tag{8}$$

eliminating the influence of the covariance matrix and the step length. For a detailed description of the parameters, see [12]. The length of the path in (8) is important. In the case of short path lengths, several movement of the centers counteract each other which is an indication that the step-size is too large and should be reduced. If on the other hand, the ES takes several consecutive steps in approximately the same direction, progress and algorithm speed would be improved, if larger changes were possible. Long path lengths, therefore, are an indicator for a required increase of the step length. Ideally, the CSA should result in uncorrelated steps.

After some calculations, see [12], the ideal situation is revealed as standard normally distributed steps, which leads to

$$\ln(\sigma^{(g+1)}) = \ln(\sigma^{(g)}) + \frac{c_\sigma}{d_\sigma}\Big(\frac{\|\mathbf{p}_\sigma^{(g+1)}\| - \mu_{\chi_n}}{\mu_{\chi_n}}\Big) \tag{9}$$

as the CSA-rule. The change is multiplicative in order to avoid numerical problems and results in non-negative scaling parameters. The parameter μ_{χ_n} in (9) stands for the mean of the χ-distribution with n degrees of freedom. If a random variable follows a χ_n^2 distribution, its square root is χ-distributed. The degrees of freedom coincide with the search space dimension. The CSA-rule works well in many application cases. It can be shown, however, that the original CSA encounter problems in large noise regimes resulting in a loss of step-size control and premature convergence. Therefore, uncertainty handling procedures and other safeguards are advisable.

An alternative approach for adapting the step-size is *self-adaptation* first introduced in [20] and developed further in [22]. It subjects the strategy parameters of the mutation to evolution. In other words, the scaling parameter or in its full form, the whole covariance matrix, undergoes recombination, mutation, and indirect selection processes. The working principle is based on an indirect stochastic linkage between good individuals and appropriate parameters: On average good parameters should lead to better offspring than too large or too small values or misleading directions. Although self-adaptation has been developed to adapt the whole covariance matrix, it is used nowadays mainly to adapt the step-size or a diagonal covariance matrix. In the case of the mutation strength, usually a log-normal distribution

$$\sigma_l^{(g)} = \sigma_{\text{base}}\exp(\tau\mathcal{N}(0,1)) \tag{10}$$

is used for mutation. The parameter τ is called the *learning rate* and is usually chosen to scale with $1/\sqrt{2N}$. The variable σ_{base} is either the parental scale factor or the result of recombination. For the step-size, it is possible to apply the same type of recombination as for the positions although different forms – for instance a multiplicative combination – could be used instead. The self-adaptation of the step-size is referred to as σ-*self-adaptation* (σSA) in the remainder of this paper.

The newly created mutation strength is then directly used in the mutation of the offspring. If the resulting offspring is sufficiently good, the scale factor is passed to the next generation. The baseline σ_{base} is either the mutation strength of the parent or if recombination is used the recombination result. Self-adaptation with recombination has been shown to be "robust" against noise [3] and is used in the CMSA-ES as update rule for the scaling factor. In [5] it was found that the CMSA-ES performs comparably to the CMA-ES for smaller populations but is less computational expensive for larger population sizes.

2 Concerning the Covariance Estimator

The covariance matrix $\mathbf{C}_\mu$ which appears in (2) and (7) can be interpreted as the sample covariance matrix with sample size μ. Two differences are present. The first using μ instead of $\mu-1$ can be explained by using the known mean instead of an estimate. The second lies in the non-identically distributed random variables of the population since order statistics appear. We will disregard that problem for the time being.

In the case of identically independently distributed random variables, the estimate converges almost surely towards the "true" covariance Σ for $\mu \to \infty$. In addition, the sample covariance matrix is related (in our case equal) to the maximum likelihood (ML) estimator of Σ. Both facts serve a justification to take $\mathbf{C}_\mu$ as the substitute for the unknown true covariance for large μ. However, the quality of the estimate can be quite poor if $\mu < N$ or even $\mu \approx N$.

This was first discovered by Stein [23,24]. Stein's phenomenon states that while the ML estimate is often seen as the best possible guess, its quality may be poor and can be improved in many cases. This holds especially for high-dimensional spaces. The same problem transfers to covariance matrix estimation, see [21]. Also recognized by Stein, in case of small ratios μ/N the eigenstructure of $\mathbf{C}_\mu$ may not agree well with the true eigenstructure of Σ. As stated in [17], the largest eigenvalue has a tendency towards too large values, whereas the smallest shows the opposite behavior. This results in a larger spectrum of the sample covariance matrix with respect to the true covariance for $N/\mu \not\to 0$ for $\mu, N \to \infty$ [2]. As found by Huber [16], a heavy tail distribution leads also to a distortion of the sample covariance.

In statistics, considerable efforts have been made to find more reliable and robust estimates. Owing to the great inportance of the covariance matrix in data mining and other statistical analyses, work is still ongoing. The following section provides a short introduction before focussing on the approach used for evolution strategies.

3 Approaches for Estimating the Covariance

As stated above, the estimation of high-dimensional covariance matrices has received a lot of attention, see e.g. [6]. Several types have been introduced, for example: shrinkage estimators, banding and tapering estimators, sparse matrix transform estimators, and the graphical Lasso estimator. This paper concentrates on shrinkage estimators and on an idea inspired by a maximum entropy approach. Both classes can be computed comparatively efficiently. Future research will consider other classes of estimators.

3.1 Shrinkage Estimators

Most (linear) shrinkage estimators use the convex combination

$$\mathbf{S}_{\text{est}}(\rho) = \rho\mathbf{F} + (1-\rho)\mathbf{C}_\mu \tag{11}$$

with $\mathbf{F}$ the *target* to correct the estimate provided by the sample covariance. The parameter $\rho \in]0,1[$ is called the *shrinkage intensity.* Equation (11) is used to shrink the eigenvalues of $\mathbf{C}_\mu$ towards the eigenvalues of $\mathbf{F}$. The shrinkage intensity ρ should be chosen to minimize

$$\mathrm{E}\Big(\|\mathbf{S}_{\text{est}}(\rho) - \Sigma\|_F^2\Big) \tag{12}$$

with $\|\cdot\|_F^2$ denoting the squared Frobenius norm with

$$\|\mathbf{A}\|_F^2 = \frac{1}{N}\mathrm{Tr}\Big[\mathbf{A}\mathbf{A}^\mathrm{T}\Big], \tag{13}$$

see [17]. To solve this problem, knowledge of the true covariance Σ would be required which is usually unobtainable.

Starting from (12), Ledoit and Wolf obtained an analytical expression for the optimal shrinkage intensity for the target $\mathbf{F} = \mathrm{Tr}(\mathbf{C}_\mu)/N\,\mathbf{I}$. The result does not make assumptions on the underlying distribution. In the case of $\mu \approx N$ or vastly different eigenvalues, the shrinkage estimator does not differ much from the sample covariance matrix, however.

Other authors introduced different estimators, see e.g. [7] or [6]). Ledoit and Wolfe themselves considered non-linear shrinkage estimators [18]. Most of the approaches require larger computational efforts. In the case of the non-linear shrinkage, for example, the authors are faced with a non-linear, non-convex optimization problem, which they solve by using sequential linear programming [18]. A general analytical expression is unobtainable, however.

Shrinkage estimators and other estimators aside from the standard case have not been used in in evolution strategies before. A literature review resulted in one application in the case of Gaussian based estimation of distribution algorithms albeit with quite a different goal [9]. There, the learning of the covariance matrix during the run lead to non positive definite matrices. A shrinkage procedure was applied to "repair" the covariance matrix towards the required structure. The authors used a similar approach as in [17] but made the shrinkage intensity adaptable.

Interestingly, (3), (6), and (7) of the ES algorithm can be interpreted as a special case of shrinkage. In the case of the CMSA-ES, for example, the estimate is shrunk towards the old covariance matrix. The shrinkage intensity is determined by

$$c_\tau = 1 + \frac{N(N+1)}{2\mu} \tag{14}$$

as $\rho = 1 - 1/c_\tau$. As long as the increase of μ with the dimensionality N is below $\mathcal{O}(N^2)$, the coefficient (14) approaches infinity for $N \to \infty$. Since the contribution of the sample covariance to the new covariance in (7) is weighted with $1/c_\tau$, its influence fades out for increasing dimensions. It is the aim of the paper to investigate whether a further shrinkage can improve the result.

Transferring shrinkage estimators to ESs must take the situation in which the estimation occurs into account since it differs from the assumptions in statistical literature. The covariance matrix $\Sigma = \mathbf{C}^{g-1}$ that was used to create the offspring is known. The sample is based on rank-based selection, however, which differs from the iid case usually considered. Only if there were no selection pressure, the sample $\mathbf{x}_1, \ldots, \mathbf{x}_\mu$ would represent normally distributed random variables. In this context, it is interesting to note that the argumentation in [12] with respect to the setting of the CMA-ES parameter argues to choose the parameter so that the distribution of the random variables remains unchained as long as no selection pressure occurs. In other words, if $\mathbf{p}^{(g)} \sim \mathcal{N}(\mathbf{0}, \mathbf{C}^{(g)})$ then also $\mathbf{p}^{(g+1)}$ for both evolution paths, (4) and (8). However, due to sampling and using the covariance estimate, larger deviations may occur. Applying shrinkage could improve the situation. However, the choice of the target remains. Most shrinkage approaches consider diagonal matrices as shrinkage targets. If we were following that approach, we could choose the matrix $\mathbf{F} = \text{diag}(\mathbf{C}_\mu)$. This would leave the diagonal elements of the sample covariance matrix unchanged decreasing only the off-diagonal entries. However, a shrinkage towards a diagonal does not appear to be a good idea for optimizing functions that are not oriented towards the coordinate system.

3.2 A Maximum Entropy Covariance Estimation

Therefore, we make use of another concept following [25]. Confronted with the problem of determining a reliable covariance matrix by combining a sample covariance matrix with a pooled variance matrix, the authors introduced a *maximum entropy covariance selection principle.* Since a combination of covariance matrices also appears in evolution strategies, a closer look at their approach is interesting. Defining a population matrix $\mathbf{C}_p$ and the sample covariance matrix $\mathbf{S}_i$, the mixture

$$\mathbf{S}_{mix}(\eta) = \eta \mathbf{C}_p + (1 - \eta)\mathbf{S}_i \tag{15}$$

was considered. In departure from usual approaches, focus lay on the combination of the two matrixes that maximizes the entropy. To this end, the coordinate system was changed to the eigenspace of $\mathbf{S}_{mix} = \mathbf{C}_p + (1 - \eta)\mathbf{S}_i$. Let $\mathbf{M}_S$ denote the (normalized) eigenvectors of the mixture matrix. The representations of $\mathbf{C}_p$ and $\mathbf{S}_i$ in this coordination system read

$$\begin{aligned} \Phi^C &= \mathbf{M}_S^{\mathrm{T}} \mathbf{C}_p \mathbf{M}_S \\ \Phi^S &= \mathbf{M}_S^{\mathrm{T}} \mathbf{S}_i \mathbf{M}_S. \end{aligned} \tag{16}$$

Both matrices are usually not diagonal. To construct the new estimate for the covariance matrix,

$$\Lambda^C = \mathrm{diag}(\Phi^C)$$
$$\Lambda^S = \mathrm{diag}(\Phi^S) \tag{17}$$

were determined. By taking $\lambda_i = \max(\lambda_i^C, \lambda_i^S)$, a covariance matrix estimate could finally be constructed via $\mathbf{M}_S \Lambda \mathbf{M}_S^{\mathrm{T}}$. The approach maximizes the possible contributions to the principal direction of the mixture matrix and is based on a maximum entropy derivation for the estimation.

3.3 New Covariance Estimators

This paper proposes a combination of a shrinkage estimator and the basis transformation introduced [25] for a use in evolution strategies. This paper focuses on the CMSA-ES. The aim is to switch towards a suitable coordinate system and then either to discard the contributions of the sample covariance that are not properly aligned or to shrink the off-diagonal components. Two choices for the mixture matrix represent themselves. The first

$$\mathbf{S}_{mix} = \mathbf{C}^g + \mathbf{C}_\mu \tag{18}$$

is be chosen in accordance to [25]. The second takes the covariance result that would have been used in the original CMSA-ES

$$\mathbf{S}_{mix} = (1 - c_\tau)\mathbf{C}^g + c_\tau \mathbf{C}_\mu \tag{19}$$

and introduces a single step recursion which may be more appropriate for small population sizes. Both choices will be investigated in this paper. They in turn can be coupled with several further ways to proceed and to construct the new covariance matrix. Switching towards the eigenspace of $\mathbf{S}_{mix}$, results in the covariance matrix representations $\Phi_\mu := \mathbf{M}_S^{\mathrm{T}} \mathbf{C}_\mu \mathbf{M}_S$ and $\Phi_\Sigma := \mathbf{M}_S^{\mathrm{T}} \mathbf{C}^g \mathbf{M}_S$.

1. The first approach for constructing a new estimate of the sample covariance is to apply the principle of maximal contribution to the axes from [25] and to determine
$$\Lambda_\mu = \max\Big(\mathrm{diag}(\Phi_\mu), \mathrm{diag}(\Phi_\Sigma)\Big) \tag{20}$$
The sample covariance matrix can then be computed as $\mathbf{C}'_\mu = \mathbf{M}_S \Lambda_\mu \mathbf{M}_S^{\mathrm{T}}$.
2. Another approach would be to discard all entries of Φ_μ except the diagonal
$$\Lambda_\mu = \mathrm{diag}(\Phi_\mu) \tag{21}$$
3. A third approach consists of applying a shrinkage estimator like
$$\Phi_\mu^S = (1 - \rho)\Phi_\mu + \rho \mathrm{diag}(\Phi_\mu). \tag{22}$$
This approach does not discard the off-diagonal entries completely. The shrinkage intensity ρ remains to be determined.

4 Experimental Evaluation

This section describes the experiments that were performed to explore the new approaches. For our investigation, the CMSA-ES version is considered since it operates just with the population covariance matrix and effects from changing the estimate should be easier to discerned. The competitors consist of algorithms which use shrinkage estimators as defined in (18) to (22). This code is not optimized for performance with respect to absolute computing time, since this paper aims at a proof of concept. The experiments are performed for the search space dimensions $N = 2, 5, 10$, and 20. The maximal number of fitness evaluations is $\mathrm{FE}_{\max} = 2 \times 10^4 N$. The CMSA-ES versions use $\lambda = \lfloor \log(3N) + 8 \rfloor$ offspring and $\mu = \lceil \lambda/4 \rceil$ parents. The start position of the algorithms is randomly chosen from a normal distribution with mean zero and standard deviation of 0.5. A run terminates prematurely if the difference between the best value obtained so far and the optimal fitness value $|f_{\text{best}} - f_{\text{opt}}|$ is below a predefined precision set to 10^{-8}. For each fitness function and dimension, 15 runs are used.

4.1 Test Suite

The experiments are performed with the black box optimization benchmarking (BBOB) software framework and the test suite introduced for the black box optimization workshops, see [13]. The aim of the workshop is to benchmark and compare metaheuristics and other direct search methods for continuous optimization. The framework allows the plug-in of algorithms adhering to a common interface and provides a comfortable way of generating the results in form of tables and figures.

The test suite contains noisy and noise-less functions with the position of the optimum changing randomly from run to run. This paper focuses on the 24 noiseless functions [11]. They can be divided into four classes: separable functions (function ids 1–5), functions with low/moderate conditioning (ids 6–9), functions with high conditioning (ids 10–14), and two groups of multimodal functions (ids 15–24).

4.2 Performance Measure

The following performance measure is used in accordance to [13]. The expected running time (ERT) gives the expected value of the function evaluations (f-evaluations) the algorithm needs to reach the target value with the required precision for the first time, see [13]. In this paper, we use

$$\mathrm{ERT} = \frac{\#(FEs(f_{\text{best}} \geq f_{\text{target}}))}{\#succ} \tag{23}$$

as an estimate by summing up the fitness evaluations $FEs(f_{\text{best}} \geq f_{\text{target}})$ of each run until the fitness of the best individual is smaller than the target value, divided by all successfull runs (Fig. 1).

4.3 Results and Discussion

Due to space restrictions, Figure 3 and Table 1 and Fig. 2 show only the results from the best experiments which were achieved for the variant which used (22) together with (19) as the transformation matrix (called CMSA-shr-ES in the following). First of all, it should be noted that there is no significant advantage to either algorithm for the test suite functions. Table 1 and Fig. 2 show the ERT loss ratio with respect to the best result from the BBOB 2009 workshop for predefined budgets given in the first column. The median performance of both algorithms improves with the dimension until the budget of 10^3 – which is interesting. An increase of the budget goes along with a decreased performance which is less pronounced for the CMSA-shr-ES in the case of the larger dimensional space. This indicates that the CMSA-shr-ES may perform more favorable in larger search spaces as envisioned. Further experiments which a larger maximal number of fitness evaluations and larger dimensional spaces will be conducted which should shed more light on the behavior. Furthermore, the decrease in performance with the budget hints at a search stagnation probably due to convergence into local optima. Restart strategies may be beneficial, but since they have to be fitted to the algorithms, we do not apply them in the present paper.

Figure 3 shows the expected running time for reaching the precision of 10^{-8} for all 24 functions and search space dimensionalities. In the case of the separable functions (1–5), both algorithms show a very similar behavior, succeeding in optimizing the first two functions and exhibiting difficulties in the case of the difficult rastrigin variants. On the linear slope, the original CMSA-ES shows fewer expected function evaluations for smaller dimensions which starts to change when the dimensionality is increased. For the functions with ids 6–9, with moderate condition numbers, there are advantages to the CMSA-shr-ES, with the exception of the rotated rosenbrock (9). Most of the functions with high conditioning, ids 10–12, and 14, can be solved by both variants with slightly better results for the CMSA-ES. The sharp ridge (id 13) appears as problematic, with the CMSA-shr-ES showing fewer fitness evaluations for hitting the various precisions goals in Table 1.

Interestingly, the CMSA-shr variant seems to perform better for the difficult multimodal functions, e.g., Gallaghers 101 peak function, a finding which should be explored in more detail. The results for the last two multimodal functions can be explained in part in that the computing resources were insufficient for the optimization. Even the best performing algorithms from the BBOB workshop needed more resources than we used in our experiments.

Further experiments will be conducted in order to shed more light on the behavior. Special attention will be given to the choice of the shrinkage factor, since its setting is unlikely to be optimal and may have influenced the outcome strongly. Furthermore, the question remains whether the population size should be increased for the self-adaptation process. Also, larger search space dimensionalities than $N = 20$ are of interest.

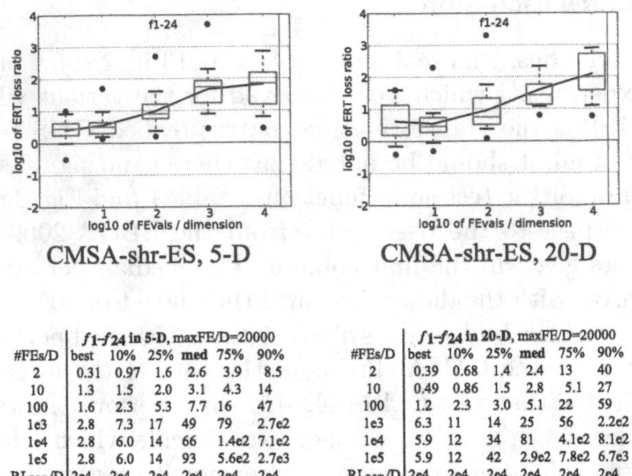

#FEs/D	*f*1–*f*24 in 5-D, maxFE/D=20000 best	10%	25%	med	75%	90%
2	0.31	0.97	1.6	2.6	3.9	8.5
10	1.3	1.5	2.0	3.1	4.3	14
100	1.6	2.3	5.3	7.7	16	47
1e3	2.8	7.3	17	49	79	2.7e2
1e4	2.8	6.0	14	66	1.4e2	7.1e2
1e5	2.8	6.0	14	93	5.6e2	2.7e3
RL_{US}/D	2e4	2e4	2e4	2e4	2e4	2e4

#FEs/D	*f*1–*f*24 in 20-D, maxFE/D=20000 best	10%	25%	med	75%	90%
2	0.39	0.68	1.4	2.4	13	40
10	0.49	0.86	1.5	2.8	5.1	27
100	1.2	2.3	3.0	5.1	22	59
1e3	6.3	11	14	25	56	2.2e2
1e4	5.9	12	34	81	4.1e2	8.1e2
1e5	5.9	12	42	2.9e2	7.8e2	6.7e3
RL_{US}/D	2e4	2e4	2e4	2e4	2e4	2e4

Fig. 1. The CMSA-shr-ES. ERT loss ratio (in number of f-evaluations divided by dimension) divided by the best ERT seen in GECCO-BBOB-2009 for the target ftarget, or, if the best algorithm reached a better target within the budget, the budget divided by the best ERT. Line: geometric mean. Box-Whisker error bar: 25–75 %-ile with median (box), 10–90 %-ile (caps), and minimum and maximum ERT loss ratio (points). The vertical line gives the maximal number of function evaluations in a single trial in this function subset.

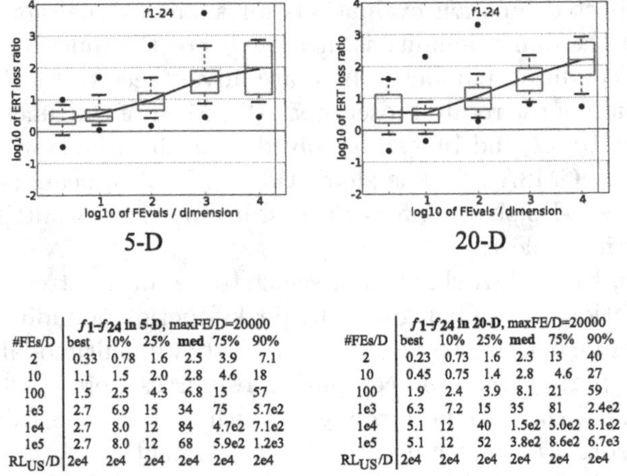

#FEs/D	*f*1–*f*24 in 5-D, maxFE/D=20000 best	10%	25%	med	75%	90%
2	0.33	0.78	1.6	2.5	3.9	7.1
10	1.1	1.5	2.0	2.8	4.6	18
100	1.5	2.5	4.3	6.8	15	57
1e3	2.7	6.9	15	34	75	5.7e2
1e4	2.7	8.0	12	84	4.7e2	7.1e2
1e5	2.7	8.0	12	68	5.9e2	1.2e3
RL_{US}/D	2e4	2e4	2e4	2e4	2e4	2e4

#FEs/D	*f*1–*f*24 in 20-D, maxFE/D=20000 best	10%	25%	med	75%	90%
2	0.23	0.73	1.6	2.3	13	40
10	0.45	0.75	1.4	2.8	4.6	27
100	1.9	2.4	3.9	8.1	21	59
1e3	6.3	7.2	15	35	81	2.4e2
1e4	5.1	12	40	1.5e2	5.0e2	8.1e2
1e5	5.1	12	52	3.8e2	8.6e2	6.7e3
RL_{US}/D	2e4	2e4	2e4	2e4	2e4	2e4

Fig. 2. The CMSA-ES. ERT loss ratio (in number of f-evaluations divided by dimension) divided by the best ERT seen in GECCO-BBOB-2009 for the target ftarget, or, if the best algorithm reached a better target within the budget, the budget divided by the best ERT. Line: geometric mean. Box-Whisker error bar: 25–75 %-ile with median (box), 10–90 %-ile (caps), and minimum and maximum ERT loss ratio (points). The vertical line gives the maximal number of function evaluations in a single trial in this function subset.

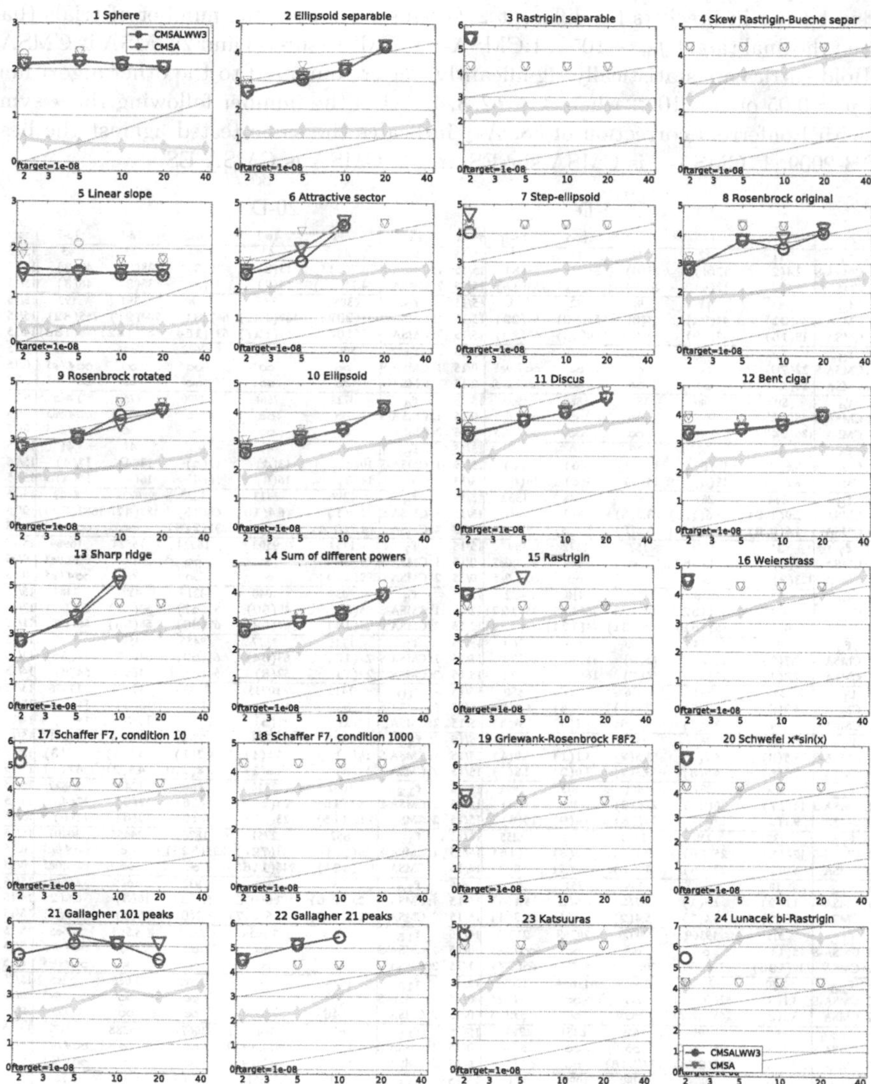

Fig. 3. Expected running time ERT in number of f-evaluations) divided by dimension for target function value as $\log_{10}$ values versus dimension. Different symbols correspond to different algorithms given in the legend of f_1 and f_{24}. Light symbols give the maximum number of function evaluations from the longest trial divided by dimension. Horizontal lines give linear scaling, slanted dotted lines give quadratic scaling. Black stars indicate statistically better result compared to all other algorithms with $p < 0.01$ and Bonferroni correction number of dimensions (six). Legend: .1: CMSA-S is CMSA-shr-ES and 2: CMSA is CMSA-ES.

Table 1. ERT in number of function evaluations divided by the best ERT measured during BBOB-2009 given in the respective first row with the central 80 % range divided by two in brackets for different Δf values. #succ is the number of trials that reached the final target $f_{\text{opt}}+10^{-8}$. 1:CMSA-S is CMSA-shr-ES and 2:CMSA is CMSA-ES. Bold entries are statistically significantly better compared to the other algorithm, with $p = 0.05$ or $p = 10^{-k}$ where $k \in \{2, 3, 4, \ldots\}$ is the number following the $\star$ symbol, with Bonferroni correction of 48. A $\downarrow$ indicates the same tested against the best BBOB-2009. 1: CMSA-S is CMSA-shr-ES and 2: CMSA is CMSA-ES.

5-D

Δf	1e+1	1e-1	1e-3	1e-5	1e-7	#succ
f_1	11	12	12	12	12	15/15
1: CMSA-S	2.4(2)	12(4)	25(6)	38(8)	53(8)	15/15
2: CMSA	2.4(2)	12(4)	26(7)	40(8)	54(15)	15/15
f_2	83	88	90	92	94	15/15
1: CMSA-S	26(21)	44(24)	53(30)	56(30)	57(29)	15/15
2: CMSA	19(15)	47(19)	54(23)	60(23)	63(30)	15/15
f_3	716	1637	1646	1650	1654	15/15
1: CMSA-S	22(70)	∞	∞	∞	∞ *1.0e5*	0/15
2: CMSA	94(140)	∞	∞	∞	∞ *1.0e5*	0/15
f_4	809	1688	1817	1886	1903	15/15
1: CMSA-S	46(62)	∞	∞	∞	∞ *1.0e5*	0/15
2: CMSA	186(247)	∞	∞	∞	∞ *1.0e5*	0/15
f_5	10	10	10	10	10	15/15
1: CMSA-S	8.6(3)	16(6)	17(6)	17(6)	17(6)	15/15
2: CMSA	8.2(3)	13(6)	14(6)	14(6)	14(6)	15/15
f_6	114	281	580	1038	1332	15/15
1: CMSA-S	2.0(1.0)	2.6(1.0)	3.2(3)	2.5(1)	3.2(3)	15/15
2: CMSA	1.6(0.9)	2.2(0.9)	3.2(3)	3.7(3)	5.8(4)	15/15
f_7	24	1171	1572	1572	1597	15/15
1: CMSA-S	57(5)	1217(1281)	∞	∞	∞ *1.0e5*	0/15
2: CMSA	123(3)	∞	∞	∞	∞ *1.0e5*	0/15
f_8	73	336	391	410	422	15/15
1: CMSA-S	2.6(1)	88(152)	79(133)	76(125)	75(122)	12/15
2: CMSA	3.0(1)	85(151)	77(131)	74(124)	72(122)	12/15
f_9	35	214	300	335	369	15/15
1: CMSA-S	2.7(2)	15(9)	15(8)	15(7)	15(7)	15/15
2: CMSA	2.9(1)	17(9)	16(7)	16(7)	15(7)	15/15
f_{10}	349	574	626	829	880	15/15
1: CMSA-S	6.7(6)	6.7(5)	7.6(5)	6.3(4)	6.3(3)	15/15
2: CMSA	7.2(4)	7.4(5)	8.3(5)	7.3(5)	7.4(4)	15/15
f_{11}	143	763	1177	1467	1673	15/15
1: CMSA-S	8.8(8)	4.7(2)	3.5(2)	3.1(1)	2.9(1)	15/15
2: CMSA	7.3(6)	4.0(3)	3.3(2)	3.0(2)	2.8(1)	15/15
f_{12}	108	371	461	1303	1494	15/15
1: CMSA-S	14(17)	20(22)	22(19)	10(9)	11(9)	15/15
2: CMSA	9.1(2)	17(21)	19(18)	9.4(9)	10(8)	15/15
f_{13}	132	250	1310	1752	2255	15/15
1: CMSA-S	12(13)	25(16)	7.1(4)	8.2(3)	13(16)	14/15
2: CMSA	5.0(7)	27(25)	7.2(5)	7.2(4)	7.0(3)	14/15
f_{14}	10	58	139	251	476	15/15
1: CMSA-S	1.6(2)	2.9(1)	5.7(2)	8.6(3)	8.1(4)	15/15
2: CMSA	1.3(2)	3.4(1)	5.4(2)	8.7(4)	8.3(4)	15/15
f_{15}	511	19369	20073	20769	21359	14/15
1: CMSA-S	15(1)	∞	∞	∞	∞ *1.0e5*	0/15
2: CMSA	132(196)	72(80)	70(77)	68(77)	66(77)	1/15
f_{16}	120	2662	10449	11644	12095	15/15
1: CMSA-S	3.1(3)	526(620)	∞	∞	∞ *1.0e5*	0/15
2: CMSA	3.3(3)	245(282)	∞	∞	∞ *1.0e5*	0/15
f_{17}	5.2	899	3669	6351	7934	15/15
1: CMSA-S	3.3(5)	98(167)	∞	∞	∞ *1.0e5*	0/15
2: CMSA	2.6(3)	223(278)	177(218)	∞	∞ *1.0e5*	0/15
f_{18}	103	3968	9280	10905	12469	15/15
1: CMSA-S	70(0.7)	101(126)	∞	∞	∞ *1.0e5*	0/15
2: CMSA	0.90(0.7)	22(38)	151(172)	∞	∞ *1.0e5*	0/15
f_{19}	1	242	1.2e5	1.2e5	1.2e5	15/15
1: CMSA-S	3.1(2)	2730(3097)	∞	∞	∞ *1.0e5*	0/15
2: CMSA	3.0(3)	1186(1448)	∞	∞	∞ *1.0e5*	0/15
f_{20}	16	38111	54470	54861	55313	14/15
1: CMSA-S	1.6(1)	∞	∞	∞	∞ *1.0e5*	0/15
2: CMSA	1.9(1)	∞	∞	∞	∞ *1.0e5*	0/15
f_{21}	41	1674	1705	1729	1757	14/15
1: CMSA-S	377(1220)	388(448)	382(469)	376(434)	370(455)	2/15
2: CMSA	612(1220)	836(970)	821(909)	810(882)	797(939)	1/15
f_{22}	71	938	1008	1040	1068	14/15
1: CMSA-S	514(705)	693(853)	645(793)	626(673)	610(773)	2/15
2: CMSA	513(705)	694(852)	646(794)	626(721)	610(703)	2/15
f_{23}	3.0	14249	31654	33030	34256	15/15
1: CMSA-S	2.0(2)	∞	∞	∞	∞ *1.0e5*	0/15
2: CMSA	3.2(3)	∞	∞	∞	∞ *1.0e5*	0/15
f_{24}	1622	6.4e6	9.6e6	1.3e7	1.3e7	3/15
1: CMSA-S	33(62)	∞	∞	∞	∞ *1.0e5*	0/15
2: CMSA	17(31)	∞	∞	∞	∞ *1.0e5*	0/15

20-D

Δf	1e+1	1e-1	1e-3	1e-5	1e-7	#succ
f_1	43	43	43	43	43	15/15
1: CMSA-S	4.7(1)	15(3)	27(3)	38(2)	48(3)	15/15
2: CMSA	4.7(0.9)	14(2)	24(2)	35(3)	46(3)	15/15
f_2	385	387	390	391	393	15/15
1: CMSA-S	331(232)	567(121)	647(81)	704(91)	755(82)	15/15
2: CMSA	332(106)	561(127)	637(154)	687(182)	786(186)	14/15
f_3	5066	7635	7643	7646	7651	15/15
1: CMSA-S	∞	∞	∞	∞	∞ *4.0e5*	0/15
2: CMSA	∞	∞	∞	∞	∞ *4.0e5*	0/15
f_4	4722	7666	7700	7758	1.4e5	9/15
1: CMSA-S	∞	∞	∞	∞	∞ *4.0e5*	0/15
2: CMSA	∞	∞	∞	∞	∞ *4.0e5*	0/15
f_5	41	41	41	41	41	15/15
1: CMSA-S	10(3)	13(4)	13(4)	13(4)	13(4)	15/15
2: CMSA	11(3)	16(4)	16(4)	16(4)	16(4)	15/15
f_6	1296	3413	5220	6728	8409	15/15
1: CMSA-S	1.6(1)	6.4(10)	63(78)	388(475)	682(690)	0/15
2: CMSA	1.7(0.4)	7.0(6)	231(278)	∞	∞ *4.0e5*	0/15
f_7	1351	9503	16524	16524	16969	15/15
1: CMSA-S	649(862)	∞	∞	∞	∞ *4.0e5*	0/15
2: CMSA	299(444)	∞	∞	∞	∞ *4.0e5*	0/15
f_8	2039	4040	4219	4371	4484	15/15
1: CMSA-S	25(9)	49(50)	50(47)	50(49)	50(51)	13/15
2: CMSA	23(14)	67(62)	67(60)	66(58)	66(57)	11/15
f_9	1716	3277	3455	3594	3727	15/15
1: CMSA-S	27(10)	63(64)	64(60)	64(57)	64(54)	13/15
2: CMSA	27(7)	52(8)	54(8)	54(9)	54(8)	14/15
f_{10}	7413	10735	14920	17073	17476	15/15
1: CMSA-S	17(6)	19(5)	15(4)	15(3)	15(3)	15/15
2: CMSA	17(10)	21(5)	17(5)	16(5)	18(5)	14/15
f_{11}	1002	6278	9762	12285	14831	15/15
1: CMSA-S	14(4)	4.1(1)	4.2(1)	5.1(3)	5.5(3)	15/15
2: CMSA	13(5)	3.6(1)	3.8(1.0)	4.6(2)	4.8(1)	15/15
f_{12}	1042	2740	4140	12407	13827	15/15
1: CMSA-S	8.5(16)	23(22)	26(15)	12(5)	12(4)	15/15
2: CMSA	5.8(15)	23(25)	26(16)	12(6)	13(6)	15/15
f_{13}	652	2751	18749	24455	30201	15/15
1: CMSA-S	46(0.8)	583(727)	299(325)	∞	∞ *4.0e5*	0/15
2: CMSA	704(921)	946(1163)	∞	∞	∞ *4.0e5*	0/15
f_{14}	75	304	932	1648	15661	15/15
1: CMSA-S	2.0(0.6)	2.6(0.5)	5.9(2)	16(5)	6.8(2)	14/15
2: CMSA	2.0(0.6)	2.6(0.7)	6.2(0.8)	16(5)	6.0(2)	15/15
f_{15}	30378	3.1e5	3.2e5	4.5e5	4.6e5	15/15
1: CMSA-S	∞	∞	∞	∞	∞ *4.0e5*	0/15
2: CMSA	∞	∞	∞	∞	∞ *4.0e5*	0/15
f_{16}	1384	77015	1.9e5	2.0e5	2.2e5	15/15
1: CMSA-S	2.9(3)	∞	∞	∞	∞ *4.0e5*	0/15
2: CMSA	107(146)	∞	∞	∞	∞ *4.0e5*	0/15
f_{17}	63	4005	30677	56288	80472	15/15
1: CMSA-S	1.1(1)	1399(1498)	∞	∞	∞ *4.0e5*	0/15
2: CMSA	1.1(1)	650(749)	∞	∞	∞ *4.0e5*	0/15
f_{18}	621	19561	67569	1.3e5	1.5e5	15/15
1: CMSA-S	0.96(0.5)	∞	∞	∞	∞ *4.0e5*	0/15
2: CMSA	1.0(0.4)	∞	∞	∞	∞ *4.0e5*	0/15
f_{19}	1	3.4e5	6.2e6	6.7e6	6.7e6	15/15
1: CMSA-S	4.8(4)	∞	∞	∞	∞ *4.0e5*	0/15
2: CMSA	6.4(9)	∞	∞	∞	∞ *4.0e5*	0/15
f_{20}	82	3.1e6	5.5e6	5.6e6	5.6e6	14/15
1: CMSA-S	2.0(0.6)	∞	∞	∞	∞ *4.0e5*	0/15
2: CMSA	2.0(0.9)	∞	∞	∞	∞ *4.0e5*	0/15
f_{21}	561	14103	14643	15567	17589	15/15
1: CMSA-S	51(1.0)	43(57)	41(55)	39(51)	34(45)	6/15
2: CMSA	179(356)	184(227)	178(191)	167(193)	148(171)	2/15
f_{22}	467	23491	24948	26847	1.3e5	12/15
1: CMSA-S	430(857)	∞	∞	∞	∞ *4.0e5*	0/15
2: CMSA	429(857)	∞	∞	∞	∞ *4.0e5*	0/15
f_{23}	3.2	67457	4.9e5	8.1e5	8.4e5	15/15
1: CMSA-S	2.1(2)	∞	∞	∞	∞ *4.0e5*	0/15
2: CMSA	2.3(3)	∞	∞	∞	∞ *4.0e5*	0/15
f_{24}	1.3e6	5.2e7	5.2e7	5.2e7	5.2e7	3/15
1: CMSA-S	∞	∞	∞	∞	∞ *4.0e5*	0/15
2: CMSA	∞	∞	∞	∞	∞ *4.0e5*	0/15

5 Conclusions

Evolution strategies are well performing variants of evolutionary algorithms used in continuous optimization. They ultilize normally distributed mutations as their main search procedure. Their performance depends on the control of the mutation process which is governed by adapting step-sizes and covariance matrices. One possible improvement concerns the covariance matrix adaptation which makes use of the sample covariance matrix. In statistical research, this estimate has been identified as not agreeing well with the true covariance for the case of large dimensional spaces and small sample sizes, or more correctly for sample sizes that do not increase sufficiently fast with the dimensionality.

While modern approaches for covariance matrix adaptation correct the estimate, the question arises whether the performance of these evolutionary algorithms may be further improved by applying other estimators for the covariance.

This paper took a closer look at covariance estimation in evolution strategies and provided a comparison with approaches in modern statistics. Furthermore, it introduced and discussed new adaptation schemes for use in optimization. In cases, where the fitness function requires highly different eigenvalues and a rotation other than the cartesian coordinate system. Therefore, a switch towards the eigenspace of the covariance matrix was proposed in this paper and investigated in experiments on the BBOB test suite. While work remains to be done, this paper provided an important first step on the way.

References

1. Audet, C.: A survey on direct search methods for blackbox optimization and their applications. In: Pardalos, P.M., Rassias, T.M. (eds.) Mathematics without Boundaries: Surveys in Interdisciplinary Research. Springer, New York (2013). Also Les Cahiers du GERAD G-2012-53 (2012)
2. Bai, Z.D., Silverstein, J.W.: No eigenvalues outside the support of the limiting spectral distribution of large-dimensional sample covariance matrices. Ann. Probab. **26**(1), 316–345 (1998)
3. Beyer, H.G., Meyer-Nieberg, S.: Self-adaptation of evolution strategies under noisy fitness evaluations. Genet. Program. Evolvable Mach. **7**(4), 295–328 (2006)
4. Beyer, H.G., Schwefel, H.P.: Evolution strategies: a comprehensive introduction. Nat. Comput. **1**(1), 3–52 (2002)
5. Beyer, H.-G., Sendhoff, B.: Covariance matrix adaptation revisited – the CMSA evolution strategy –. In: Rudolph, G., Jansen, T., Lucas, S., Poloni, C., Beume, N. (eds.) PPSN 2008. LNCS, vol. 5199, pp. 123–132. Springer, Heidelberg (2008)
6. Chen, X., Wang, Z., McKeown, M.: Shrinkage-to-tapering estimation of large covariance matrices. IEEE Trans. Signal Process **60**(11), 5640–5656 (2012)
7. Chen, Y., Wiesel, A., Eldar, Y.C., Hero, A.O.: Shrinkage algorithms for MMSE covariance estimation. IEEE Trans. Signal Process. **58**(10), 5016–5029 (2010)
8. Sarkar, M., Theuwissen, A.: Introduction. In: Sarkar, M., Theuwissen, A. (eds.) A Biologically Inspired CMOS Image Sensor. SCI, vol. 461, pp. 1–14. Springer, Heidelberg (2013)
9. Dong, W., Yao, X.: Covariance matrix repairing in gaussian based EDAs. In: IEEE Congress on, Evolutionary Computation, CEC 2007, pp. 415–422 (2007)

10. Eiben, A.E., Smith, J.E.: Introduction to Evolutionary Computing. Natural Computing Series. Springer, Berlin (2003)
11. Finck, S., Hansen, N., Ros, R., Auger, A.: Real-parameter black-box optimization benchmarking 2010: presentation of the noiseless functions. Technical report, Institute National de Recherche en Informatique et Automatique (2010) 2009/22
12. Hansen, N.: The CMA evolution strategy: a comparing review. In: Lozano J., et al. (eds.) Towards a new evolutionary computation. Advances in estimation of distribution algorithms, pp. 75–102. Springer (2006)
13. Hansen, N., Auger, A., Finck, S., Ros, R.: Real-parameter black-box optimization benchmarking 2012: experimental setup. Technical report, INRIA (2012). http://coco.gforge.inria.fr/bbob2012-downloads
14. Hansen, N., Ostermeier, A.: Completely derandomized self-adaptation in evolution strategies. Evolutionary Computation **9**(2), 159–195 (2001)
15. Hansen, N., Auger, A., Ros, R., Finck, S., Pošík, P.: Comparing results of 31 algorithms from the black-box optimization benchmarking BBOB-2009. In: Proceedings of the 12th Annual Conference Companion on Genetic and Evolutionary Computation, GECCO 2010, pp. 1689–1696. ACM, New York (2010). http://doi.acm.org/10.1145/1830761.1830790
16. Huber, P.J.: Robust Statistics. Wiley, New York (1981)
17. Ledoit, O., Wolf, M.: A well-conditioned estimator for large dimensional covariance matrices. J. Multivar. Anal. Arch. **88**(2), 265–411 (2004)
18. Ledoit, O., Wolf, M.: Non-linear shrinkage estimation of large dimensional covariance matrices. Ann. Stat. **40**(2), 1024–1060 (2012)
19. Meyer-Nieberg, S., Beyer, H.G.: Self-adaptation in evolutionary algorithms. In: Lobo, F., Lima, C., Michalewicz, Z. (eds.) Parameter Setting in Evolutionary Algorithms, pp. 47–76. Springer Verlag, Heidelberg (2007)
20. Rechenberg, I.: Evolutionsstrategie: Optimierung technischer Systeme nach Prinzipien der biologischen Evolution. Frommann-Holzboog Verlag, Stuttgart (1973)
21. Schäffer, J., Strimmer, K.: A shrinkage approach to large-scale covariance matrix estimation and implications for functional genomics. Stat. Appl. Genet. Mol. Biol. **4**(1), Article 32 (2005)
22. Schwefel, H.P.: Numerical Optimization of Computer Models. Wiley, Chichester (1981)
23. Stein, C.: Inadmissibility of the usual estimator for the mean of a multivariate distribution. In: Proceedings 3rd Berkeley Symposium Mathematical Statistics and Probability, vol. 1, pp. 197–206. Berkeley, CA (1956)
24. Stein, C.: Estimation of a covariance matrix. In: Rietz Lecture, 39th Annual Meeting. IMS, Atlanta, GA (1975)
25. Thomaz, C.E., Gillies, D., Feitosa, R.: A new covariance estimate for bayesian classifiers in biometric recognition. IEEE Trans. Circuits Syst. Video Technol. **14**(2), 214–223 (2004)

A Relax and Fix Approach to Solve the Fixed Charge Network Design Problem with User-Optimal Flow

Pedro Henrique González[1,2(✉)], Luidi Gelabert Simonetti[1], Carlos Alberto de Jesus Martinhon[1], Edcarllos Santos[1], and Philippe Yves Paul Michelon[2]

[1] Instituto de Computação, Universidade Federal Fluminense, Niterói, Brazil
pedro.gonzalez@alumni.univ-avignon.fr

[2] Laboratoire d'Informatique d'Avignon, Université d'Avignon et des Pays de Vaucluse, Avignon, France

Abstract. Due to the constant development of society, increasing quantities of commodities have to be transported in large urban centers. Therefore, network planning problems arise as tools to support decision-making, aiming to meet the need of finding efficient ways to perform such transportations. This paper reviews a bi-level formulation, a one level formulation obtained by applying the complementary slackness theorem, Bellman's optimality conditions, Big-M Linearization and also presents a heuristic procedure, through combining a randomized constructive algorithm with an improved Relax-and-Fix heuristic, so high quality solutions could be found.

Keywords: Network design problem · Dynamic programming · Relax-and-Fix · Bi-level problem

1 Introduction

The Fixed Charge Network Design Problem (FCNDP) involves selecting a subset of edges from a graph, in such a way that a given set of commodities can be transported from their origins to their destinations. The problem consists in minimizing the sum of the fixed costs (due to selected edges) and variable costs (depending on the flow of goods on the edges). Fixed and variable costs can be represented by linear functions and arcs are not capacitated. The FCNDP belongs to a large class of network design problems [1]. In the literature, one can find several variations of FCNDP [2] such as shortest path problem, minimum spanning tree problem, vehicle routing problem, traveling salesman problem and network Steiner problem [1]. Moreover, as illustrated by several books and papers [2]–[4], generic network design problem has numerous applications. Mathematical formulations for FCNDP not only represent the FCNDP, but also problems of communication, transportation, sewage systems and resource planning. It also appears in other contexts, such as flexible production systems [5] and automated

E. Pinson et al. (Eds.): ICORES 2014, CCIS 509, pp. 173–185, 2015.
DOI: 10.1007/978-3-319-17509-6_12

manufacturing systems [6]. Finally, network design problems arise in many vehicle fleet applications that do not involve the construction of physical facilities, but rather model decision problems such as sending a vehicle through a road or not [7,8].

In network planning problems, not only the simplest versions are NP-Hard [9,10], but also the task of finding feasible solutions (for problems with budget constraint on the fixed cost) is extremely complex [11]. Due to the natural difficulties of the problem, heuristics methods are presented as a good alternative in the search for quality solutions.

In thc paper, we intend to address a specific variation of FCNDP. The Fixed-Charge Uncapacited Network Design Problem with User-optimal Flows (FCNDP-UOF), which consists of adding multiple shortest path problems to the original problem. The FCNDP-UOF can be modeled as a bilevel discrete linear programming problem. This type of problem involves two distinct agents acting simultaneously rather than sequentially when making decisions. On the upper level, the leader (1st agent) is in charge of choosing a subset of edges to be opened in order to minimize the sum of fixed and variable costs. In response, on the lower level, the follower (2nd agent) must choose a set of shortest paths in the network, resulting in the paths through which each commodity will be sent. The effect of an agent on the other is indirect: the decision of the followers is affected by the network designed on the upper level, while the leader's decision is affected by variable costs imposed by the routes setted in the lower level.

The inclusion of shortest path problem constraints in a mixed integer linear programming is not straightforward. Difficulties arise both in modeling and designing efficient methods. As far as we know, there are few works done on FCNDP-UOF in the literature, and most of them address to a particular variant. This problem or its variants could be seen on [12]–[18] and has been treated as part of larger problems in some applications on [19].

The FCNDP-UOF problem appears in the design of a road network for hazardous materials transportation [13,14,16,17]. During the solution of this problem the government defines a selection of road segments to be opened/closed to the transportation of hazardous materials assuming that hazmat shipments in the resulting network will be done along shortest paths. There are no costs associated with the selection of roads to compose the network but the government wants to minimize the population exposure in case of an incident during a dangerous-goods transportation. This is a particular case of the FCNDP-UOF problem where the fixed costs are equal to zero.

It is interesting to specify the contributions of each work cited above. Billheimer and Gray [12] present and formally define the FCNDP-UOF. Kara and Verter [13] and [14] works focus on exact methods, presenting a mathematical formulation and several metrics for the hazardous materials transportation problem. Mauttone et al. [15] not only presented a different model, but also presented a Tabu Search for the FCNDP-UOF. Both, [16] and [17] presented heuristic approaches to tackle the hazardous materials transportation problem. At last, [18], presented a extension of the model proposed by Kara and Verter and also a GRASP metaheuristic.

This text is organized as follows. In Sect. 2, we start by describing the problem followed by a bi-level and a one-level formulation, presented by [15]. Then in Sect. 3 we present our solution approach. Section 4 reports on our computational results and compare our results with the mathematical formulation and with heuristic results found in the literature. At last, in Sect. 5 the conclusion and future works are presented.

2 General Description of the FCNDP-UOF

In this section we describe the problem and present a bilevel and a one-level formulation for the FCNDP-UOF proposed respectively by [20] and [15] for the FCNDP-UOF, which we address as MLF Model.

Since the structure of the problem can be easily represented by a graph, the basic structures to create a network are a set of nodes V that represents the facilities and a set of uncapacited and undirected edges E representing the connection between installations. Furthermore, the set K is the set of commodities to be transported over the network, and these commodities may represent physical goods as raw material for industry, hazardous material or even people. Each commodity $k \in K$, has a flow to be delivered through a shortest path between its source $o(k)$ and its destination $d(k)$. The formulation presented here works with variants presenting commodities with multiple origins and destinations, and for treating such a case, it is sufficient to consider that for each pair $(o(k), d(k))$, there is a new commodity resulting from the dissociation of one into several commodities.

2.1 Mathematical Formulation

This subsection shows a small review of FCNDP-UOF in order to exemplify the characteristics and make easier the understanding of it.

The model for FCNDP-UOF has two types of variables, one for the construction of the network and another related to representing the flow. Let y_{ij} be a binary variable, we have that $y_{ij} = 1$ if the edge $[i, j]$ is chosen as part of the network and $y_{ij} = 0$ otherwise.

List of Symbol

V Set of nodes.
E Set of admissible edges.
K Set of commodities.
A^E Set of arcs obtained by bi-directing the edges in E.
G Associated graph $G(V, E)$.
δ_i^+ Set of all arcs leaving node i.
δ_i^- Set of all arcs arriving at node i.
c_a Length of the arc a.
$e(a)$ Edge e related to the arc a.

$o(k)$ Origin node for commodity k.
$d(k)$ Destiny node for commodity k.
g_{ij}^k Variable cost of transporting commodity k through the arc $(i,j) \in E$.
f_{ij} Fixed cost of opening the edge $[i,j] \in E$.
y_{ij} Indicates whether edge $[i,j]$ belongs in the solution.
x_{ij}^k Indicates whether commodity k passes through the arc (i,j).

Bi-level formulation. FCNDP-UOF is a variation of the FCNDP where each $k \in K$ has to be transported through a shortest path between its origin $o(k)$ and its destination $d(k)$. This change entails adding new constraints to the general problem. In FCNDP-UOF, besides selecting a subset of E whose sum of fixed and variable costs is minimal (leading problem), each commodity $k \in K$ must be transported through the shortest path between $o(k)$ and $d(k)$ (follower problem). The FCNDP-UOF belongs to the class of NP-Hard problems and can be modeled as a bi-level discrete integer programming problem [20], as follows:

Bi-level formulation:

$$\min \sum_{(i,j)\in E} f_{ij}y_{ij} + \sum_{k\in K}\sum_{(i,j)\in A^E} g_{ij}^k x_{ij}^k$$

$$\text{s.t.} \quad y_{ij} \in \{0,1\}, \qquad \forall (i,j) \in E, \tag{1}$$

where x_{ij}^k is a solution of the problem:

$$\min \sum_{k\in K}\sum_{(i,j)\in A^E} c_{ij}x_{ij}^k$$

$$\text{s.t.} \quad \sum_{j\in\delta^+(i)} x_{ij}^k - \sum_{(j\in\delta^-(i)} x_{ij}^k = b_i^k, \qquad \forall i \in V, \forall k \in K, \tag{2}$$

$$x_{ij}^k + x_{ji}^k \leq y_{ij}, \qquad \forall (i,j) \in E, \forall k \in K, \tag{3}$$

$$x_{ij}^k \geq 0, \qquad \forall (i,j) \in A^E, \forall k \in K. \tag{4}$$

where:

$$b_i^k = \begin{cases} -1 \text{ if } i = d(k), \\ 1 \text{ if } i = o(k), \\ 0 \text{ otherwise.} \end{cases}$$

Analyzing the model described by constraints (1)–(4), we can see that the set of constraints (1) ensures that y_e assume only binary values. In (2), we have flow constraints. Constraints (3) do not allow flow into arcs whose corresponding edges are closed. Finally, (4) imposes the non-negativity restriction of the variables x_{ij}^k. An interesting remark is that solving the follower problem is equivalent to solving $|K|$ shortest paths problems independently.

One-Level Formulation. The FCNDP-UOF can be formulated as an one-level integer programming problem replacing the objective function and the constraints defined by (2), (3) and (4) of the follower problem for its optimality conditions [15]. This could be done by applying the fundamental theorem of duality and the complementary slackness theorem [21]. However, optimality conditions for the problem in the lower level are, in fact, the optimality conditions of the shortest path problem and they could be expressed in a more compact and efficient way if we consider the Bellman's optimality conditions for the shortest path problem [22] and using a simple lifting process [23].

Unfortunately this new formulation loses the interesting feature of being linear. To bypass this problem a Big-M linearization is applied. After these modifications, one can write the model as an one-level mixed integer linear programming problem, as follows:

$$\min \sum_{(i,j)]\in E} f_{ij}y_{ij} + \sum_{k\in K}\sum_{(i,j)\in A^E} g_{ij}^k x_{ij}^k$$

$$\text{s.t.} \quad \sum_{j\in\delta^+(i)} x_{ij}^k - \sum_{j\in\delta^-(i)} x_{ji}^k = b_i^k, \qquad \forall i\in V, \forall k\in K, \tag{5}$$

$$x_{ij}^k + x_{ji}^k \leq y_{ij}, \qquad \forall e=[i,j]\in E, \forall k\in K, \tag{6}$$

$$\pi_i^k - \pi_j^k \leq M_{e(a)} - y_{e(a)}(M_{e(a)} - c_a) - 2c_a x_{ji}^k, \quad \forall a=(i,j)\in A^E, k\in K, \tag{7}$$

$$\pi_{d(k)}^k = 0, \qquad \forall k\in K, \tag{8}$$

$$\pi_i^k \geq 0, \qquad \forall i\in \backslash\{d(k)\}, \forall k\in K, \tag{9}$$

$$x_{ij}^k \in \{0,1\}, \qquad \forall (i,j)\in A^E, \forall k\in K, \tag{10}$$

$$y_e \in \{0,1\}, \qquad \forall e\in E. \tag{11}$$

where:

$$b_i^k = \begin{cases} -1 \text{ if } i=d(k), \\ 1 \text{ if } i=o(k), \\ 0 \text{ otherwise.} \end{cases}$$

The variables π_i^k, $k\in K$, $i\in V$, represent the shortest distance between vertex i and vertex $d(k)$. Then we define that $\pi_{d(k)}^k$ will always be equal zero. Assuming that constraints (6), (10) and (11) are satisfied, it is easy to see that constraints (7) are equivalent to Bellman"s optimality conditions for $|K|$ pairs $(o(k), d(k))$.

3 Solution Approach

We address this section to present and explain the Partial Decoupling Heuristic and the Relax and Fix Heuristics. Before explaining the improved Relax-and-Fix heuristic, called PDRF, a small review of the Relax-and-Fix heuristic is presented.

3.1 Partial Decoupling Heuristic

A total decoupling heuristic for the FCNDP-UOF, is based on the idea of dissociating the problem of building a network from the shortest path problem. However, as discussed in [16], the decoupling of the original problem can provide worst results than when addressing both problems simultaneously. Therefore, this algorithm proposes what we call partial decoupling, where certain aspects of the follower problem are considered when trying to build a solution to the leading problem. So in order to build the network the following cost is used: $(f_{ij} \times (1 - y_{ij})) + (\alpha \times g_{ij}^{k''} + (1 - \alpha) \times c_a)$, which means that we consider whether the is edge open or not, plus a linear combination of the variable cost and the length of the edge. The α factor works as a mediator of the importance of the $g_{ij}^{k''}$ and c_e values. In the beginning of the iterations α prioritizes the variable cost ($g_{ij}^{k''}$), while in the end it prioritizes the edge length (c_a).After building the network, a shortest path algorithm is applied to take every product from its origin $o(k)$ to its destination $d(k)$, considering c_a as the edge cost. It is important to note that $g_{ij}^{k} = q^k \beta_{ij}$, where q^k represents the amount of commodity k and β_{ij} represents the shipping cost through the arc (i, j).

The algorithm presented here is a small variation of the Partial Decoupling Heuristic presented in [18]. The procedure is further explained on Algorithm 1. The partial decoupling heuristic consists in using the *Dijkstra* algorithm for the shortest path problem. Procedures *DijkstraLeader* and *DijkstraFollower*, sequentially solve the problem of network construction, followed by the shortest path problem for each commodity $k \in K$, so that in the end of the procedure, all commodities have been transported from its origin to its destination. The *DLCost* and *DSCost* are respectively *DijkstraLeader* and *DijkstraFollower* procedures costs. The notation $s \leftarrow \langle y, x \rangle$ represents that the solution s is storing the values of the variables y and x that were just defined by *DijkstraLeader* and *DijkstraFollower*. The function *CloseEdge* closes all the edges that at the end of the *DijkstraFollower* procedure are open and do not have flow. The random function returns a random element from the set passed as a parameter. In order to choose the insertion order of $|K|$ commodities, the procedure uses a candidate list consisting of a subset of products not yet routed, whose amount is greater than or equal to $\gamma\%$ times the largest amount of commodity not routed. The function Rearm(K) adds all commodities to set K and makes all variables return to its initial state.

Algorithm 1. Partial Decoupling Heuristic.

```
Input: G;K;γ
Data: MinCost ← ∞, α ← 1, y ← 0, x ← 0;
begin
    K̂ ← K;
    for numIterDP in 1...MaxIterDP do
        while K ≠ ∅ do
            K̄ ← CandidateList(K, γ);
            k' ← Random(K̄);
            for each e = (i, j) ∈ E do
                DLCost(e, k') ← (f_e × (1 − y_e)) + (α × g_ij^k' + (1 − α) × c_e);
            y ← DijkstraLeader(DLCost, k');
            K ← K\{k'};
        for each e = (i, j) ∈ E do
            DSCost(e) ← c_e;
        for k ∈ K̂ do
            x ← DijkstraFollower(DSCost, k);
        s ← ⟨y, x⟩;
        CloseEdge(s);
        if Cost(s) < MinCost then
            s_best ← s;
            MinCost ← Cost(s_best);
        α ← α − 1/MaxIterDP;
        Rearm(K);
    return s_best
```

3.2 Relax and Fix Heuristic

Given a mixed integer programming formulation:

$$\begin{cases} \min & c^1z^1 + c^2z^2; \\ \text{s.t.} & A^1z^1 + A^2z^2 = b; \\ & z^1 \in \mathbb{Z}_+^{n_1}, z^2 \in \mathbb{Z}_+^{n_2}; \end{cases} \tag{12}$$

without loss of generality, let's suppose that the variables z_j^1 for $j \in N_1$ are more important than the variables z_j^2 for $j \in N_2$, with $n_i = |N_i|$ for $i = 1, 2$.

The idea of the Relax and Fix, consists in solving two (or more) easier LPs or MIPs. The first one allows us to fix (i.e., $z_j^i = w$, $w \in \mathbb{Z}_+^{n_i}$) or limit the range of more important variables, while the second allows us to choose good values for other variables z^2.

In order to do so, first it is necessary to solve a relaxation like:

$$\begin{cases} \min c^1z^1 + c^2z^2; \\ \text{s.t. } A^1z^1 + A^2z^2 = b; \\ \quad\; z^1 \in \mathbb{Z}_+^{n_1}, z^2 \in \mathbb{R}_+^{n_2}; \end{cases} \tag{13}$$

in which the integrality of z^2 variables is dropped. Let $(\bar{z}^1, \bar{z}^2)$ be the corresponding solution. Thenceforth, fix the most important variables, according to criterias based on the problem peculiarity, and solve the new problem. After that, $(\bar{z}^1, \bar{z}^2)$ becomes the corresponding solution if the solution of the relaxed model is feasible. At last, the algorithm returns $z^H = (\bar{z}^1, \bar{z}^2)$.

In terms of algorithm, the Relax and Fix procedure can be seen as:

Algorithm 2. Relax and Fix Heuristic.

```
Input: n1, n2, N1, N2
Data: MinCost ← ∞
begin
    for i = 1...2 do
        for j ∈ N2 do
            z_i^j ∈ {0,1};
        s ← SolveLR(N1, N2);
        for j ∈ N1 do
            if z_i^j = w then
                z_i^j ← w;
        if Cost(s) < MinCost and Feas(s) = TRUE then
            s_best ← s ;
            MinCost ← Cost(s_best) ;
    return s_best
```

The function SolveLR(N_1, N_2) solves the linear relaxation of the Generalized Model for the sets N_1 and N_2. The function $Feas(s)$ returns true if the solution s passed as parameter is a feasible solution to the problem and returns false otherwise.

3.3 PDRF

In order to adapt the Relax and Fix for the FCNDP-UOF, we separate the set of variables x_{ij}^k, $(i,j) \in A^E$, $k \in K$, in $|K|$ disjoint sets, where $|K|$ is the number of commodities to be transported. At each iteration, the variables $x_{ij}^k \in Q_k$ are defined as binary. After solving the relaxed model, if it returns a feasible solution, we fix the variables y_e, that are both zero and attend to the reduced cost criterion for variable fixing, as zero.

To choose the order of x_{ij}^k variables to become binary, the procedure uses a candidate list. In order to choose a commodity, a candidate list consisting of the commodities whose amount is greater than or equal to $\gamma\%$ times the largest amount of the commodity whose variables are not set as binary.

The function $SolveLR(V, E, K, MinCost)$ solves the linear relaxation of the MLF Model for the sets V, E and K, taking into consideration the primal bound $MinCost$. The RCVF(y_{ij}) function returns TRUE if the value of the linear relaxation plus the Reduced Cost of y_{ij} is greater than the current solution. Since y_{ij} and x_{ij}^k are decision variables in the integer programming model.

Algorithm 3. PDRF.

```
Input: G;K;γ
Data: MinCost ← ∞
begin
    s ← PartialDecoupling(γ);
    MinCost ← Cost(s) ;
    K̄ ← K;
    while K̄ ≠ ∅ do
        k ← CandidateList(K̄, γ);
        for e ∈ E do
            x_e^k ∈ {0, 1};
        s ← SolveLR(V, E, K, MinCost);
        for e ∈ E do
            if y_e = 0 and RCVF(y_e) = TRUE then
                y_e ← 0;
        if Cost(s) < MinCost and Feas(s) = TRUE then
            s_best ← s ;
            MinCost ← Cost(s_best) ;
    return s_best
```

Since the Partial Decoupling Heuristic provides a feasible solution, no recovery strategy was developed in case the current fixing of the variables turns out to be infeasible.

4 Computational Results

In this section we present computational results for the one-level model and for the Relax-and-Fix presented in the previous section.

The algorithms were coded in Xpress Mosel using FICO Xpress Optimization Suite, on an Intel®Core TM 2 CPU 6400@2.13 GHz computer with 2 GB of RAM. Computing times are reported in seconds. In order to test not only the performance of the one-level model, but also the performance of the presented heuristic, we used networks data obtained through communication with the authors of [15].

The instances are grouped according to the number of nodes in the graph (10, 20, 30), followed by the graph density (0.3, 0.5, 0.8) and finally the amount of different commodities to be transported. For the presented tables, we report the optimum value found by exact model (*Opt*), the best solution(*Best Sol*) and best time(*Best Time*) reached by selected approach, and the gap value between best solution and the solution found by the heuristic(*GAP*). We also reported the average values for time(*Avg Time*) and for solutions(*Avg Sol*). Finally, reported standard deviation values for time(*Dev Time*) and solution(*Dev Sol*). In all three tables the results in bold represent the best solution found, while the underlined ones represent that the optimum has been found.

In Tables 1 and 2, we present the results reached for the instances generated by [15]. For these five instances, three heuristics were compared: the Tabu

Table 1. Computational results for Tabu Search and GRASP approach.

	Opt	Tabu Search MLF			GRASP						
		Best Sol	Best Time	GAP	Avg Sol	Avg Time	Dev Sol	Dev Time	Best Sol	Best Time	GAP
30-0.8-30-001	4830	4927	1110	0.020	4871	332.144	0	9.227	**4871**	330.908	0.008
30-0.8-30-002	6989	7322	93	0.048	7122.2	328.295	182.39	4.115	**<u>6989</u>**	325.357	0.000
30-0.8-30-003	7746	8142	565	0.051	8124	337.191	16.43	33.634	**8112**	321.838	0.047
30-0.8-30-004	8384	8828	1287	0.053	8384	318.062	0	26.091	**<u>8384</u>**	338.249	0.000
30-0.8-30-005	7428	7502	794	0.010	7442.8	321.434	33.09	17.889	**<u>7428</u>**	344.367	0.000
Avg			769.8	0.04		327.42	46.38	18.18		332.14	0.01

Table 2. Computational results for Tabu Search and PDRF approach.

	Opt	Tabu Search MLF			PDRF					
		Best Sol	Best Time	GAP	Avg Sol	Avg Time	Dev Time	Best Sol	Best Time	GAP
30-0.800000-30-001	4830	4927	1110	0.02	4830	9.28	0.04	**<u>4830</u>**	8.36	0
30-0.800000-30-002	6989	**7322**	93	0.048	7322	29.49	0.02	**7322**	26.78	0.047
30-0.800000-30-003	7746	8142	565	0.051	8112	29.00	0.03	**8112**	26.35	0.04
30-0.800000-30-004	8384	**8828**	1287	0.053	8828	60.51	0.19	**8828**	54.86	0.05
30-0.800000-30-005	7428	**7502**	794	0.010	7585	13.07	0.03	7585	11.85	0.02
Avg			769.8	0.03		28.27	0.06		25.64	0.03

Search heuristic proposed by [15], the GRASP heuristic of [18] and the PDRF algorithm. For the Tabu Search, the average time was high and no optimum solution was found. When observing the gap value, the table shows that the GRASP heuristic obtained best solutions in general, however the computational time is very high in comparison with the PDRF heuristic. Moreover, the standard deviation obtained by GRASP presented high values suggesting the algorithm has a irregular behavior and for the PDRF algorithm all standard deviation values for solutions were 0. Although for those instances GRASP outperform the PDRF in solution quality, when looking the best solutions obtained (3 out of 5), Table 2 shows that PDRF outperform the Tabu Search presented by [15].

In Table 3 were used another 45 instances generated by Mautonne, Labbé and Figueiredo, whose results were not published by them. For this group of instances, the computational results suggest the efficiency of PDRF heuristic. On average, the PDRF was 20 times faster than GRASP. Also, PDRF found 29 optimal solutions, while GRASP found only 7 optimal solutions. Besides that, the PDRF also improved or equaled GRASP results for 40 (36 improvements) out of 45 instances.

4.1 Statistical Analysis

In order to verify whether or not the differences of mean values obtained by the evaluated strategies shown in Table 3 are statistically significant, we employed the Wilcoxon-Mann-Whitney test technique [24]. This test could be applied to

Table 3. Computational results for GRASP and PDRF approach.

		GRASP							PDRF					
	Opt	Avg Sol	Avg Time	Dev Sol	Dev Time	Best Sol	Best Time	GAP	Avg Sol	Avg Time	Dev Time	Best Sol	Best Time	GAP
20-0.300000-10-001	5978	6513.58	15.65	136.48	0.34	6411	15.50	0.07	5978.00	0.15	0.0424	**5978**	0.12	0.00
20-0.300000-10-002	10469	10813.30	16.57	185.69	0.58	**10664**	16.38	0.02	10724.00	0.58	0.0011	10724	0.48	0.02
20-0.300000-10-003	7020	7286.40	15.99	132.14	0.34	7200	15.67	0.03	7020.00	0.86	0.0199	**7020**	0.69	0.00
20-0.300000-10-004	5484	5754.74	15.84	116.73	0.33	5598	15.71	0.02	5543.00	1.73	0.0143	**5543**	1.43	0.01
20-0.300000-10-005	7932	8322.00	16.04	0.00	0.40	8322	16.01	0.05	8070.00	0.37	0.0011	**8070**	0.31	0.02
20-0.300000-20-001	9488	9488.00	32.10	0.00	1.36	**9488**	31.84	0.00	9488.00	0.47	0.0004	**9488**	0.39	0.00
20-0.300000-20-002	11521	11699.86	31.64	201.31	0.91	11607	30.94	0.01	11522.00	1.06	0.0031	**11522**	0.88	0.00
20-0.300000-20-003	8270	8670.82	32.57	222.90	0.72	8568	32.44	0.04	8270.00	1.39	0.0044	**8270**	1.15	0.00
20-0.300000-20-004	11901	12320.58	31.94	300.06	1.07	**11985**	31.62	0.01	12400.00	1.43	0.0024	12400	1.19	0.04
20-0.300000-20-005	9656	10379.38	32.12	178.59	0.46	10297	31.93	0.07	9656.00	1.11	0.0008	**9656**	0.93	0.00
20-0.300000-30-001	12510	13244.00	49.28	0.00	0.76	13244	48.69	0.06	12510.00	0.96	0.0016	**12510**	0.80	0.00
20-0.300000-30-002	14216	14854.90	49.81	364.81	1.76	14737	49.41	0.04	14216.00	1.03	0.0045	**14216**	0.85	0.00
20-0.300000-30-003	13393	14687.52	48.18	577.28	1.41	14629	47.79	0.09	13393.00	3.44	0.0036	**13393**	2.86	0.00
20-0.300000-30-004	14452	15420.97	48.62	327.77	0.63	15329	48.32	0.06	14452.00	2.02	0.0034	**14452**	1.68	0.00
20-0.300000-30-005	11419	12599.00	51.32	0.00	1.08	12599	51.02	0.10	11419.00	1.14	0.0018	**11419**	0.95	0.00
20-0.500000-10-001	4784	4784.00	21.56	0.00	0.83	**4784**	21.43	0.00	4932.00	0.98	0.0038	4932	0.82	0.03
20-0.500000-10-002	7689	7689.00	21.86	0.00	0.57	**7689**	21.73	0.00	7689.00	0.58	0.0025	**7689**	0.48	0.00
20-0.500000-10-003	6184	6184.00	22.68	0.00	0.47	**6184**	22.45	0.00	6237.00	0.50	0.0005	6237	0.41	0.01
20-0.500000-10-004	5189	5532.91	22.41	95.20	0.29	5489	22.19	0.06	5444.00	0.17	0.0004	**5444**	0.14	0.05
20-0.500000-10-005	6051	6233.72	22.78	80.47	0.59	6172	22.74	0.02	6051.00	1.50	0.0077	**6051**	1.25	0.00
20-0.500000-20-001	8816	9964.00	46.50	0.00	0.95	9964	45.85	0.13	8816.00	1.13	0.0015	**8816**	0.94	0.00
20-0.500000-20-002	8584	8721.34	47.45	150.45	1.83	8584	46.89	0.00	8584.00	0.81	0.0005	**8584**	0.67	0.00
20-0.500000-20-003	7560	8354.83	45.72	214.84	0.92	8305	44.65	0.10	7560.00	1.79	0.0057	**7560**	1.49	0.00
20-0.500000-20-004	7634	7750.74	45.28	100.06	0.84	7674	44.92	0.01	7634.00	0.78	0.0008	**7634**	0.65	0.00
20-0.500000-20-005	8270	8636.00	44.86	0.00	1.12	8636	44.77	0.04	8270.00	2.16	0.0042	**8270**	1.79	0.00
20-0.500000-30-001	10156	12600.00	67.99	0.00	2.34	12600	67.99	0.24	10156.00	1.51	0.0005	**10156**	1.26	0.00
20-0.500000-30-002	11403	12932.00	68.66	0.00	1.91	12932	68.66	0.13	11403.00	3.14	0.0026	**11403**	2.61	0.00
20-0.500000-30-003	11600	13021.40	73.29	334.74	1.35	12867	71.57	0.11	11671.00	7.72	0.0328	**11671**	6.42	0.01
20-0.500000-30-004	11785	12333.56	70.88	317.15	1.32	12260	68.82	0.04	11978.00	2.41	0.0015	**11978**	2.01	0.02
20-0.500000-30-005	9559	10989.00	69.47	0.00	1.82	10989	69.33	0.15	9559.00	3.06	0.0029	**9559**	2.55	0.00
20-0.800000-10-001	3947	4120.80	34.32	105.35	0.90	4040	34.32	0.02	3947.00	0.24	0.0004	**3947**	0.20	0.00
20-0.800000-10-002	3743	3915.00	34.51	0.00	1.13	3915	34.02	0.05	3809.00	1.57	0.0150	**3809**	1.30	0.02
20-0.800000-10-003	3412	3480.24	34.81	74.75	0.58	**3412**	34.39	0.00	3412.00	0.06	0.0004	**3412**	0.05	0.00
20-0.800000-10-004	4086	4209.00	35.27	0.00	0.80	4209	34.99	0.03	4086.00	0.81	0.0008	**4086**	0.67	0.00
20-0.800000-10-005	4498	4542.98	35.64	97.51	0.77	**4498**	35.28	0.00	4574.00	0.68	0.0008	4574	0.56	0.02
20-0.800000-20-001	5796	6909.00	70.88	0.00	1.73	6909	69.22	0.19	5992.00	1.60	0.0099	**5992**	1.33	0.03
20-0.800000-20-002	7037	7635.54	71.48	187.03	1.02	7590	70.34	0.08	7321.00	13.91	0.0134	**7321**	11.58	0.04
20-0.800000-20-003	4596	6251.89	69.00	89.48	1.84	5422	68.18	0.18	4596.00	3.88	0.0008	**4596**	3.23	0.00
20-0.800000-20-004	4851	5187.00	70.26	69.01	2.45	5250	69.98	0.08	4851.00	1.24	0.0019	**4851**	1.03	0.00
20-0.800000-20-005	6086	6855.53	72.13	86.23	1.93	6267	71.42	0.03	6086.00	4.12	0.0110	**6086**	3.42	0.00
20-0.800000-30-001	7769	9425.00	105.01	0.00	2.17	9425	101.23	0.21	7769.00	1.94	0.0037	**7769**	1.61	0.00
20-0.800000-30-002	7681	8735.33	110.77	126.42	1.98	8666	109.89	0.13	7681.00	4.32	0.0131	**7681**	3.59	0.00
20-0.800000-30-003	5144	5947.89	107.30	201.43	2.67	5889	106.24	0.14	5709.00	3.09	0.0268	**5709**	2.56	0.11
20-0.800000-30-004	7188	8768.08	104.77	177.53	3.74	8630	104.56	0.20	7387.00	19.36	0.0154	**7387**	16.11	0.03
20-0.800000-30-005	7374	8175.16	108.08	127.82	1.46	7942	108.08	0.08	7374.00	4.30	0.0050	**7374**	3.58	0.00
AVG		8709.77	49.85	119.53	1.21		49.32	0.07	8116.4	2.38	0.006		1.98	0.01

compare algorithms with some random features and identify if the difference of performance between them is due to randomness.

According to [24], this statistical test is used when two independent samples are compared and whenever it is necessary to have a statistical test to reject the null hypothesis, with a significance θ level (i.e., it is possible to reject the null hypothesis with the probability of $(1 - \theta \times 100\,\%)$). For the sake of this analysis we considered $\theta = 0.01$. The hypotheses considered in this test are:

- Null Hypothesis (H0): there are no significant differences between the solutions found by PDRF and the original method;
- Alternative Hypothesis (H1): there are significant differences (bilateral alternative) between the solutions found by PDRF and the GRASP.

In the 50 instances tested, GRASP found 4 better solution then PDRF, while PDRF found 39. Concerning the statistical significance, the null hypotesis was reject for all 4 wons of the GRASP and for 35 of the 39 better solutions found by

PDRF. We notice that almost all differences of performance (86 % of the tests) are statistically significant. We can also observe that in 77 % of tests the PDRF obtained statisticaly significant results. These results indicate the superiority of the proposed strategy.

5 Conclusion and Future Works

We proposed a new algorithm for a variant of the fixed-charge uncapacitated network design problem where multiple shortest path problems are added to the original problem. In the first phase of the algorithm, the Partial Decoupling heuristic is used to build a initial solution. In the second phase, a Relax and Fix heuristic is applied to improve the solution cost.

The proposed approach was tested on a set of instances grouped by graph density, number of nodes and commodities. Our results have shown the efficiency of PDRF in comparison with a GRASP and Tabu Search heuristic, once that the proposed algorithm presented best average time for all instances, often reaching optimum solutions. In a few cases, GRASP reached best solution values, however the computational time spend was not good when compared with PDRF.

As future work, we intend to work on exact approaches as Benders decomposition and Lagrangian relaxation since both are very effective for similar problems, as could be seen in [25,26].

Acknowledgements. This work was supported by CAPES (Process Number: BEX 9877/13-4) and by Laboratoire d'Informatique d'Avignon, Université d'Avignon et des Pays de Vaucluse, Avignon, France. Luidi Simonetti is partially funded by CNPq grants 304793/2011-6.

References

1. Magnanti, T.L., Wong, R.T.: Network Design and Transportation Planning: Models and Algorithms. Transp. Sci. **18**, 1–55 (1984)
2. Boesch, F.T.: Large-scale Networks: Theory and Design. 1 edn. IEEE Press selected reprint series (1976)
3. Boyce, D., Janson, B.: A discrete transportation network design problem with combined trip distribution and assignment. Transp. Res. Part B: Methodol. **14**, 147–154 (1980)
4. Mandl, C.E.: A survey of mathematical optimization models and algorithms for designing and extending irrigation and wastewater networks. Water Resour. Res. **17**, 769–775 (1981)
5. Kimemia, J., Gershwin, S.: Network flow optimization in flexible manufacturing systems. In: 1978 IEEE Conference on Decision and Control including the 17th Symposium on Adaptive Processes, pp. 633–639. IEEE (1978)
6. Graves, S.C., Lamar, B.W.: An integer programming procedure for assembly system design problems. Oper. Res. **31**, 522–545 (1983)
7. Simpson, R.W.: Scheduling and Routing Models for Airline Systems. Massachusetts Institute of Technology, Flight Transportation Laboratory, Cambridge (1969)

8. Magnanti, T.L.: Combinatorial optimization and vehicle fleet planning: perspectives and prospects. Networks **11**, 179–213 (1981)
9. Johnson, D.S., Lenstra, J.K., Kan, A.H.G.R.: The complexity of the network design problem. Networks **8**, 279–285 (1978)
10. Wong, R.T.: Accelerating Benders decomposition for network design. Ph.D. Thesis, Massachusetts Institute of Technology (1978)
11. Wong, R.T.: Worst-case analysis of network design problem heuristics. SIAM J. Algebraic Discrete Methods **1**, 51–63 (1980)
12. Billheimer, J.W., Gray, P.: Network design with fixed and variable cost elements. Transp. Sci. **7**, 49–74 (1973)
13. Kara, B.Y., Verter, V.: Designing a road network for hazardous materials transportation. Transp. Sci. **38**, 188–196 (2004)
14. Erkut, E., Tjandra, S.A., Verter, V.: Hazardous materials transportation. Handb. Oper. Res. Manage. Sci. **14**, 539–621 (2007)
15. Mauttone, A., Labbé, M., Figueiredo, R.M.V.: A Tabu Search approach to solve a network design problem with user-optimal flows. In: V ALIO/EURO Conference on Combinatorial Optimization, pp. 1–6. Buenos Aires (2008)
16. Erkut, E., Gzara, F.: Solving the hazmat transport network design problem. Comput. Oper. Res. **35**, 2234–2247 (2008)
17. Amaldi, E., Bruglieri, M., Fortz, B.: On the hazmat transport network design problem. In: Pahl, J., Reiners, T., Voß, S. (eds.) INOC 2011. LNCS, vol. 6701, pp. 327–338. Springer, Heidelberg (2011)
18. González, P.H., Martinhon, C.A.D.J., Simonetti, L.G., Santos, E., Michelon, P.Y.P.: Uma Metaheurística GRASP para o Problema de Planejamento de Redes com Rotas Ótimas para o Usuário. In: XLV Simpósio Brasileiro de Pesquisa Operacional, Natal (2013)
19. Holmberg, K., Yuan, D.: Optimization of internet protocol network design and routing. Networks **43**, 39–53 (2004)
20. Colson, B., Marcotte, P., Savard, G.: Bilevel programming: a survey. 4OR **3**, 87–107 (2005)
21. Bazaraa, M.S., Jarvis, J.J., Sherali, H.D.: Linear Programming and Network Flows. Wiley-Interscience, New York (2004)
22. Ahuja, R.K., Magnanti, T.L., Orlin, J.B.: Network flows: theory, algorithms, and applications. Prentice-Hall Inc, Upper Saddle River, NJ, USA (1993)
23. De Giovanni, L.: The internet protocol network design problem with reliability and routing constraints. Ph.D. Thesis, Politecnico di Torino (2004)
24. Hettmansperger, T.P., McKean, J.W.: Robust Nonparametric Statistical Methods. CRC Press, Boca Raton (1998)
25. Bektas, T., Crainic, T. G., Gendron, B.: Lagrangean decomposition for the multicommodity capacitated network design problem (2007)
26. Costa, A.M., Cordeau, J.F., Gendron, B.: Benders, metric and cutset inequalities for multicommodity capacitated network design. Comput. Optim. Appl. **42**, 371–392 (2007)

Analysis of a Downward Substitution Strategy in a Manufacturing/Remanufacturing System

Fethullah Gocer[1], S. Sebnem Ahiska[1(✉)], and Russell E. King[2]

[1] Industrial Engineering Department, Galatasaray University, 34349 Ortakoy, Istanbul, Turkey
{fgocer, sahiska}@gsu.edu.tr

[2] Edward P. Fitts Department of Industrial and Systems Engineering, North Carolina State University, Raleigh, NC, USA
king@ncsu.edu

Abstract. A hybrid production system is considered where both manufacture of new product and remanufacture of returned items is performed. Due to consumer perception, new and remanufactured products are treated as different products with different costs and selling prices as well as separate demand streams. Remanufactured products have a higher stock out risk because the remanufacturing capacity is limited by the amount of returns available for remanufacture. One way to cope with this risk is to use a downward substitution strategy, i.e. a higher valued manufactured product is substituted for an out of stock lower valued remanufactured product. We formulate this control problem as an infinite-horizon hybrid manufacturing/remanufacturing system with product substitution under stochastic demand and returns. We model it as a Markov Decision Process in order to determine the optimal manufacturing and remanufacturing decisions under product substitution. The effects of stochastic demand/return distributions on the profitability of the substitution strategy are investigated through numerical experimentation.

Keywords: Product substitution · Remanufacturing · Manufacturing · Inventory control · Markov decision process

1 Introduction

In traditional production systems, manufacturers use virgin raw materials and parts during the manufacturing process. Once the products' ownership is transferred to the customers, it is usually the customer's responsibility to return or dispose of the products at the end of their usable life. However, nowadays, more manufacturers are collecting back their products from customers after usage or at the end of their life due to both environmental regulations and concerns as well as the potential economic benefits of product recovery. Product recovery in the form of remanufacturing can substantially reduce the resource consumption and waste disposal resulting in savings in material, energy and disposal costs.

Cost savings was the primary driver for manufacturers as they began to consider remanufacturing operations. With increased product returns, the profitability of these systems increase [1]. However, as governments tighten environmental laws and

E. Pinson et al. (Eds.): ICORES 2014, CCIS 509, pp. 186–198, 2015.
DOI: 10.1007/978-3-319-17509-6_13

regulations, many manufacturers are required to incorporate product recovery activities where a significant portion of production uses recovered material. While manufacturers often consider remanufacturing as an obligation forced by government regulations, in recent years, they have also realized that customers may also prefer remanufactured products for the price advantage as well as environmental consciousness.

In this study, we consider inventory control of a hybrid manufacturing/remanufacturing system, which has two modes of supply in order to satisfy customer demand: manufacture of new items and remanufacture of returned items. Production planning and control focuses on the effective utilization of resources in order to satisfy customer demand in a cost-efficient manner. In a hybrid system where the new and remanufactured items are viewed as not having the same value, there are typically three types of inventories: manufactured items, returned items and remanufactured items. Here, we consider product substitution among manufactured and remanufactured items to mitigate lost sales (backorders) in a cost effective way.

Typically in hybrid systems studies, the manufactured and remanufactured items are assumed to be alike; therefore they are stored in the same serviceable inventory and used to satisfy a common demand stream. In some cases though, customers may perceive a lower quality in a remanufactured item and expect to pay less than for a new item, resulting in a segmented market among the items. When manufactured and remanufactured items are non-identical, product substitution may be used in case of a stock-out. Under 'upward substitution' a customer demanding a new item agrees to accept a remanufactured one. This is related to customer-driven substitution such that when a customer's first-choice product is out-of-stock, he/she buys a similar product within same category [2]. Alternatively, 'downward substitution' (also called one-way or firm-driven substitution [2]) is when a higher-value item is substituted for a stocked-out lower-value item. This strategy is commonly used by automotive spare part manufacturers, e.g. for parts such as injectors and engine starters [3].

In this study we determine the optimal manufacturing and remanufacturing decisions for a periodically reviewed stochastic hybrid manufacturing/remanufacturing system under downward substitution, extending earlier research of Ahiska and Kurtul [3] by numerically investigating how the profitability is affected by the characteristics of the demand/return distributions.

2 Literature Review

Many producers are taking back used items after usage and recovering them through product recovery options such as remanufacturing due to environmental concerns as well as regulatory obligations and economic benefits. Hybrid manufacturing and remanufacturing systems are more difficult to control than the traditional manufacturing systems due to many factors. First, the flow of product returns in terms of quantity and timing is uncertain. Second, the manufacturing and remanufacturing processes are usually interrelated either through sharing common production resources (such as storage area, production line or workforce) or production of substitutable products. Hence, for efficient control of manufacturing and remanufacturing systems, the coordination between them is essential.

Inventory management of hybrid production systems has received significant attention in the literature over the last couple of decades. However, the studies that specifically analyze the use of product substitution strategies in these systems are scarce. Most of these studies consider a single-period setting. Inderfurth [4] investigates analytically the structure of optimal inventory policy for a hybrid system under one-way product substitution in a single-period setting. Kaya [5] considers partial substitution of manufactured and remanufactured products in a single-period newsvendor setting. Jin et al. [6] use a threshold level to control when to offer new products as substitutes for remanufactured products in a single-period setting. Bayindir et al. [7] use a continuous-review inventory policy to control the hybrid system, and they determine whether the remanufacturing option is profitable under one-way substitution. Their study is extended by adding a capacity constraint for the single-period version of the problem, and they investigate the effect of substitution on the optimal utilization of remanufacturing [8].

Some work on hybrid systems with substitution assumes deterministic demand and returns. Pineyro and Viera [9] formulate an economic lot-sizing problem where new items can substitute for remanufactured items. They find optimal or near optimal solutions using a Tabu-search procedure. Li et al. [10] propose a dynamic program to minimize manufacturing, remanufacturing, holding and substitution costs for an uncapacitated multi-product production planning problem with time-varying demands in a finite time horizon with no disposal or backlog. In another study by Li et al. [11], the finite-horizon multi-period two-product capacitated dynamic lot sizing problem is analyzed for deterministic time-varying demands. They develop a dynamic programming approach to provide the optimal solution to the capacitated production planning model with remanufacturing and substitution.

Inventory models with two-way substitution is another stream of research that enable consumers to substitute products within the same category. Korugan and Gupta [12] is among the earliest work on product substitution in a stochastic hybrid system. They study a system where the demand for a product is satisfied with either new items or remanufactured items. In a later work, Korugan [13] considers alternative substitution policies for hybrid manufacturing/remanufacturing system using an MDP.

Recently, Ahiska and Kurtul [3] consider a multi-period periodic-review inventory control problem for a hybrid manufacturing/remanufacturing system with product substitution and find the optimal inventory policies both with and without one-way product substitution using discrete-time MDPs under stochastic demands and returns. In this paper, we extend the work of Ahiska and Kurtul [3] by analyzing the profitability of the downward substitution strategy under different demand and return settings.

3 Problem Description

We consider a recoverable system with two production processes: manufacturing and remanufacturing. Manufacturing produces new items using externally supplied virgin materials while remanufacturing uses returned items to produce remanufactured items. Remanufactured items are viewed as having an inferior value by customers and

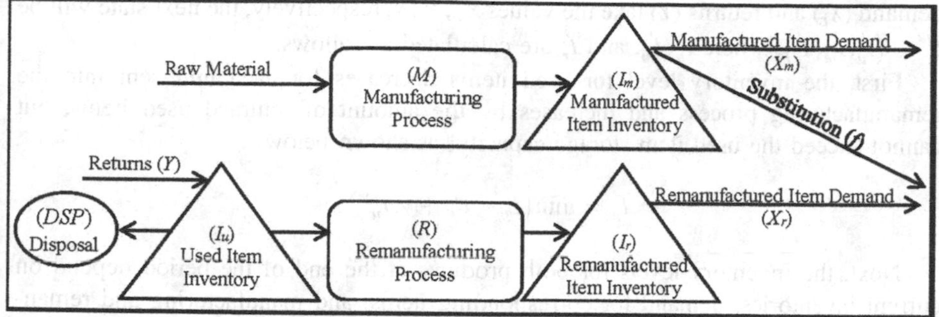

Fig. 1. Hybrid manufacturing/remanufacturing system under downward substitution.

therefore sold for a lower price than new items and have a different demand profile. Hence, there is a segmented market for manufactured and remanufactured items.

Demand is stochastic, which may cause excessive inventory to build up or lost sales to occur if poor production decisions are made. Downward substitution is considered to reduce the lost sales risk for remanufactured products such that when the remanufactured item is out of stock, a new item is sold to the customer at the remanufactured item price (i.e. the discounted price). No explicit cost associated with substitution is considered other than the opportunity cost of selling the manufactured item at the discounted price.

Figure 1 illustrates the hybrid system. There are three stocking points: recoverable inventory that includes returned used items; remanufactured inventory; and manufactured inventory. We assume 'undesirable' returns are first culled from the good quality returned items before coming into the model. The returned items are disposed if the recovered inventory is full, otherwise they are stored for later remanufacture. After manufacturing and remanufacturing operations, the resulting items are stored in their respective inventories. During each period, demand for the manufactured and remanufactured items reduce the corresponding inventory levels. At the beginning of every period, the quantities to manufacture and remanufacture must be determined.

This problem was formulated by Ahiska and Kurtul [3] as a discrete-time MDP to find the optimal manufacturing and remanufacturing decisions. The MDP model formulation is briefly described below.

The state of the system in a period, denoted by S, is represented by three variables I_u, I_r and I_m which are the inventory levels of used (i.e. recoverable), remanufactured and manufactured items, respectively. These inventory levels are bounded as $I_m^{\min} \leq I_m \leq I_m^{\max}$, $I_r^{\min} \leq I_r \leq I_r^{\max}$ and $0 \leq I_u \leq I_u^{\max}$. If $I_j^{\min} < 0$ then backordering of the demand is allowed up to $-I_j^{\min}$ for $j = r, m$.

In this system we have to make the decisions of how many units to manufacture (d_m), and to remanufacture (d_r). For each system state, we find the feasible values for (d_m, d_r) decisions considering the production and storage capacities.

Given that the current state is $S = (I_u, I_m, I_r)$, the manufacturing and remanufacturing decisions are (d_m, d_r), and manufactured item demand (X_m), remanufactured item demand (X_r) and returns (Y) take the values x_m, x_r, y, respectively, the next state will be $S' = (I'_u, I'_m, I'_r)$, where I'_u, I'_m, and I'_r are calculated as follows.

First, the inventory level for used items decreases for each unit sent into the remanufacturing process and increases by the amount of returned used items, but cannot exceed the used item storage capacity, as shown below.

$$I'_u = \min\left(I_u - d_r + y, I_u^{\max}\right)$$

Next, the inventory levels for both products at the end of the period depend on current inventories, demand for corresponding items, and manufacturing and remanufacturing decisions, as well as on the product substitution strategy such that unfulfilled remanufactured item demand is met from the manufactured item stock if stock is available after first satisfying the demand for manufactured items. The amount of remanufactured item demand satisfied from new item stock, i.e. substitution, f, is computed as follows.

Clearly, if $I_r > x_r$ (no shortage for remanufactured items) or if $I_m < x_m$ (no manufactured items left in stock after satisfying manufactured item demand), no product substitution will occur ($f = 0$). In this case, the amount of remanufactured item demand that remains unsatisfied, denoted as l, is $l = \max\ (x_r - I_r, 0)$. If $I_r < x_r$ (i.e. there is a shortage of $x_r - I_r$ remanufactured items) and if $I_m > x_m$, then there are $I_m - x_m$ items left in manufactured item stock that can be used to satisfy the remanufactured item shortage. In this case, the amount of substitution is $f = \min\ (I_m - x_m, x_r - I_r)$ and the amount of remanufactured item demand that remains unsatisfied after product substitution occurs is $l = \max\ (x_r - I_r - f, 0)$. General formulations for f and l that cover all the 'if' conditions defined in this paragraph are: $f = [\ \min\ (I_m - x_m, x_r - I_r)]^+$ and $l = [x_r - I_r - f]^+$ where $[a]^+ = \max\ (a, 0)$.

Given the substitution amount f and unsatisfied remanufactured item demand l, the inventory levels for manufactured and remanufactured items at the beginning of next period are:

$$I'_m = \max\left(I_m - x_m - f, I_m^{\min}\right) + d_m$$
$$I'_r = \max\left(I_r - x_r, -l, I_r^{\min}\right) + d_r$$

The state transitions under a no substitution strategy can be simply obtained by setting $f = 0$ in the formulations above.

The transition probability from S to S' under decision (d_m, d_r), represented by $P[S, S', (d_m, d_r)]$ equals the sum of the probabilities of occurrence for demands and returns, (x_m, x_r, y), that lead to transition from S to S' under decision (d_m, d_r). The objective is to maximize the expected profit per period defined as the total revenue from sales minus the total cost including manufacturing and remanufacturing cost, holding costs for different stocking points, backordering cost, lost sales cost and disposal cost.

The following notation is used.

p_m: unit price for manufactured product
p_r: unit price for remanufactured product

s_m: setup cost for manufacturing
s_r: setup cost for remanufacturing
c_m: unit manufacturing cost
c_r: unit remanufacturing cost
h_m: unit holding cost per period for manufactured product
h_r: unit holding cost per period for remanufactured product
h_u: unit holding cost per period for used (returned) product
b_m: unit backorder cost per period for manufactured product
b_r: unit backorder cost per period for remanufactured product
l_m: unit lost sales cost for manufactured products
l_r: unit lost sales cost for remanufactured products
k: unit disposal cost for used products
DSP: disposal amount for the current period

$$DSP = \left[I_u - d_r + y - I_u^{\max}\right]^+$$

LS_m: manufactured item lost sales in the current period

$$LS_m = \left[I_m^{\min} - (I_m - x_m)\right]^+$$

LS_r: remanufactured item lost sales in the current period

$$LS_r = \left[I_r^{\min} + l\right]^+$$

BO_m: backordered manufactured item demand in the current period

$$BO_m = \begin{cases} -\max\left(I_m - x_m, I_m^{\min}\right) & \textit{if } I_m < x_m \\ 0 & \textit{otherwise} \end{cases}$$

BO_r: backordered remanufactured item demand in the current period

$$BO_r = \min\left(l, -I_r^{\min}\right)$$

Q_m: number of manufactured items sold in the current period

$$Q_m = \begin{cases} x_m & \textit{if } x_m < I_m \\ \max(I_m, 0) & \textit{otherwise} \end{cases}$$

Q_r: number of remanufactured items sold in the current period

$$Q_r = \begin{cases} x_r & \textit{if } x_r < I_r \\ \max(I_r, 0) & \textit{otherwise} \end{cases}$$

$\delta_m(d_m)$: cost of manufacturing d_m units

$$\delta_m(d_m) = \begin{cases} s_m + c_m d_m & d_m > 0 \\ 0 & d_m = 0 \end{cases}$$

$\delta_r(d_r)$: cost of remanufacturing d_r units

$$\delta_r(d_r) = \begin{cases} s_r + c_r d_r & d_r > 0 \\ 0 & d_r = 0 \end{cases}$$

Given that the system state is S, demand is x_m and x_r, y units of return occur, and decisions d_r and d_m are made, the profit is calculated as:

$$\begin{aligned} \text{Profit}(S,(d_m,d_r),(x_m,x_r,y)) = {} & p_r(Q_r+f)+p_m Q_m \\ & -[\delta(d_m)+\delta(d_r)] - \left[h_m\left[I'_m\right]^+ + h_r\left[I'_r\right]^+ + h_u I'_u\right] \\ & -[b_m BO_m + b_r BO_r] - [l_m LS_m + l_r LS_r] - kDSP. \end{aligned}$$

Then the expected profit in a given period is calculated as:

$$\text{E}[\text{Profit}(S,(d_m,d_r))] = \sum_{x_m,x_r,y} P[x_m,x_r,y]\text{Profit}(S,(d_m,d_r),(x_m,x_r,y))$$

where $P[x_m, x_r, y]$ represents the joint probability mass function for the random variables X_m, X_r, and Y.

The formulation is solved with a variant of Howard's [14] policy iteration method using the fixed policy successive approximation method by Morton [15] for computational efficiency.

4 Numerical Experiments and Results

In this section, we analyze the profitability of using the downward substitution strategy under different demand/return distributions. For the experimentation, we consider a product produced by an international automotive spare part manufacturer. Due to privacy concerns, the data is scaled and the identity of the firm is kept anonymous. Due to the vigorous competition in the sector, over the last few years the firm noticed that the lost sales due to stock-outs of remanufactured products were resulting in losses of customers and damage to the image of the firm in the market. Hence, customer satisfaction is very important, and in order to guarantee a high level of customer satisfaction, the company is considering a substitution strategy. The product considered is an 'engine starter' which is a type of electric motor. This product family was among the firm's first production, and a better service level for this product is considered to be prestigious by the manufacturer [3].

The unit selling prices for the manufactured and remanufactured engine starter are 68.39€ and 51.85€, respectively, and the unit manufacturing and remanufacturing costs are 22.74€ and 17.46€. The manufacturer tolerates the backordering of manufactured item demand up to a certain level (i.e. $I_m^{min} < 0$) while backordering of remanufactured item demand is not allowed (i.e. $I_r^{min} = 0$) due to the risks associated with receiving returns. If some remanufactured item demand remains unsatisfied after substitution, then it is lost. Unit backordering cost/period for the manufactured item is calculated as 20 % of its unit price while the lost sales cost for both items are calculated as 25 % of the corresponding unit price. The annual holding costs for both items are calculated as 20 % of the corresponding unit cost, and the holding cost for a used item is considered to be half of the holding cost for a remanufactured item. The lead times for manufacturing and remanufacturing are both one period. No set up cost exists for either production option.

The first set of experiments is designed to investigate how profitability of product substitution is affected by the means of the demand and return distributions. We use bounded discrete stochastic distributions with three different shapes (Uniform, Normal, and right skewed) for manufactured and remanufactured item demands and used item returns. The mean of each different-shape distribution is assigned three different values: low, medium and high, as shown in Table 1.

Table 1. Mean values for different distributions.

	Mean		
Distribution shape	High	Medium	Low
Uniform (Uni)	2.00	1.50	1.00
Normal (Nrm)	2.51	2.00	1.50
Right skewed (RS)	1.20	1.05	0.54

In all, 27 combinations of the means are created by assigning the different levels of the mean of the distribution for manufactured item demand ($E[X_m]$), remanufactured item demand ($E[X_r]$) and used item returns ($E[Y]$). These 27 combinations coupled with the three distribution shapes yield a total of 81 scenarios. For each scenario, the optimal expected profit per period for the hybrid system under substitution and no substitution strategies is determined by solving the MDP as defined in the previous section.

The improvements in profit gained by substitution vs. no substitution are reported in Table 2. We make the following observations: When the mean of remanufactured item demand is at least as much as the mean of returns ($E[Xr] \geq E[Y]$), the substitution strategy results in additional profit for the manufacturer. Among the 54 scenarios where $E[Xr] \geq E[Y]$, the highest improvement in profit was 85 %. When returns are substantially higher than the remanufactured item demand (i.e. $E[Xr] < E[Y]$), the use of substitution is not economically justified. It caused loss of profit but only up to 3 % among the 27 scenarios we considered (see Table 2). Further experimentation (not shown here) reveals that if the average returns exceed the demand but at a lower level than the amounts shown in Table 1, substitution is still profitable.

It is worth noting that the mean of manufactured item demand does not affect the amount of change in profit by substitution. However because the profit of manufacturing is lower for lower manufactured item demand, a same amount of change in profit by substitution corresponds to a higher percent change of profit over no substitution

Table 2. The improvement in profit by substitution for different combinations of E[X_m], E[X_r] and E[Y] under different-shape distributions.

Means			The improvement in profit by substitution (in absolute value-AV and %)					
E[X_m]	E[X_r]	E[Y]	Uniform		Normal		Right skewed	
			AV	%	AV	%	AV	%
high	High	High	2.63	1.75	−0.01	0.00	2.00	2.29
high	High	Med	20.26	15.25	20.90	12.00	6.41	7.63
high	High	Low	40.77	37.11	41.64	27.61	27.06	43.84
high	Med	High	−1.43	−1.06	−1.36	−0.77	−0.44	−0.53
high	Med	Med	1.57	1.16	0.01	0.00	1.80	2.18
high	Med	Low	20.33	17.49	20.49	13.02	20.86	32.81
high	Low	High	−1.45	−1.24	−1.27	−0.80	−1.06	−1.63
high	Low	Med	−1.45	−1.23	−1.26	−0.79	−1.07	−1.64
high	Low	Low	0.57	0.47	0.03	0.02	0.74	1.11
med	High	High	2.61	2.03	−0.01	0.00	1.99	2.47
med	High	Med	20.17	18.23	20.93	13.86	6.40	8.30
med	High	Low	40.66	46.42	41.68	32.63	27.07	49.31
med	Med	High	−1.43	−1.27	−1.36	−0.89	−0.45	−0.59
med	Med	Med	1.56	1.38	0.01	0.01	1.79	2.37
med	Med	Low	20.27	21.58	20.52	15.28	20.85	36.74
med	Low	High	−1.45	−1.53	−1.28	−0.95	−1.08	−1.86
med	Low	Med	−1.45	−1.51	−1.27	−0.94	−1.09	−1.87
med	Low	Low	0.56	0.57	0.03	0.02	0.73	1.22
low	High	High	2.60	2.44	0.00	0.00	2.00	3.47
low	High	Med	20.09	22.73	21.08	16.43	6.42	11.84
low	High	Low	40.57	62.10	41.85	39.87	27.11	84.82
low	Med	High	−1.43	−1.58	−1.35	−1.04	−0.44	−0.82
low	Med	Med	1.54	1.70	0.02	0.02	1.81	3.42
low	Med	Low	20.17	28.14	20.68	18.54	20.90	61.82
low	Low	High	−1.45	−2.00	−1.27	−1.13	−1.06	−3.03
low	Low	Med	−1.45	−1.97	−1.27	−1.12	−1.07	−3.02
low	Low	Low	0.55	0.73	0.04	0.03	0.74	2.01

case as the mean of manufactured item demand decreases. In short, the profitability of product substitution strategy is mainly dependent on the size of remanufactured item demand relative to that of returns.

Clearly, substitution results in a higher improvement in profit when the expected remanufactured item demand gets higher and/or the expected return gets lower. For representative results supporting this comment, see Fig. 2, which plots the % improvements in profit by substitution for nine scenarios with the low level of mean manufactured item demand and the Normal shaped distribution, and the mean of remanufactured item demand and returns as low, medium and high. As the ratio of the

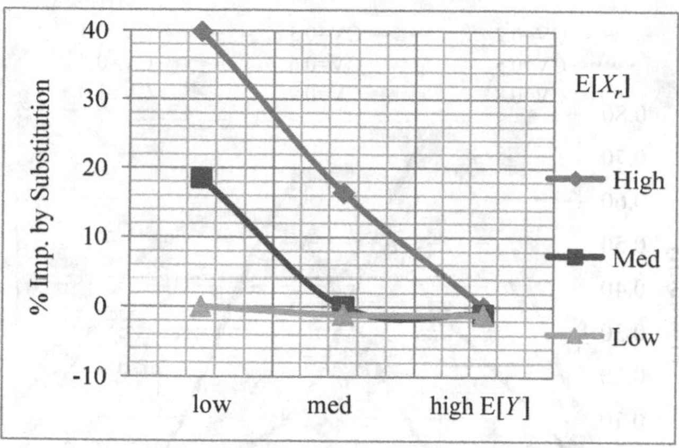

Fig. 2. % improvement in profit as E[X_r] and E[Y] change (for Normal-shape distribution and low E[X_m]).

mean remanufactured item demand to the mean returns increases from lowest ($E[X_r]$ = low, $E[Y]$ = high) to highest ($E[X_r]$ = high, $E[Y]$ = low), the percent change in profit when using product substitution increases from −1.1 % to 39.9 %.

We performed a second set of experiments in order to clearly see how the economic attractiveness of the substitution varies as the return distribution changes. For this purpose, nine different return distributions are created with different coefficients of variations (CVs) ranging from 0.2 to 1.0 with an increment of 0.1, which are plotted in Fig. 3. All the distributions have a standard deviation of 0.5, hence they differ only by their mean, which ranges from 2.5 to 0.5 as CV changes from 0.2 to 1.0. The stochastic distribution with CV of 0.6 is used for the demand distributions for remanufactured and manufactured items in this set of experiments.

Figure 4 shows how the expected profit for the hybrid system with/without product substitution changes as the mean of the return distribution decreases from 2.5 to 0.5 (i.e. CV increases from 0.2 to 1). The expected profit is also plotted separately for the no substitution case.

The following observations are made: Recall that the CV of remanufactured item demand distribution was set 0.6. Hence, in all the scenarios with return distribution's CV < 0.6, the mean of return is higher than the mean of remanufactured item demand (E[Y] > E[X_r]). When CV < 0.6, the use of substitution does not provide substantial additional profit over no substitution case (only around 0.2 %) since the amount of returns available are typically sufficient to meet remanufactured item demand. However when CV exceeds 0.6 (i.e. E[Y] goes below E[X_r]), a decrease in returns increases the economic attractiveness of product substitution from 0.6 % to nearly 28 %.

Another observation is that when CV < 0.6, an increase in CV (i.e. decrease in expected returns) results in an increase in remanufacturing profit while the effect is opposite for CV > 0.6. This can be explained as follows: For CV < 0.6, the expected remanufacturing amount (consequently, the sales revenue for remanufactured items and the remanufacturing cost) remains unchanged as expected returns decrease because the

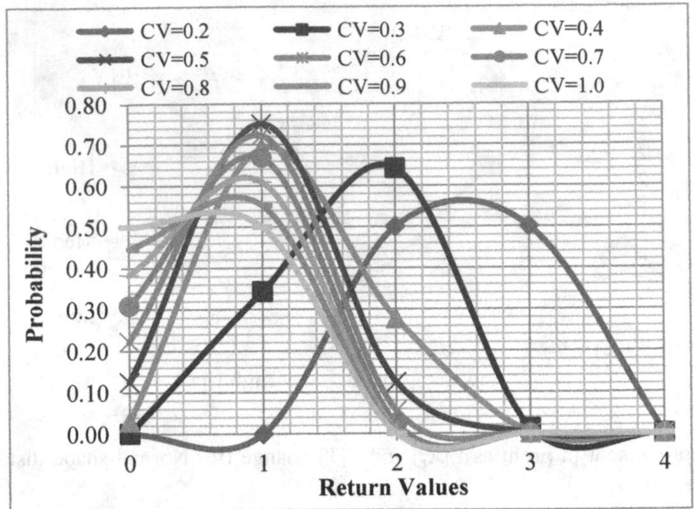

Fig. 3. The return distributions with different coefficient of variations (CVs).

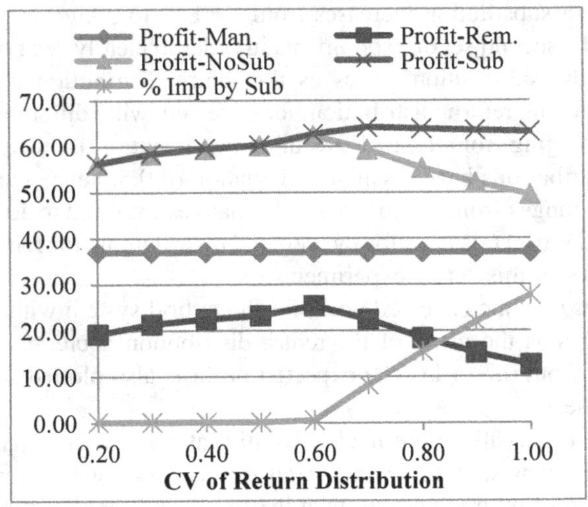

Fig. 4. The expected profits under different CVs.

returns are sufficient to meet the remanufactured item demand and the expected remanufacturing amount is just as much as remanufactured item demand. In this case the increase in profit for remanufacturing is explained by the significant amount of savings obtained in disposal cost since less disposal is needed as returns get lower (see Fig. 5). For CV > 0.6 (i.e. returns are not sufficient to meet all remanufactured item demand), a decrease in expected returns decreases the profit for remanufacturing due to a decrease in revenue from remanufactured items and an increase in the lost sales cost (see Fig. 5).

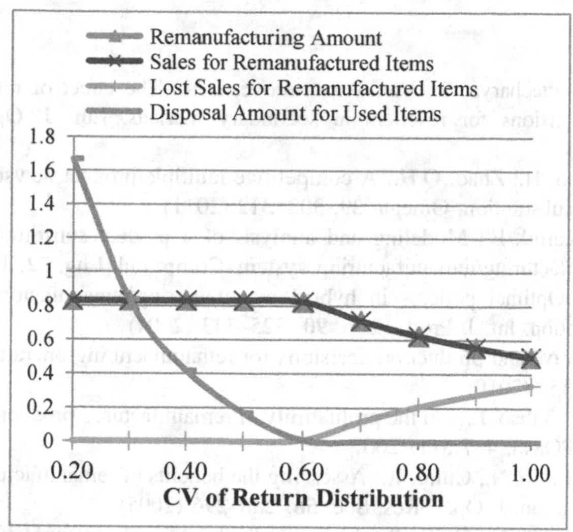

Fig. 5. Expected values for remanufacturing amount, sales/lost sales for remanufactured items and disposal amount for used items for the no substitution case under different return CVs.

5 Conclusions

We analyze a periodically reviewed stochastic manufacturing/remanufacturing system where the remanufactured items have an inferior value from customers' point of view compared to newly manufactured items. A downward product substitution strategy is employed in case of a stock-out for remanufactured items. The problem is formulated as a discrete-time MDP in order to find the optimal inventory policies for both with and without product substitution. Through a numerical study based on real data for a product produced by an automotive spare part manufacturer, the profitability of substitution is investigated under different demand and return distributions. The results show that the substitution strategy is economically attractive when the expected demand for remanufactured items is at least as much of expected returns, and the improvement in profit by substitution increases significantly as the size of returns decreases relative to the size of remanufactured item demand. These results should encourage the manufacturers operating hybrid systems to use the product substitution strategy since it may increase significantly their profit along with improving the service level by reducing the expected lost sales for remanufactured parts.

Acknowledgements. This work has been financially supported by Galatasaray University research fund grant no. 13.402.002 and the Center for Additive Manufacturing and Logistics at N.C. State University. The authors would like to thank Emre Kurtul for his help in collecting data used in this study.

References

1. Robotis, A., Bhattacharya, S., Van Wassenhove, L.N.: The effect of remanufacturing on procurement decisions for resellers in secondary markets. Eur. J. Oper. Res. **163**(3), 688–705 (2005)
2. Huang, D., Zhou, H., Zhao, Q.H.: A competitive multiple product newsboy problem with partial product substitution. Omega. **39**, 302–312 (2011)
3. Ahiska, S.S., Kurtul, E.: Modeling and analysis of a product substitution strategy for a stochastic manufacturing/remanufacturing system. Comp. Ind. Eng. **72**, 1–11 (2014)
4. Inderfurth, K.: Optimal policies in hybrid manufacturing/remanufacturing systems with product substitution. Int. J. Prod. Econ. **90**, 325–343 (2004)
5. Kaya, O.: Incentive and production decisions for remanufacturing operations. Eur. J. Oper. Res. **201**, 442–453 (2010)
6. Jin, Y., Ana, M., Yihao, L.: On the profitability of remanufactured products. In: 18th Annual Conference of POMS, 4–7 May 2007
7. Bayindir, Z.P., Erkip, N., Gullu, R.: Assessing the benefits of remanufacturing option under one-way substitution. J. Oper. Res. Soc. **56**, 286–296 (2005)
8. Bayindir, Z.P., Erkip, N., Gullu, R.: Assessing the benefits of remanufacturing option under one-way substitution and capacity constraint. Comp. Oper. Res. **34**, 487–514 (2007)
9. Pineyro, P., Viera, O.: The economic lot-sizing problem with remanufacturing and one-way substitution. Int. J. Prod. Econ. **124**, 482–488 (2010)
10. Li, Y., Chen, J., Cai, X.: Uncapacitated production planning with multiple product types, returned product remanufacturing, and demand substitution. OR Spectr. **28**, 101–125 (2006)
11. Li, Y., Chen, J., Cai, X.: Heuristic genetic algorithm for capacitated production planning problems with batch processing and remanufacturing. Int. J. Prod. Econ. **105**, 301–317 (2007)
12. Korugan, A., Gupta, S.M.: Substitution policies for a hybrid system. SPIE Proc. **4193**, 1–6 (2001)
13. Korugan, A.: The effect of product substitution at non-boundary inventory states. In: Optics East. International Society for Optics and Photonics, pp. 224–233 (2004)
14. Howard, R.A.: Dynamic Programming and Markov Processes. The MIT Press, Cambridge (1960)
15. Morton, T.E.: Asymptotic convergence rate of cost differences for Markovian decision processes. Oper. Res. **19**, 244–248 (1971)

Risk Tolerance Evaluation for an Oil and Gas Company Using a Multi-criteria Approach

António Quintino[1,2(✉)], João Carlos Lourenço[1], and Margarida Catalão-Lopes[1]

[1] CEG-IST, Instituto Superior Técnico, Universidade de Lisboa, Lisbon, Portugal
{antonio.quintino,joao.lourenco, mcatalao}@tecnico.ulisboa.pt

[2] Refining and Distribution Risk Department, Galp Energia, Lisbon, Portugal

Abstract. Oil and gas companies' earnings are heavily affected by prices fluctuations of crude oil, refined products and natural gas. The use of hedging strategies should take into account the company's risk tolerance, which assessment has no consensual technique. The present research evaluates the risk tolerance of an oil and gas company with four approaches: Howard's, Delquie's, CAPM and a risk assessment questionnaire. Monte Carlo simulation with a Copula-GARCH prices modeling and stochastic optimization are used to find optimal derivatives portfolios according to the risk tolerances previously obtained. The hedging results are then evaluated with a multi-criteria model showing how this analysis can have a decisive role in the final hedging recommendation.

Keywords: Portfolio hedging · Risk tolerance · Multi-criteria decision analysis

1 Introduction

Deregulation of the United States energy markets in the 1970's provided the ingredients for the steady growth of derivatives in the energy markets [1]. Several studies have focused on the pros-and-cons of hedging practices in oil and gas (O&G) companies, but in general, there exists a common agreement on lower earnings unpredictability [2]. The introduction of the decision-maker utility as a decision criterion [3] assured the foundations for risk and return concepts across economic thinking, including the early use of utility functions in portfolio optimization [4].

This research evaluates the hedging options based on risk tolerance parameters and confronts the results with a multi-criteria evaluation model to answer the following question: which amount (if any) should an O&G company hedge with derivatives?

The remainder of this paper is organized as follows. Section 2 describes the problem formulation, Sect. 3 presents the price variables stochastic modeling and correlation fitting, Sect. 4 describes the risk tolerance evaluation, Sect. 5 shows the results obtained by stochastic optimization with different risk tolerances, Sect. 6 presents a multi-criteria evaluation of eight hedging options against four criteria (payout exposure, downside gains, upside gains, and risk premium), and Sect. 7 presents the conclusions.

E. Pinson et al. (Eds.): ICORES 2014, CCIS 509, pp. 199–214, 2015.
DOI: 10.1007/978-3-319-17509-6_14

2 Problem Formulation

The case study O&G company is organized in three business units: the Exploration and Production unit (E&P) produces only a partial amount of the crude oil the Refining and Distribution (R&D) unit needs (crude oil buying is regular), and the Natural Gas unit (NG) imports natural gas from foreign suppliers and sells it to final consumers. A corporate risk measure to align hedging operations with the company supposed risk preferences do not exist. Since this research is focused on commodities price risk, we take as reference the company's revenues affected in first instance by price fluctuations, i.e. the gross margin, calculated as the difference between the value of the goods sold (crude oil, refined products and natural gas) and the value of acquired goods (crude oil and natural gas). The hedging horizon considered is one year in line with next year budget.

2.1 Physical Earnings Formulation

Crude oil production in oilfields of the Exploration and Production (E&P) business unit takes place under the two most applied agreements, which regulate the division of profits between O&G companies and host governments [5]: "Production Sharing Contracts" (PSC) and "Concessions" (CON). PSC are common in African and non-OECD countries and under this regime the O&G company receives a defined share of the production remaining after cost recovery, the Entitled Production quantity e_p (in barrels of crude oil, bbl). In CON regimes, the O&G company receives an earning e resulting from a defined percentage of the crude oil market price p affecting the produced quantities. The general formula for the E&P earnings for both regimes m_e (in $) is:

$$m_e = e_p p + e \tag{1}$$

The crude oil price has two major world reference indexes: the Brent price in Europe and the Western Texas Intermediate price (WTI) in the U.S.A. The Refining and Distribution (R&D) business unit is composed by the refining industrial complex and the distribution network (wholesale and retail). Since any change in the refined products price is quickly transferred to the final consumer, the price risk affects essentially the refining business, which operates with very narrow gross margins, resulting from the difference between the price's outputs (refined products) and inputs (crude oil). Due to the high prices volatility, negative refining margins can occur for long periods, especially in older and less complex refineries, explaining why some of them are being shut down. This makes it difficult to anticipate the yearly earnings of a refinery and justifies why hedging is a common practice [6]. The refining gross margin m_r (in $/bbl) is given by:

$$m_r = \left(\sum_{i=1}^{n} y_i x_i - p \right) q_r \tag{2}$$

where y_i is the yield (the percentage of each i refined product taken from a unit of crude oil), x_i is the unitary price of each refined product i, p is the unitary price of crude oil and q_r is the yearly crude oil quantity refined (in tonnes). The Natural Gas (NG) business unit buys natural gas from other countries, based on long-term contracts with price formulas indexed to the prices of crude oil and refined products baskets. The selling price formulas are diversified according to consumer's types (households, power plants and industrial consumers). The NG gross margin m_g (in \$) is given by:

$$m_g = \left(\sum_{i=1}^{n} z_i s_i - \sum_{j=1}^{n} w_j b_j \right) q_g \qquad (3)$$

where s_i and b_j are respectively the selling and buying price indexes, z_i and w_j are respectively the selling and buying weights, and q_g is the yearly total quantity of natural gas (measured in m^3 or kWh).

2.2 Derivatives Payout Formulation

The goal underneath this research is one year term hedging, thus we will choose the most common and tradable derivatives for each business unit, which include swaps and european options priced in the OTC ("over the counter") market through large banks, and Brent crude oil futures (ICE Brent) priced in the ICE exchange (a NYSE company). For the E&P business unit we will consider selling crude oil futures. The unitary payout d_e (in \$/bbl) is given by:

$$d_e = f - p_t \qquad (4)$$

where f is the future agreed price for selling the Brent crude oil (\$/bbl), and p_t is the Brent crude oil spot price at future maturity time t (f at maturity time p_t is lower than the f pre-agreed price, E&P receives the difference between these two prices, otherwise it pays the difference.) For Refining we will choose the following derivatives: selling swaps, which allows protection from lower margins (even losing the potential benefit of higher margins), and collars (i.e. selling calls and buying puts), since they provide a price band to benefit from price movements without incurring in costs.

These derivatives have a simplified refining margin underlying (also known as crack spread), based on the refined products with most traded forward prices. We will name this simplified refining margin the "hedge margin" m_h (in \$):

$$m_h = \left(\sum_{i=1}^{5} y_i x_i - p \right) \qquad (5)$$

where y_i is the yield of product i entering the "hedge margin" (only 5 of the 18 products from the production of the refinery have enough forward price liquidity to enter a hedge basket), x_i is the market price of product i and p is the Brent crude oil price.

The hedge margin swap is a derivative based on a fixed hedge margin price where the swap seller (the company) receives or pays the price difference between the fixed agreed price f_s and the spot price p_h at each future t fixed time lags, usually monthly until the end of the contract. The swap payout definition for the swap hedge margin d_s (in \$/bbl) is given by:

$$d_s = f_s - p_h \tag{6}$$

The collar is a derivative instrument resulting from buying a put and selling a call. If the spot price at maturity time t is lower than the floor price f_c^f the company receives the difference from the counterparty, and if the spot price at maturity time t is higher than the cap price f_c^c the company pays the difference. The collar payout d_c (in \$/bbl) is given by:

$$d_c = \min(f_c^f - p_h; 0) + \max(f_c^c - p_h; 0) \tag{7}$$

where p_h is the hedge margin spot price at maturity month t.

The NG business unit acts as an importer and distributor and is concerned with natural gas prices increases that may not transfer to clients. With the same logic of the refining margin, selling swaps of the natural gas margin allows protection from lower natural gas margins. The monthly swap payout definition d_g (in \$/kWh) is given by:

$$d_g = f_g - p_g \tag{8}$$

where f_g is the initial fixed agreed price for the natural gas margin, usually the average forward natural gas margin m_g for contract duration, and p_g is the natural gas margin spot price at each future maturity month t, until the end of the contract.

2.3 Company Earnings Formulation

The sum of the total derivatives payout d (where q_{eh}, q_{sh}, q_{ch} and q_{gh} are the hedged quantities) with the physical margin of each business unit (m_e, m_r and m_g) defines the company's gross margin m (in \$):

$$m = \underbrace{d_e q_{eh} + d_s q_{sh} + d_c q_{ch} + d_g q_{gh}}_{d} + m_e + m_r + m_g \tag{9}$$

The option to include all physical earnings and derivatives payouts to evaluate the company's risk reduction instead of doing it separately by business unit is based on previous analyses [7].

3 Prices Modeling

Our method can be synthesized in three steps: first, modeling the price of each product i (crude oil, refined products, natural gas) with a GARCH model; second, finding the best copula function to correlate each price standardized returns residuals z_{it} from the GARCH model; and third, insert the copula function in the GARCH model.

For this research we follow the main historic pricing reference [8] for energy markets quoted for the Northwest Europe (a.k.a. Rotterdam prices) from 2006 to 2012. For the OTC forward prices we follow the quoted monthly prices [9] for the Northwest Europe in 2013 and the ICE Brent for future prices.

Historic prices are modeled by their monthly price returns for each product i. The price returns r_{it} in month t are given by:

$$r_{it} = \ln \frac{p_{it}}{p_{i(t-1)}} \tag{10}$$

where p_{it} is the average price i in month t and $p_{i(t-1)}$ is the average price i in month $t-1$. The Generalized Autoregressive Conditional Heteroscedasticity model (GARCH) proposed by Bollerslev [10] achieved the best fit for each of the prices returns (using the SIC-Schwarz information criterion and the AIC-Akaike information criterion as goodness of fit measures). For each price i in month t, the monthly prices returns r_{it} for a GARCH (1,1) are given by:

$$r_{it} = \mu_i + \sigma_{it} z_{it} \tag{11}$$

with

$$\sigma_{it}^2 = \omega_i + \alpha_i (r_{i(t-1)} - \mu_i)^2 + \beta_i \sigma_{i(t-1)}^2 \tag{12}$$

where μ_i is the series trend (in our case the forward curve for price i), z_{it} are independent variables that follow a Normal distribution $\mathcal{N}(0,1)$, and the conditional variance σ_{it}^2 assumes an autoregressive moving average process (ARMA), with α_i weighing the moving average part and β_i affecting the auto-regressive part. The absence of autocorrelation was confirmed by the Ljung-Box statistic.

Modeling correlation between the different products prices, assuring nonlinear and complex interdependencies, leads us to copula's functions. The Sklar theorem [11] provides the theoretical foundation for the application of copulas' functions. It assumes a stochastic multi-variable vector X_n (in our case, the price return for each product n) with continuous marginals and cumulative density function:

$$F_n(x_n) = P(X_n \leq x_n) \tag{13}$$

Applying the probability integral transform to each component, gives

$$[U_1, U_2, \ldots U_n] = [F_1(X_1), F_2(X_2), \ldots, F_n(X_n)] \tag{14}$$

The copula function C is defined as the joint cumulative distribution function

$$C[u_1, u_2, \ldots u_n] = P[U_1 \leq u_1, U_2 \leq u_2, \ldots, U_n \leq u_n] \tag{15}$$

The copula C contains all information on the dependence structure between the components of $(X_1, X_2, \ldots X_n)$, whereas the marginal cumulative distribution functions F_n contain all information on the marginal distributions, making them especially adequate to portfolio risk analysis [12]. Applying the SIC and the AIC criteria we obtain the Student's copula (*t*-copula) as the best copula function to model the correlation of the *n* prices returns residuals. The cumulative distribution function for the Student's t-copula is defined by:

$$C(u_1;\, u_n; \rho, d) = \Gamma_{d,\rho}\left\lfloor t_d^{-1}(u_1),\, t_d^{-1}(u_n)\right\rfloor \tag{16}$$

where Γ is the *t*-copula with *d* degrees of freedom and correlation matrix ρ, t^{-1} is the inverse Student's distribution with *d* degrees-of-freedom, and u_n are the marginal distributions of the standardized price returns residuals, z_i in our case, as shown:

$$u_i = z_i = \frac{r_i - \mu_i}{\sigma_i} \tag{17}$$

The t-copulas have the advantage of preserving the tail dependence in extreme events, and are steadily used in advanced hedging strategies [13]. The Copula-GARCH model is given by:

$$r_{it} = \left[\omega_i + \alpha_i\left(r_{i(t-1)} - \mu_i\right)^2 + \beta_i\sigma^2_{i(t-1)}\right]^{1/2} \cdot \Gamma_{d,\rho}\left[t_d^{-1}(z_{it})\right] \tag{18}$$

4 Risk Modelling

4.1 Risk Measures

The axioms necessary and sufficient for a risk measure be coherent where defined by Artzner [14]. The Conditional Value-at-Risk (CVaR) as a coherent risk measure was proposed by Rockafellar and Uryasev [15] and is given by:

$$CVaR_{1-\alpha} = E(X|X_\alpha \leq VaR_{1-\alpha}) \tag{19}$$

where X_α is the value defined for having Value-at-Risk (VaR) for a significance level of α. CVaR allows to measure the average losses (gains) below (above) a pre-defined value of VaR. For our study we assumed the CVaR (left and right tail) for the gains' probability density function.

4.2 Risk Tolerance

The selection of the optimal derivatives portfolio is influenced by the decision-maker's attitudes towards financial risk. This is the point where utility theory commands the selection of the optimal portfolio, assessing the decision maker's risk tolerance. The exponential utility function is one of the most widely used, and is well tested on portfolio risk management in the oil industry [16]:

$$u(x) = 1 - e^{-\frac{x}{\rho}} \tag{20}$$

Its single parameter (the risk tolerance ρ), no initial wealth dependence and constant absolute risk aversion [17] explain the exponential utility function wide use. In a lottery game, the risk tolerance value ρ is the value that the decision maker is willing to accept in order to play a game where there are only two outcomes: winning the amount ρ with a 50 % probability or lose $\rho/2$ with 50 % probability. The exponential utility function performs better than other utility functions [18] but advises a post sensitivity analysis to assure the results coherence. The certainty equivalent is the amount that the decision agent accepts in order to avoid the uncertain outcome. The exponential utility function certainty equivalent *CE* for outcomes with normal distributions (which is our case, after a K-S test) can be simplified to [19]:

$$CE \approx \mu(x) - \frac{\sigma_x^2}{2 \cdot \rho} \tag{21}$$

where $\mu(x)$ is the yearly average gross margin for the company according to expression (9), σ_x^2 is the gross margin variance, ρ is the company's risk tolerance considered. The second right-hand side term (i.e. the ratio between the variance and two times the risk tolerance) corresponds to the risk premium.

4.3 Risk Tolerance Evaluation Methods

The methods for company risk tolerance assessment were chosen according to their past wide application in oil and gas companies, with the exception of the Delquie method with an analytical formula, applicable to any business.

Howard. The most referred research for corporate values of risk tolerance suggests setting the risk tolerance ρ at 6 % of sales, 1 to 1.5 times the yearly net income, or 1/6 of equity in the "O&G" companies [20]. We set it to the yearly net income (closer to 6% of sales) since the equity ratio did not adhere to current oil business reality.

Delquie. A more analytic approach [21] proposes the risk tolerance to be set to a fraction of the maximum acceptable loss that the company can afford for a given p significance level, which can be considered a proxy for the Value-at-Risk (VaR_{1-p}):

$$\rho = \frac{\text{VaR}(p)}{-\ln p} \tag{22}$$

With a significance level $p = 5\%$ and the company's one year gross margin $\text{VaR}_{95\%}$ evaluated at $\$505 \times 10^6$, we computed a risk tolerance of $\$166 \times 10^6$.

Questionnaire. Another common way to estimate corporate risk tolerance is through a questionnaire [22] answered by a decision-making group (DM), representing the company's point of view. We confronted the DM group with a set of questions to assess the amount of money for which they were indifferent, as a company, in order to have a 50–50 chance of winning that sum or losing half of it. A complementary set of questions was made on the risk premium that they were willing to pay in order to receive, with certainty, the average gross margin estimated for next year's budget. We achieved a risk tolerance mean of $\$180 \times 10^6$ with a standard deviation of $\$42 \times 10^6$.

CAPM. Another risk tolerance method estimation, derived from the Capital Asset Pricing Method-CAPM [23], is to assume the *CE* as the effective cash-flow when each year t nominal cash-flow CF_t is discounted through the ratio of the risk free rate r_f to the rate that the company demands for investments, the Weighted Average Cost of Capital (*WACC*). We evaluated all the forecasted project cash-flows 10 years ahead (essentially E&P based) achieving a risk tolerance of $\$220 \times 10^6$.

$$CE_t = CF_t \left[\frac{(1 + r_f)^t}{(1 + WACC)^t} \right] \tag{23}$$

The four risk tolerance methods, the respective reference measures and values, and the risk tolerance results are presented in Table 1.

Table 1. Risk Tolerance results (in $\$10^6$).

Method	Measure	Measure Value	Risk Tolerance ρ
Howard	Net Income	317	317
Delquie	$VaR_{95\%}$	505	166
CAPM	CE	370	220
Questionnaire	Gross Margin	760	180

Delquie's method has the most conservative risk tolerance, while Howard's method estimated the highest value. The CAPM and the Questionnaire method have intermediate risk tolerance values.

5 Optimization Results

In order to evaluate the consequences of the risk tolerance estimates in Table 1, we run stochastic optimizations with Optquest [24] for eight risk tolerance values, including the four presented in Table 1, maximizing the company's utility (which is equivalent to maximizing the corresponding certainty equivalent):

$$\max CE \approx \max \left\{ m(d, m_e, m_r, m_g) - \frac{\sigma_m^2}{2\rho} \right\} \tag{24}$$

The notional quantities q_{eh}, q_{sh}, q_{ch} and q_{gh} from expression (9) are the dependent variables. The stochastic price p_t of each product is embedded in the gross margin of each business unit, m_e, m_r, m_g, and in the derivatives payout d, at the same time.

After having achieved the optimal solution for each of the eight risk tolerance values, we ran a Monte Carlo simulation using Model Risk [25]. Figure 1 shows the risk tolerance impact in the company's certainty equivalent and in the "left tail" gains (minimum gains), measured by the risk measure $CVaR_{95\%}$.

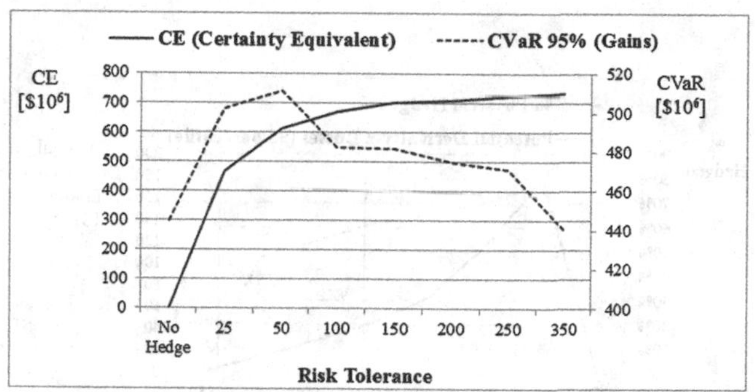

Fig. 1. CE and $CVaR_{95\%}$ as a function of risk tolerance.

As risk tolerance increases, the certainty equivalent increases. However, after a risk tolerance level of $\$50 \times 10^6$, we see a drop in the company's CVaR. Looking at Fig. 2, the decrease in CVaR is explained by the decreasing amount of derivatives d in the optimal solutions, which allows greater potential upside gains, but also greater potential downside losses. We define w_h, the *"% Physical Hedged"*, as the ratio between the notional quantities of derivatives contracts and the total physical company's production.

$$w_h = \frac{q_{eh} + q_{sh} + q_{ch} + q_{gh}}{q_e + q_r + q_g} \tag{25}$$

Less hedging means that the minimum gains are lower. Looking at the risk tolerance vertical lines, the Delquie method recommends about 20% hedging, the risk questionnaire about 15%, CAPM about 7% hedging and the Howard's method only 3%

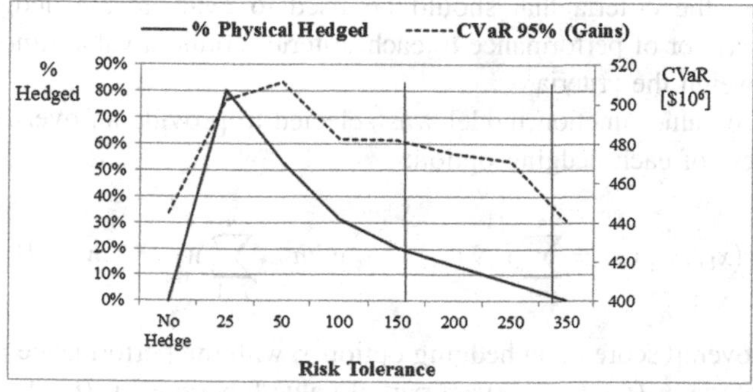

Fig. 2. CVaR95% and % Physical hedged as a function of risk tolerance ($\$10^6$).

hedging. Since different risk tolerances imply significant differences in terms of potential derivatives losses, as shown in Fig. 3, which is then the "real" company's risk tolerance? Yearly potential derivatives losses may vary from \$20 × 10^6 to \$140 × 10^6, which can have a heavy impact in the Mark-to-Market (MTM) company's quarterly financial statements and inherently in the company value, when compared with peer companies.

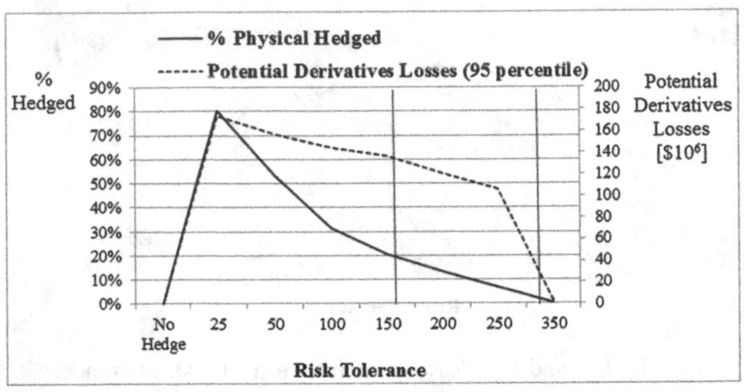

Fig. 3. % Physical hedged and potential derivatives losses as a function of risk tolerance.

6 Multi-criteria Evaluation

As we observe in the results presented in Sect. 5, the risk tolerance estimation widely affects the hedging optimal solutions. In addition, it is not clear if the in-house risk assessment questionnaire accurately defined the company's risk tolerance. Therefore, we will test in what extent the questionnaire reflects the decision maker's risk preferences.

The company is interested in selecting the most attractive hedging strategy from the set of eight options previously built. However, the company's decision-making group (DM) is not sure about which one to select. In fact, the DM members suspect that there is no option that is the best according to all points of view that came to their mind. To help the DM, we developed a multi-criteria evaluation model [26] using the MACBETH approach [27, 28], which required the group to: discuss their points of view and select the criteria that should be used to evaluate the hedging options; associate a descriptor of performance to each criterion; build a value function for each criterion; and weigh the criteria.

The additive value function model was selected to provide an overall measure of the attractiveness of each hedging option:

$$v(x_1, \ldots, x_n) = \sum_{i=1}^{n} w_i v_i(x_i) \qquad with \quad \sum_{i=1}^{n} w_i = 1, w_i > 0 \tag{26}$$

where v is the overall score of an hedging option x_i with the performance profile (x_1,..., x_n) on the n criteria, v_i (i = 1, …, n) are partial value functions, w_i (i = 1, …, n) are the criteria weights. Note that we are assuming, as working hypothesis, that the criteria are compensatory, i.e., a bad performance of an option on a criterion may be compensated by a good performance of that option in other criterion. However, this hypothesis must be validated.

The DM members discussed the points of view they considered relevant for evaluating hedging options, having in mind the next year gross margin budget as overall objective. After discussion, four evaluation criteria were selected: (1) downside gains (the bellow budget possible gains), (2) upside gains (the above budget possible gains), (3) payout exposure (the possible cash-out to pay derivatives) and (4) risk premium (the difference between the expected earnings and the certainty equivalent, as described in Eq. 21). Since all these criteria have monotonically increasing value functions, where "more" is always preferred to "less" within the same criterion, no matter the values in the other criteria, we assumed that they are mutually preferential independent, which is a necessary condition to use an additive value function model [29].

A "good" and a "neutral" reference levels (which the DM members considered to have a "very positive attractiveness" and "neither positive nor negative attractiveness", respectively) were defined in each descriptor of performance. The 5th and 95th percentiles from the Monte Carlo simulation results were used to define the upper and lower performance levels, respectively, on each descriptor of performance. Other intermediate levels, between the upper and the lower performance levels, were also created on each descriptor of performance, resulting in five levels per descriptor. For example, Fig. 4 presents the five performance levels defined on criterion "payout exposure", where 0 and 150 were defined as "good" and "neutral" reference levels, respectively. The performances of the hedging alternatives in all the criteria are expressed in $\$10^6$.

Performance levels :

- +	Quantitative level
1	0
2	50
3	100
4	150
5	200

Fig. 4. Performance levels for the "payout exposure" criterion (in $\$10^6$).

A value function was built for each of the four criteria using the MACBETH method and software (www.m-macbeth.com) [27, 28] fixing 100 and 0 as the value scores of the "good" and "neutral" reference levels, respectively, on all four criteria. According to the MACBETH questioning protocol, the decision-makers had to judge the difference in attractiveness between each two levels of the descriptor of performance using the semantic scale: very weak, weak, moderate, strong, very strong or extreme. For example, in the matrix of judgments for criterion "payout exposure" presented in Fig. 5 the decision-makers considered the difference in attractiveness between 0 and 150 to be very strong ("v. strong", in Fig. 5). After, M-MACBETH proposed a value function scale compatible with all the judgments inputted in the matrix of judgments, using the linear programming procedure presented by Bana e Costa et al. [28].

	0	50	100	150	200
0	no	weak	strong	v. strong	extreme
50		no	moderate	strong	v. strong
100			no	strong	v. strong
150				no	moderate
200					no

Fig. 5. MACBETH matrix of judgments for the "payout exposure" criterion.

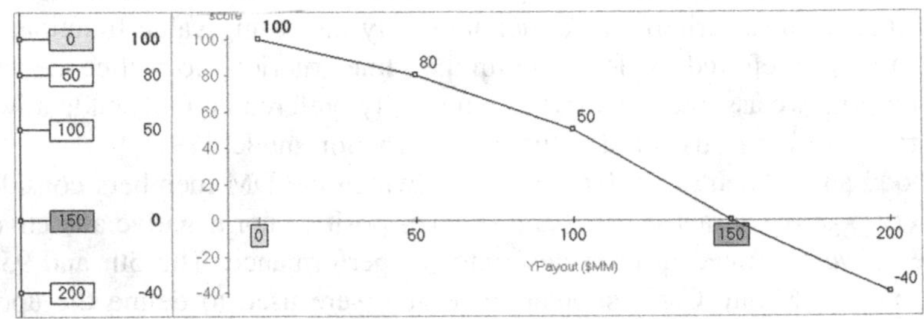

Fig. 6. Value function for the "payout exposure" criterion (performances in $\$10^6$).

The decision-makers were then asked to validate the proposed scale in terms of the proportions between the resulting scale intervals, and adjust them, if needed. Figure 6 shows the final (validated) value function for the "payout exposure" criterion.

The following step consisted in eliciting the criteria weights. For this purpose the DM first ranked the "neutral to good" performance improvements on the four criteria. In Fig. 7 we see (in the "overall references" column) that the DM considered the "neutral to good" (500 to 600) performance improvement in criterion "Dwn Gains" as the most attractive. The performance improvement from "neutral to good" (300 to 100) in criterion "Risk Premium" was ranked second. The "neutral to good" improvement on criterion "Up Gains" was ranked third and the "neutral to good" improvement on criterion "PayoutExp" was ranked fourth.

Overall references	PayoutExp	Dwn Gains	Up Gains	Risk Premium
	0	600	1150	0
[Dwn Gains]	50	550	1100	100
[Risk Premium]	100	500	1050	200
[Up Gains]	150	450	1000	300
[PayoutExp]	200	400	950	400

Fig. 7. Performance levels on the four criteria (in $\$10^6$).

Then the DM judged the attractiveness of each neutral to good improvement, which allowed filling in the column "neutral all over" in the matrix presented in Fig. 8. For example, the neutral to good improvement on criterion "Dwn Gains" was considered to be "very strong" ("v. strong" in the first row of column "Neutral All Over" in Fig. 8). Thereafter, the DM judged the differences in attractiveness between each two pairs of neutral to good improvements. For example, the difference in attractiveness from increasing the performance on criterion "Dwn Gains" from neutral to good (500 to 600) against increasing the performance on criterion "Up Gains" from neutral to good (950 to 1100) was deemed "moderate" (see the highlighted cell in Fig. 8). The answers of the DM allowed filling in the MACBETH weighting judgments matrix presented in Fig. 8. Note that by accepting to make this trade-offs between criteria, the DM is validating our hypothesis of compensation between criteria.

	[Dwn Gains]	[Risk Premium]	[Up Gains]	[PayoutExp]	Neutral All Over	Current scale
[Dwn Gains]	no	weak	moderate	strong	v. strong	37.03
[Risk Premium]		no	very weak	weak	strong	25.93
[Up Gains]			no	weak	strong	22.22
[PayoutExp]				no	moderate	14.82
Neutral All Over					no	0.00

Fig. 8. MACBETH weighting judgments.

M-MACBETH then generated the criteria weights by linear programming, which were shown to the group for validation and possible adjustment. Column "Current scale", in Fig. 8, shows the resulting criteria weights.

In the last step, the performances on the four criteria of the eight hedging options resulting from the eight risk tolerance levels – from A (no hedge) until H (risk tolerance of $\$350 \times 10^6$) – were inputted in M-MACBETH (see Fig. 9). (Note that the performances of the options are the results generated for each of the eight risk tolerance scenarios in Sect. 5.)

With these data inputted the partial value scores (on each criterion) of the hedging options and their overall value scores were calculated by M-MACBETH (see Fig. 10).

In Fig. 10 (column "Overall") we see that the most overall attractive option, considering the expressed preferences of the decision-makers, is option A (No hedge). Option H (RT = 350), which corresponds to the highest risk tolerance ($\rho = \$350 \times 10^6$), is ranked second, whereas B (RT = 25), which corresponds to the lowest risk tolerance option ($\rho = \$25 \times 10^6$), is the least preferred.

Options	PayoutExp	Dwn Gains	Up Gains	Risk Premium
A (No hedge)	0	445	1080	0
B (RT=25)	174	502	1012	293
C (RT=50)	157	511	1032	155
D (RT=100)	144	482	1036	90
E (RT=150)	137	485	1025	59
F (RT=200)	120	475	1030	51
G (RT=250)	106	471	1043	38
H (RT=350)	30	441	1089	33

Fig. 9. Performances of the eight options on the four criteria (in $\$10^6$).

Options	Overall	PayoutExp	Dwn Gains	Up Gains	Risk Premium
Good All Over	100.00	100.00	100.00	100.00	100.00
A (No hedge)	51.68	100.00	-57.50	86.67	150.00
H (RT=350)	44.73	88.00	-63.50	92.67	133.50
G (RT=250)	43.53	44.00	-29.00	62.00	131.00
F (RT=200)	39.32	30.00	-25.00	53.33	124.50
E (RT=150)	38.73	13.00	-15.00	50.00	120.50
D (RT=100)	34.19	6.00	-18.00	57.33	105.00
C (RT=50)	34.19	-5.60	11.00	54.67	72.50
B (RT=25)	7.99	-19.20	2.00	41.33	3.50
Neutral All Over	0.00	0.00	0.00	0.00	0.00
Scaling constants :		0.1482	0.3703	0.2222	0.2593

Fig. 10. Overall and partial value scores of the options and criteria weights.

7 Conclusions

The multi-criteria evaluation of the hedging options using the judgments of the decision-makers, who also answered to the risk tolerance questionnaire, gave us different results in terms of preferred hedging options. Since the multi-criteria evaluation takes into account the intra-criterion and inter-criteria preference judgments of the DM, the results obtained with the model reflect the overall attractiveness of the options for the DM. The most preferred hedging option "A" (No hedge) is closer to the Howard risk estimation ($\rho \approx \$350 \times 10^6$) and confirms Smith [30] finding that "large companies with reasonably diversified shareholders should have risk tolerances that are much larger than those typically suggested in the decision analysis literature" (p. 114), including the risk tolerance levels initially estimated through the questionnaire.

With this research we show that it is possible to perform a structured approach to model the entire O&G company business model and evaluate price risk management in an integrated way. Gross margins from the three business units and a basket of derivatives may be considered together in a certainty equivalent maximization problem, and it becomes clear how the hedging solutions vary with different risk tolerances.

Defining a preliminary risk tolerance level for the company through a tailored risk assessment questionnaire and comparing it with other methods for risk tolerance estimation allows achieving preliminary solutions based on stochastic portfolio optimization for each risk tolerance. However, a multi-criteria final assessment should be done, using the Monte Carlo simulation results, in order to ascertain how decision-makers valuate the underneath multiple consequences from each hedging option. This multi-criteria risk tolerance evaluation can in fact help the company in the always difficult decision "to hedge or not to hedge" and, if yes, which amount to hedge.

Finally, it is important to note that these results were obtained with data and preference judgments concerning a specific moment in time. Using data from a few months before or later, with different crude oil and refined products forward prices, may lead to different results under this approach. Therefore, further research should be done to evaluate the results of the model in different price conditions and involving other decision makers, preferably also including board members.

References

1. Administration, E.I.: Derivatives and Risk Management in the Petroleum, Natural Gas, and Electricity Industries, U.S.D. Energy, Energy Information Administration. Energy Information Administration, Washington (2002)
2. Haushalter, D.: Finance policy, basis risk and corporate hedging: evidence from oil and gas producers. J. Financ. **55**(1), 107–152 (2000)
3. von Neumann, J., Morgenstern, O.: Theory of Games and Economic Behavior. NJ. Princeton University Press, Princeton (1944)
4. Levy, H., Markowitz, H.M.: approximating expected utility by a function of mean and variance. Am. Econ. Rev. **69**, 308–317 (1979)
5. Kretzschmar, G.L., Kirchner, A., Reusch, H.: Risk and return in oilfield asset holdings. Energy Econ. **30**(6), 3141–3155 (2008)
6. Ji, Q., Fan, Y.: A dynamic hedging approach for refineries in multiproduct oil markets. Energy **36**(2), 881–887 (2011)
7. Quintino, A., Lourenço, J.C., Catalão-Lopes, M.: An integrated risk management model for an oil and gas company. In: 2013 International Conference on Economics and Business Administration, WSEAS, Editor 2013, pp. 144–151. Europment Conferences: Rhodes, Greece (2013)
8. Platts, Market Data - Oil (2012)
9. Reuters, Commodities: Energy (2012)
10. Bollerslev, T.: Generalized autoregressive conditional heteroskedasticity. J. Econ. **31**, 307–327 (1986)
11. Sklar, A.: Fonctions de répartition à n dimensions et leurs marges. Publ. Inst. Statist. Univ. Paris **8**, 229–231 (1959)
12. Chollete, L.: Economic implications of copulas and extremes. Penger og Kreditt—Norges Bank **2**, 56–58 (2008)
13. Lu, X.F., Lai, K.K., Liang, L.: Portfolio value-at-risk estimation in energy futures markets with time-varying copula-GARCH model. Ann. Oper. Res. **219**(1), 135–153 (2011)
14. Artzner, P., Delbaen, F., Eber, J.-M., Heath, D.: Coherent measures of risk. Math. Financ. **9**(3), 203–228 (July 1999)
15. Rockafellar, R.T., Uryasev, S.: Conditional value-at-risk for general loss distributions. J. Bank & Financ **26**, 1443–1471 (2002)
16. Walls, M.R.: Corporate risk tolerance and capital allocation: a practical approach to setting and implementing an exploration risk policy. Soc. Petrol Eng. **2**, 125–132 (1994)
17. Pratt, J.W.: Risk aversion in the small and in the large. Econometrica **32**(1/2), 122–136 (1964)
18. Kirkwood, C.W.: Approximating risk aversion in decision analysis applications. Decis. Anal. **1**(1), 51–67 (2004)
19. Clemen, R.T.: Making Hard Decisions: An Introduction to Decision Analysis, 2nd edn. Duxbury Press, Belmont (1996)
20. Howard, R.A.: Decision analysis: practice and promise. Manag. Sci. **34**(6), 51–67 (1988)
21. Delquie, P.: Interpretation of the risk tolerance coefficient in terms of maximum acceptable loss. Decis. Anal. **5**(1), 5–9 (2008)
22. Walls, M.R.: Measuring and utilizing corporate risk tolerance to improve investment decision making. Eng. Econ. **50**(4), 361–376 (2005)
23. Sharpe, W.F.: Capital asset price: a thory of market equilibrium under conditions of risk. J. Financ. **19**(3), 425–442 (1964)
24. Optquest (2012). http://www.opttek.com/OptQuest

25. ModelRisk. (2012). Vose Software: http://www.vosesoftware.com/
26. Belton, V., Stewart, T.J.: Multiple Criteria Decision Analysis: an Integrated Approach. Springer-Science + Business Media B.V., (ed.) Springer, Kluwer Academic Publishers, Boston (2002)
27. Bana e Costa, C.A., Vansnick, J.C.: The MACBETH approach: Basic ideas, software, and an application. In: Meskens, N., Roubens, M.R. (eds.) Advances in Decision Analysis, pp. 131–157. Kluwer Academic Publishers, Dordrecht (1999)
28. Bana e Costa, C.A., De Corte, J.M., Vansnick, J.C.: MACBETH. Int. J. Inf. Technol. Decis. Mak **11**(2), 359–387 (2012)
29. Dyer, J.S., Sarin, R.K.: Measurable multiattribute value functions. J.Oper. Res. **27**(4), 810–822 (1979)
30. Smith, J.E.: Risk sharing, fiduciary duty, and corporate risk attitudes. Decis. Anal. **1**(2), 114–127 (2004)

A Mathematical Programming Model for the Real Time Traffic Management of Railway Networks Under Disturbances

Astrid Piconese[1], Thomas Bourdeaud'huy[2(✉)], Mariagrazia Dotoli[1], and Slim Hammadi[2]

[1] DEE, Department of Electrical Engineering and Information Science, Politecnico di Bari, Bari, Italy
piconese.as@gmail.com, dotoli@poliba.it
[2] LAGIS, Laboratoire d'Automatique, Génie Informatique et Signal, École Centrale de Lille, Lille, France
{thomas.bourdeaud_huy,slim.hammadi}@ec-lille.fr

Abstract. The real-time traffic management allows to solve unexpected disturbances that occur along a railway line during the normal development of the traffic. After a disturbance, the original timetable is restored through the rescheduling process. Despite the improvements of off-line decision support tools for trains dispatchers that enable a better use of rail infrastructure, real-time traffic management received a limited scientific attention. In this paper, we deal with the real time traffic management for regional railway networks, mainly single tracked, in which a centralized traffic control system is installed. The rescheduling problem is presented as a Mixed Integer Linear Programming Model which resolution allows to carry out the rescheduling process in a very short computational time.

Keywords: Railway systems · Real-time optimization · Regional networks · Single-tracked · Centralized traffic control · Mixed Integer Linear Programming

1 Introduction

A railway system is a complex system with many interacting processes that depend on technical devices, human behavior, external environment, and therefore contains many risks of disturbances. The usual method how railways manage their traffic performance is through a carefully designed plan of operations, defining several months in advance routes, orders and timing for all trains. This process, called *off-line timetabling*, is followed by a real-time traffic management which consists in managing disturbances that may occur during the ordinary functioning of the network.

Once a delayed train deviates from its original schedule, it may propagate its delay to other trains due to infrastructure, signaling or timing conflicts. Major

E. Pinson et al. (Eds.): ICORES 2014, CCIS 509, pp. 215–234, 2015.
DOI: 10.1007/978-3-319-17509-6_15

disturbances may influence the off-line plan of operations that should be subject to short-term adjustments in order to minimize the negative effects of the disturbances. Possible traffic control actions include changing dwell times at scheduled stops, changing train speeds along lines, or adjusting train orders at junctions, stations and passing points. Other control actions involve major modifications such as changing train routes or even canceling scheduled train journeys. The main goal of the real-time dispatching is to minimize trains delays, while satisfying the traffic regulation constraints, and ensuring compatibility with the current position of each train, see [1].

Several approaches for re-scheduling railway traffic have been suggested. They have quite different focus with respect to infrastructure characteristics, objectives and organisation. Extensive surveys of approaches for railway traffic scheduling and re-scheduling can be found in [2–5].

In this paper we deal with real-time traffic control problem for a regional single-tracked railway where an operating system called *Centralized Traffic Control* (CTC) is installed. The CTC provides a centralized control for signals and switches within a limited territory, controlled from a single control console. The command is carried out by the *Train Dispatcher* (TD).

The TD observes the status of the territory – i.e. occupation of line sections, location of trains, etc. – in a continuous manner and collects information; meanwhile, he communicates with the upper level decision-makers and the staff in the territory in order to exchange decisions taken. In case of an unplanned event and emergency he takes a decision and makes necessary actions in accordance with the rules and regulations pre-defined by the railway authority, see [6].

The TD may benefit from appropriate decision support system, such as scheduling algorithms, to perform a real-time simulation and evaluation of traffic under disturbances in order to quickly reschedule train movements and to reduce delays from a global perspective.

It is important to find a good compromise between the solution quality, the time horizon of the traffic prediction, and the computational effort. If a short time horizon is adopted, only few trains, and few conflicts, can be detected and solved with short computation times. On the other hand, a longer time horizon leads to a larger number of trains running in the system, inducing a longer computational time needed in order to eliminate completely the propagation of the disturbance. There is a tradeoff between the size of the time horizon of traffic prediction (bigger time horizon meaning better quality) and the computational time. In fact, in a small time horizon the real-time dispatching does not take into account conflicting trains outside the time horizon. On the other hand, a conflict arising far in the future may not be as relevant as a closer conflict, since other unforeseen events could still affect the further conflict, see [1].

Usually the TD reschedules the involved trains, depending on the known duration of the disturbance. He bases his decisions on his own knowledge, resolving a conflict at a time when it occurs, and then manually rebuilds the timetable, with a considerable waste of time and no certainty that his decisions will lead to an optimal solution.

Building on the formalism given in [7], we present a model that solves the rescheduling problem for regional passenger transport networks with stations of equal importance, where the CTC system is installed. We formulate the problem as a Mixed Integer Linear Programming Model (MILP).

The methodology given in [7] provides a decision support system to the train dispatcher allowing to take decisions in order to restore traffic and limit inefficiencies for passengers. In this work, the new timetable after the disturbance is obtained using an integer linear programming model, by minimizing train delays in all stations programmed in their path, while considering constraints regarding travel times, stop times at stations, safety standards and network capacity. The model is applied to a limited time horizon that is choosen by the analyst. In order to solve conflicts that may occur in the rescheduled timetable after the time horizon, an iterative heuristic algorithm is proposed. The heuristic algorithm solves a conflict at the time when it occurs. Priority is given to the train with the highest traveling time, namely the longest presence on the line. The computational time for limited time horizons is of the order of seconds, but the heuristic algorithm may require an elevated computational time that depends on the number of trains and the complexity of the railway line.

We adapt the previous methodology to regional networks mainly made of single tracks. Our proposition consists in a single integer linear programming model, taking into account the whole set of constraints imposed by the railway infrastructure and the initial schedule. The revised model solves all conflicts that arise along the railway line after the occurrence of the disturbance. The heuristic algorithm is therefore no longer applied. The rescheduled timetable is established in a shorter time, then discomfort for passengers is restricted and the quality of the transport service is increased. To show its effectiveness, we study the problem in a particular section of a railway network located in Southern Italy, *Ferrovie Del Sud Est* (FSE), see [8]. The FSE network is made of single tracks with few double-tracked segments.

The paper is organized as follows. In Sect. 2 we present the problem formalization. In Sect. 3 the mathematical model for the resolution of the problem is proposed. In Sect. 4 we present the application of the model to the case study of the FSE network. Finally, Sect. 5 contains some concluding remarks and suggestions for further research.

2 Problem Formalization

2.1 Initial Scheduling

Definition 1 (Railway Network). *A railway network is defined by a set of segments on which trains runs.*

Segment (b_i)**.** *A segment b is a railway section between two points. We define by $\mathbb{B} = \{b_1, b_2, \ldots, b_{\mathsf{B}}\} = \{b_i\}_{i \in [\![1,\mathsf{B}]\!]}$ the set of segments. B denote the cardinality of the set $\mathbb{B}$. The set of segments is partitioned into the subset $\mathbb{B}^s$ corresponding to segments into a station, and $\mathbb{B}^c$ corresponding to the subset of rail connections outside stations.*

Track (v_j). *Let b be a segment $\in \mathbb{B}$. We define by $\mathbb{V}^b = \{v_1^b, v_2^b, \dots v_{\mathsf{V}^b}^b\} = \{v_j^b\}_{j \in [\![1, \mathsf{V}^b]\!]}$ the set of parallel tracks in b. The set of all tracks in the railway network is denoted by $\mathbb{V}$. V and V^b denote respectively the cardinality of $\mathbb{V}$ and $\mathbb{V}^b$ for a given segment b. Given a track $v \in \mathbb{V}$, we denote by b^v its corresponding segment.*

Circulations in a railway network are defined by a set of trains. Train's path is made of an ordered set of movements.

Definition 2 (Trains and Movements). *We assume that all train lengths are compatible with the length of all tracks that compose the railway line. Trains are thus defined as follows.*

Train (t_k). *The set of trains using the railway network is denoted as $\mathbb{T} = \{t_1, t_2, \dots, t_{\mathsf{T}}\} = \{t_k\}_{k \in [\![1, \mathsf{T}]\!]}$. T denotes the cardinality of $\mathbb{T}$.*

Train Direction (d^t). *Each train is defined by a direction parameter refering to its destination station. Let t be a train $\in \mathbb{T}$. We denote by $d^t = 0$ the direction of a train going to the north (a.k.a. "even trains") of the railway line, and $d^t = 1$ of a train running to the south (a.k.a. "odd trains").*

Movement (μ_p). *A movement indicates the request for a track by a train. Let t be a train $\in \mathbb{T}$. We define by $\mathbb{M}^t = \{\mu_1^t, \mu_2^t, \dots, \mu_{\mathsf{M}^t}^t\} = \{\mu_p^t\}_{p \in [\![1, \mathsf{M}^t]\!]}$ the ordered set of movements of train t. M^t denotes the cardinality of $\mathbb{M}^t$. $\mu_{first}^t = \mu_1^t$ and $\mu_{last}^t = \mu_{\mathsf{M}^t}^t$ denote respectively the first and the last element in $\mathbb{M}^t$. We define by $\mathbb{M}$ the set of all movements on the railway line and by $\mathbb{M}^{last}$ the set of the last movements of each train.*

Train movements are defined by several parameters.

Movement Direction. *All movements μ of a train t share the same direction as their train, denoted as $d^\mu \in \{0, 1\}$:*

$$\forall t \in \mathbb{T}, \forall \mu \in \mathbb{M}^t, d^\mu = d^t. \tag{1}$$

Track and Segment of a Movement (b_μ, v_μ). *Each movement μ of a train is scheduled in an unique segment $b_\mu \in \mathbb{B}$. We denote by $\mathbb{M}^b$ the set of movements scheduled in the same segment $b \in \mathbb{B}$.*

Each movement $\mu \in \mathbb{M}$ must be scheduled in a track of the segment b_μ, denoted as $v_\mu \in \mathbb{V}$, according to the following constraint:

$$\forall t \in \mathbb{T}, \forall \mu \in \mathbb{M}^t, v_\mu \in \mathbb{V}^{b_\mu}. \tag{2}$$

Reference Schedule Times $(\alpha_\mu^{\text{ref}}, \delta_\mu^{\text{ref}}, \gamma_\mu^{\text{ref}})$. *Each train movement μ is associated to reference times corresponding to its initial schedule. Let $t \in \mathbb{T}$ be a train and $\mu \in \mathbb{M}^t$ one of its movements. We define three reference times $\alpha_\mu^{ref}, \delta_\mu^{ref}, \gamma_\mu^{ref} \in \mathbb{N}$, where:*

- *α_μ^{ref} is the starting time of μ as established in the initial schedule, expressed in minutes taking as reference a time $T_0 \in \mathbb{N}$.*

- δ_μ^{ref} *is the duration of* μ *(expressed in minutes) if it occurs in a rail connection, i.e. the minimum running time defined in the initial schedule. This quantity is equal to* 0 *if the movement occurs in a station. Formally:*

$$\forall t \in \mathbb{T}, \forall \mu \in \mathbb{M}^t, b_\mu \in \mathbb{B}^s \Rightarrow \delta_\mu^{ref} = 0. \tag{3}$$

- γ_μ^{ref} *is the duration of a movement* μ *(in minutes) if it occurs in a station, i.e. the minimum stopping time defined in the initial schedule. This quantity is equal to* 0 *is the movement occurs in a rail connection. Formally:*

$$\forall t \in \mathbb{T}, \forall \mu \in \mathbb{M}^t, b_\mu \in \mathbb{B}^c \Rightarrow \gamma_\mu^{ref} = 0. \tag{4}$$

Using such notations, the time interval during which a movement μ *reserves its track can be expressed as* $[\![\alpha_\mu^{ref}, \alpha_\mu^{ref} + \delta_\mu^{ref} + \gamma_\mu^{ref} = \beta_\mu^{ref}]\!]$. *Two consecutive movements must be scheduled according to these intervals, thus we have:*

$$\forall t \in \mathbb{T}, \forall p \in [\![1, \mathsf{M}^t[\![, \alpha_{\mu_{p+1}}^{ref} = \alpha_{\mu_p}^{ref} + \delta_{\mu_p}^{ref} + \gamma_{\mu_p}^{ref}. \tag{5}$$

Obviously, in order to reduce the size of the initial problem we could use only one variable $\zeta = \gamma + \delta$ to represent movements duration. However, even if such formulation would reduce the number of initial variables, the size of the problem after the presolve phase would remain unchanged, since modern solvers are able to detect such redundant variables. For clarity, we decided thus to keep using two different variables γ and δ to represent movement duration respectively in a rail connection and in a station.

Definition 3 (Safety Constraints). *Since several trains run at the same time on a railway network, several constraints must be verified to ensure the security of circulations.*

Track Occupation Constraints. *A track cannot be occupied by two trains at the same time. Such restriction can be expressed formally by constraining any pair of movements using the same track to be scheduled on disjoint timing intervals:*

$$\begin{aligned} &\forall t_1, t_2 \in \mathbb{T}, \forall \mu_i \in \mathbb{M}^{t_1}, \forall \mu_j \in \mathbb{M}^{t_2}, \\ &\qquad v_{\mu_i} = v_{\mu_j} \Rightarrow [\![\alpha_{\mu_i}^{ref}, \beta_{\mu_i}^{ref}]\!] \cap [\![\alpha_{\mu_j}^{ref}, \beta_{\mu_j}^{ref}]\!] = \varnothing. \end{aligned} \tag{6}$$

Separation Times (Δ_m, Δ_f). *The separation time is a delay that has to elapse between two movements* μ_i, μ_j *on the same track (i.e. between a train leaving one track and another one entering the same track). We denote by* $\Delta_m \in \mathbb{N}$ *the safety time required if trains meet and* $\Delta_f \in \mathbb{N}$ *if one train is following the other one.* Δ_m *and* Δ_f *are expressed in the same time units as* γ^{ref} *and* δ^{ref}. *These time delays must occur between the end of the first movement (denoted as* $\beta_{\mu_i}^{ref}$*) and the start of the other one (denoted as* $\alpha_{\mu_j}^{ref}$*). Formally, separation time constraints can be expressed as follows:*

$$\forall b \in \mathbb{B}, \forall \mu_i, \mu_j \in \mathbb{M}^b, \begin{cases} v_{\mu_i} = v_{\mu_j} \\ d^{\mu_i} = d^{\mu_j} \end{cases} \Rightarrow \begin{cases} \quad \alpha_{\mu_j}^{ref} \geq \beta_{\mu_i}^{ref} + \Delta_f \\ \vee\, \alpha_{\mu_i}^{ref} \geq \beta_{\mu_j}^{ref} + \Delta_f \end{cases}. \tag{7}$$

$$\begin{cases} v_{\mu_i} = v_{\mu_j} \\ d^{\mu_i} \neq d^{\mu_j} \end{cases} \Rightarrow \begin{cases} \quad \alpha_{\mu_j}^{ref} \geq \beta_{\mu_i}^{ref} + \Delta_m \\ \vee\, \alpha_{\mu_i}^{ref} \geq \beta_{\mu_j}^{ref} + \Delta_m \end{cases}. \tag{8}$$

Note previous equations reinforce the constraint (6) since they induce a separation delay between the time intervals of two movements on the same track. Thus, there will be no need hereafter to model constraint (6) provided contraints (7) and (8) are expressed.

A rail transport service on a railway line is graphically represented by a cartesian graph, representing the safety constraints.

Graphic Timetable. *The graphic timetable is a cartesian graph that represents the situation of a railway schedule for a day, see [9]. This diagram shows all movements, schedules and separation times.*

The time line is plotted on the x axis. The railway line (space) is plotted on the y axis. Inside stations, tracks are represented by dashed lines parallel to the x axis. Inside connection segments, trains are represented by an oblique broken line which orientation indicates train direction. For instance, train lines are oriented from bottom to up for even trains that travel from South to North (i.e. to the station on the upper end of the y axis).

(Fig. 1 - a) represents the graphic timetable of three trains t_1, t_2 and t_4. Train t_1 is directed to the South, while t_2 and t_4 travel in the opposite direction. The railway line is made of 5 segments: single-tracked stations b_1 and b_3, double-tracked station b_5 and single-tracked rail connections b_2 and b_4. Track sets for these segments are defined by: $\mathbb{V}^{b_1} = \{v_1\}$, $\mathbb{V}^{b_2} = \{v_1\}$, $\mathbb{V}^{b_3} = \{v_1, v_2\}$, $\mathbb{V}^{b_4} = \{v_1\}$, $\mathbb{V}^{b_5} = \{v_1\}$. Train's paths are defined by: $\mathbb{M}^{t_1} = \{\mu_1^{t_1}, \mu_2^{t_1}, \mu_3^{t_1}\}$, $\mathbb{M}^{t_2} = \{\mu_1^{t_2}, \mu_2^{t_2}, \mu_3^{t_2}\}$, $\mathbb{M}^{t_4} = \{\mu_1^{t_4}, \mu_2^{t_4}, \mu_3^{t_4}\}$. All trains cross station b_3 that is single-tracked. As shown, separation times are applied when two trains occupy the same track of a segment. Δ_f is the separation time that has to elapse between the end of the movement $\mu_2^{t_2}$ and the beginning of $\mu_2^{t_4}$, where t_2 and t_4 travel in the same direction. Δ_m is the separation time that has to elapse between the end of $\mu_2^{t_4}$ and the beginning of $\mu_2^{t_1}$, where t_1 and t_4 travel in opposite directions.

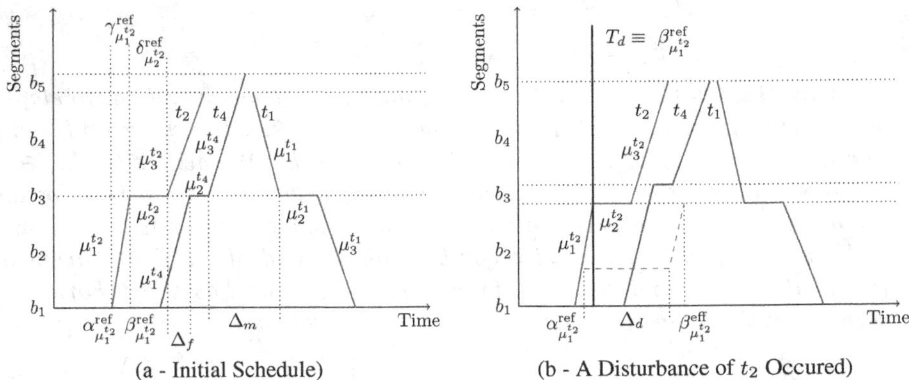

(a - Initial Schedule) (b - A Disturbance of t_2 Occured)

Fig. 1. Graphic timetables.

2.2 Disturbances Issues

When a disturbance occurs along the railway network, compromising the normal traffic operation, a rescheduling process must be accomplished taking into account time constraints imposed by the initial schedule.

Definition 4 (Disturbance). *A disturbance denotes the deviation of a train $t_d \in \mathbb{T}$ from its original schedule due to an unforeseen situation, concerning one of its movements $\mu_d \in \mathbb{M}^{t_d}$.*

Disturbance Duration and Reference Time. *When a disturbance occurs, a disturbance reference time T_d is defined as the first time on which the disturbance has an impact on the schedule of the others trains, i.e. the reference ending time of the disturbed movement. If only one disturbance affects one movement $\mu_d^{t_d} \in \mathbb{M}^{t_d}$ of a train t_d along the railway line, T_d is defined as:*

$$T_d = \beta_{\mu_d}^{ref}$$

Obvisously, such definition can be extended to several independent disturbances affecting some trains along the line. Then, T_d is defined as the minimum ending time of movements μ_i affected by the perturbation:

$$T_d = \min_{perturbed\ movements\ \mu_i} \{\beta_{\mu_i}^{ref}\}$$

We denote by Δ_d the disturbance duration expressed in minutes. The impact of a disturbance on the movement is expressed by an increase of the value of parameter δ or γ depending on the nature of the segment on which the disturbance occurs. For instance, if $b_{\mu_d} \in \mathbb{B}^c$, $\delta'_{\mu_d} = \delta_{\mu_d}^{ref} + \Delta_d$. Conversely, if $b_{\mu_d} \in \mathbb{B}^s$, $\gamma'_{\mu_d} = \gamma_{\mu_d}^{ref} + \Delta_d$. Again, in case of several disturbances, each perturbed movement μ will be associated to in own disturbance duration Δ_μ. (Fig. 1 - b) represents a railway line made of three stations (b_1, b_3 and b_5) and two single-tracked rail connections (b_2 and b_4). A disturbance occurs in b_2 and affects only the movement $\mu_1^{t_2} \in \mathbb{M}^{t_2}$. Reference time T_d is thus equal to $\beta_{\mu_1^{t_2}}^{ref}$. The dashed line shows the movement $\mu_1^{t_2}$ after the end of the rescheduling process. The recheduled crossing time $\delta_{\mu_1^{t_2}}^{eff}$ is equal to the reference time $\delta_{\mu_1^{t_2}}^{ref}$ increased by the disturbance duration Δ_d. The effective path of movement $\mu_1^{t_2}$ interferes with the path of another movement in the segment b_2 that must be consequently rescheduled.

Time Horizon (H). *The time horizon H is the time period in which the rescheduling operations are carried out. It consists of a given number of timetable minutes in which trains are scheduled on the railway line. To be considered, trains to reschedule must have all their movements inside the time horizon. Trains having some movements starting after the time horizon are not taken into account in the rescheduling process even if their first* *movements belong to the time horizon. Formally, we define the set* $\mathbb{T}^H$ *of trains that must be rescheduled in the following way:*

$$\mathbb{T}^H = \left\{ t \in \mathbb{T} \text{ s.t. } \beta^{ref}_{\mu^t_{last}} \leq T_d + H \right\} \tag{9}$$

The set $\mathbb{M}^H$ *of movements that must be rescheduled is defined accordingly, as the set of movements of trains belonging to* $\mathbb{T}^H$*:*

$$\mathbb{M}^H = \bigcup_{t \in \mathbb{T}^H} \mathbb{M}^t. \tag{10}$$

Obviously, some of these trains and movements may finish before the occurence of the first disturbance. Additional constraints are added in the next paragraph to keep them unchanged.

Effective Schedule Times. *In order to take into account the effect of the disturbance on the subsequent movements on the railway network, we introduce for any movement its effective schedule times denoted by* $\alpha^{eff}, \delta^{eff}$ *and* $\gamma^{eff} \in \mathbb{N}$*. Obviously, movements of which ending dates* β *are scheduled before* T_d *are not altered. Formally:*

$$\forall \mu \in \mathbb{M}^H \text{ s.t. } \beta^{ref}_\mu \leq T_d, \begin{cases} \alpha^{eff}_\mu = \alpha^{ref}_\mu \\ \delta^{eff}_\mu = \delta^{ref}_\mu \\ \gamma^{eff}_\mu = \gamma^{ref}_\mu \end{cases}. \tag{11}$$

Note our mathematical models will be adjusted such that one considers trains and movements before disturbance(s) only if they show effective conflicts with the circulations to reschedule, in order to reduce the number of variables. Again, the solver would have easily eliminated such variables anyway.

Effective Time Constraints. *These new scheduled times must allow to absorb the perturbation by delaying the initial reference times, according to their respective segment types, following constraints (3) and (4). Formally, these previous equations become:*

$$\forall \mu \in \mathbb{M}^H \text{ s.t. } \beta^{ref}_\mu > T_d, \qquad b_\mu \in \mathbb{B}^s \Rightarrow \begin{cases} \alpha^{eff}_\mu \geq \alpha^{ref}_\mu \\ \delta^{eff}_\mu = \delta^{ref}_\mu = 0 \\ \gamma^{eff}_\mu \geq \gamma^{ref}_\mu \end{cases}. \tag{12}$$

$$b_\mu \in \mathbb{B}^c \Rightarrow \begin{cases} \alpha^{eff}_\mu \geq \alpha^{ref}_\mu \\ \delta^{eff}_\mu \geq \delta^{ref}_\mu \\ \gamma^{eff}_\mu = \gamma^{ref}_\mu = 0 \end{cases}. \tag{13}$$

Effective schedule times of movements belonging to $\mathbb{M}^H$ *must also obviously follow the sequencing constraint (5) and safety constraints (7) and (8).*

Of course, depending on the density of traffic at the time of the disturbance, and its duration, the number of disturbed movements can vary considerably. In this paper, we consider all the scheduled movements of the day but one of our objective functions can be designed to minimize the number of rescheduled trains.

In the following section, we present the Mixed Integer Linear Programming Model that allows to solve the rescheduling problem, i.e. to give a value to each effective scheduled time while respecting the safety constraints and optimizing practical criteria.

3 Mathematical Modeling

3.1 Decision Variables

We introduce additional variables and constants used to express the problem in a linear way.

- $X_{\mu,v} \in \{0,1\}^{\mathbb{M}\times\mathbb{V}}$ is the variable that identifies the track v on which a movement μ occurs. $X_{\mu,v} = \varphi(v = v_\mu)$ where the function $\varphi(C)$ is the indicator $\varphi(C) = 1$ if the condition C is verified, 0 otherwise.
- $X^{\text{before}}_{\mu_i,\mu_j} \in \{0,1\}^{\mathbb{M}\times\mathbb{M}}$ is the variable that characterizes the chronological order of two movements $\mu_i \neq \mu_j$ if they use the same segment. $X^{\text{before}}_{\mu_i,\mu_j} = \varphi(\mu_i$ is scheduled before $\mu_j)$.
- $X^{\text{delay}}_t \in \{0,1\}^{\mathbb{T}}$ is the variable that specify if a train t deviates from its original schedule and is therefore delayed. $X^{\text{delay}}_t = \varphi(\beta^{eff}_{\mu^t_{last}} > \beta^{ref}_{\mu^t_{last}})$.
- $X^{\text{delay}}_\mu \in \{0,1\}^{\mathbb{T}}$ is the variable that specify if a movement μ deviates from its original schedule and is therefore delayed. $X^{\text{delay}}_\mu = \varphi(\beta^{eff}_\mu > \beta^{ref}_\mu)$.
- $B \in \mathbb{N}$ is a sufficiently large positive constant.
- $H \in \mathbb{N}$ is a parameter that defines the size of the time horizon.

By definition, the previous decisions variables are subject to constraints characterizing their *physical sense*.

- Any movement can only be scheduled on one track of its segment, consequently:

$$\forall \mu \in \mathbb{M}, \sum_{v \in \mathbb{V}^{b_\mu}} X_{\mu,v} = 1. \tag{14}$$

- Two movements scheduled on the same segment must be ordered:

$$\forall b \in \mathbb{B}, \forall \mu_i, \mu_j \in \mathbb{M}^b, X^{\text{before}}_{\mu_i,\mu_j} + X^{\text{before}}_{\mu_j,\mu_i} = 1. \tag{15}$$

Safety constraints presented above must be expressed using those variables in linear way.

3.2 Linearization of Safety Constraints

Safety constraints (7) and (8) express that a separation delay must elapse between two movements occupying the same track. These conditions can be expressed through the use of additional variables and constraints in order to obtain a linear formulation:

$$\forall b \in \mathbb{B}, \forall \mu_i \neq \mu_j \in \mathbb{M}^b s.t.\ d_{\mu_i} = d_{\mu_j}, \forall v \in \mathbb{V}^b, \tag{16}$$
$$\beta^{\text{eff}}_{\mu_i} - \alpha^{\text{eff}}_{\mu_j} + \Delta_f \leq B \cdot \left(3 - X^{\text{before}}_{\mu_i,\mu_j} - X_{\mu_i,v} - X_{\mu_j,v}\right).$$

$$\forall b \in \mathbb{B}, \forall \mu_i \neq \mu_j \in \mathbb{M}^b s.t.\ d_{\mu_i} \neq d_{\mu_j}, \forall v \in \mathbb{V}^b, \tag{17}$$
$$\beta^{\text{eff}}_{\mu_i} - \alpha^{\text{eff}}_{\mu_j} + \Delta_m \leq B \cdot \left(3 - X^{\text{before}}_{\mu_i,\mu_j} - X_{\mu_i,v} - X_{\mu_j,v}\right).$$

The previous equation expresses that if two movements μ_i, μ_j occurs on the same track, and if $X^{\text{before}}_{\mu_i,\mu_j} = 1$, then μ_i must end before the start of μ_j. If $X^{\text{before}}_{\mu_i,\mu_j} = 0$ or μ_i and μ_j do not occur in the same track, Eqs. (16) and (17) are trivially verified. The disjunction operator $\vee$ in Eqs. (7) and (8) is taken into account by the boolean variable $X^{\text{before}}_{\mu_i,\mu_j}$ that denotes the two possible alternatives.

Note for any pair of movements μ_i, μ_j, two instances of Eqs. (16) and (17) are considered in the mathematical model, depending on the order of movements: (μ_i, μ_j) or (μ_j, μ_i).

3.3 Objective Functions

The optimization problem compares four alternative objective functions defined as follows:

$$Obj_1 : \min \sum_{\mu_i \in \mathbb{M}^H} (\beta^{\text{eff}}_{\mu_i} - \beta^{\text{ref}}_{\mu_i}). \tag{18}$$

$$Obj_2 : \min \sum_{\mu_i \in \mathbb{M}^{last} \cap \mathbb{M}^H} (\beta^{\text{eff}}_{\mu_i} - \beta^{\text{ref}}_{\mu_i}). \tag{19}$$

$$Obj_3 : \min \sum_{t \in \mathbb{T}^H} X_t{}^{\text{delay}}. \tag{20}$$

$$Obj_4 : \min \sum_{\mu \in \mathbb{M}^H} X_\mu{}^{\text{delay}}. \tag{21}$$

- Obj_1 minimizes the delay of the traffic (i.e. the sum of the delays of all the movements trains in the railway line).
- Obj_2 minimizes the total final delay of the traffic (i.e. the final delays when trains arrive at their final destination, or rather the last stop considered within the rescheduling time horizon).
- Obj_3 minimizes the number of delayed trains.
- Obj_4 minimizes the number of delayed movements for each train.

When Obj_3 (resp. Obj_4) is used, we introduce additional constraints (22) and (23) (resp. (24) and (25)) as follows:

$$\forall t \in \mathbb{T}^H, \beta^{\text{eff}}_{\mu^t_{last}} - \beta^{\text{ref}}_{\mu^t_{last}} \leq B \cdot X_t{}^{\text{delay}}. \tag{22}$$

$$\forall t \in \mathbb{T}^H, \beta^{\text{eff}}_{\mu^t_{last}} - \beta^{\text{ref}}_{\mu^t_{last}} > B \cdot \left(X_t{}^{\text{delay}} - 1\right). \tag{23}$$

$$\forall \mu \in \mathbb{M}^H, \beta^{\text{eff}}_{\mu} - \beta^{\text{ref}}_{\mu} \leq B \cdot X_{\mu}{}^{\text{delay}}. \tag{24}$$

$$\forall \mu \in \mathbb{M}^H, \beta^{\text{eff}}_{\mu} - \beta^{\text{ref}}_{\mu} > B \cdot \left(X_{\mu}{}^{\text{delay}} - 1\right). \tag{25}$$

For instance, if the rescheduled ending time of the train coincides with its reference time, the train is not delayed and constraint (23) implies $X_t{}^{\text{delay}} = 0$. Conversly, if one movement of a train is delayed, the last movement is necessarily delayed according to Eqs. (12) and (13). Equation (22) implies then $X_t{}^{\text{delay}} = 1$.

Finally, another constraint should be added, in order to prevent movements being postponed for a long time and moving outside the considered time horizon. Indeed: constraints considered in our mathematical model are only applied to movements belonging to $\mathbb{M}^H$, within the time horizon. A movement delayed after the time horizon would thus not interfere with these movements. Then, since we do not take into account sequencing considerations between trains, a simple yet artificial way to handle disturbances and minimize the number of delayed trains could consist in rejecting disturbed trains movements *after* the time horizon.

In order to avoid such spurious solutions, we add the following constraint:

$$\forall \mu \in \mathbb{M}^H, \alpha^{\text{eff}}_{\mu} + \delta^{\text{eff}}_{\mu} + \gamma^{\text{eff}}_{\mu} \leq T_d + H. \tag{26}$$

Constraint (26) means that movements scheduled in the analyzed time horizon, after the rescheduling process, must still end within the same time horizon. Thus, we are always searching for solutions allowing to absorb the disturbance(s) during the considered period of time.

The full model is given in Fig. 2. Constraints (31) and (32) initiate the disturbance duration of the first perturbed movement. Constraint (33) specifies that each train movement is directly succeeded by the next one, that means that when a train leaves a track, it instantly begins to occupy the next one. Constraints (34) to (36) ensure that movements ending before the occurrence of the disturbance (except the disturbed movement itself) remain unchanged. Constraints (38) and (41) express that if a movement occurs in a station (resp. in a rail connection) the effective running time (resp. the effective stopping time) is null. Constraints (39) and (41) (resp. (38) to (40)) enforce the restrictions related to planned stops (resp. to planned running times) and the consequent earliest possible departure times. Constraint (42) means that each movement has to use exactly one track of its segment. Constraint (43) implies that two movements scheduled on the same segment must be ordered. Constraints (44) and (45) mean that if several movements have to use the same track of a segment, a separation time (Δ_f or Δ_m) must elapse between the end of the first movement and the beginning of the second one.

Let $(\mathbb{B}, \mathbb{V})$ be a railway network, $\mathbb{T}$ a set of trains with their schedules, (μ_d, Δ_d, T_d) the characteristics of a disturbance occuring on the system, H the considered time horizon and $(\mathbb{T}^H, \mathbb{M}^H)$ the set of trains and movements to be rescheduled. The mixed integer linear programming model $MILP$ is defined by:

$$Obj_1 : \min \sum_{\mu \in \mathbb{M}^H} \beta_\mu^{\text{eff}} - \beta_\mu^{\text{ref}} \tag{27}$$

$$Obj_2 : \min \sum_{\mu \in \mathbb{M}^{last} \cap \mathbb{M}^H} \beta_\mu^{\text{eff}} - \beta_\mu^{\text{ref}} \tag{28}$$

$$Obj_3 : \min \sum_{t \in \mathbb{T}^H} X_t^{delay} \tag{29}$$

$$Obj_4 : \min \sum_{\mu \in \mathbb{M}^H} X_\mu^{delay} \tag{30}$$

Such that:

$$\text{if } \mu_d \in \mathbb{B}^s, \quad \gamma_{\mu_d}^{\text{eff}} = \gamma_{\mu_d}^{\text{ref}} + \Delta_d \tag{31}$$

$$\text{if } \mu_d \in \mathbb{B}^c, \quad \delta_{\mu_d}^{\text{eff}} = \delta_{\mu_d}^{\text{ref}} + \Delta_d \tag{32}$$

$$\forall t \in \mathbb{T}^H, \forall \mu \in \mathbb{M}^t, \quad \alpha_{\mu_{p+1}}^{\text{eff}} - \alpha_{\mu_p}^{\text{eff}} - \delta_{\mu_p}^{\text{eff}} - \gamma_{\mu_p}^{\text{eff}} = 0 \tag{33}$$

$$\forall t \in \mathbb{T}^H, \forall \mu \in \mathbb{M}^t \ s.t.\ \beta_\mu^{\text{ref}} \leq T_d, \quad \alpha_\mu^{\text{eff}} = \alpha_\mu^{\text{ref}} \tag{34}$$

$$\forall t \in \mathbb{T}^H, \forall \mu \neq \mu_d \in \mathbb{M}^t \ s.t.\ \beta_\mu^{\text{ref}} \leq T_d, \quad \delta_\mu^{\text{eff}} = \delta_\mu^{\text{ref}} \tag{35}$$

$$\gamma_\mu^{\text{eff}} = \gamma_\mu^{\text{ref}} \tag{36}$$

$$\forall t \in \mathbb{T}^H, \forall \mu \in \mathbb{M}^t \ s.t.\ \beta_\mu^{\text{ref}} > T_d, \quad \alpha_\mu^{\text{eff}} \geq \alpha_\mu^{\text{ref}} \tag{37}$$

$$\forall t \in \mathbb{T}^H, \quad \delta_\mu^{\text{eff}} = \delta_\mu^{\text{ref}} (= 0) \tag{38}$$

$$\forall \mu \in \mathbb{M}^t \ s.t.\ \beta_\mu^{\text{ref}} > T_d \wedge b_\mu \in \mathbb{B}^s, \quad \gamma_\mu^{\text{eff}} \geq \gamma_\mu^{\text{ref}} \tag{39}$$

$$\forall t \in \mathbb{T}^H, \quad \delta_\mu^{\text{eff}} \geq \delta_\mu^{\text{ref}} \tag{40}$$

$$\forall \mu \in \mathbb{M}^t \ s.t.\ \beta_\mu^{\text{ref}} > T_d \wedge b_\mu \in \mathbb{B}^c, \quad \gamma_\mu^{\text{eff}} = \gamma_\mu^{\text{ref}} (= 0) \tag{41}$$

$$\forall \mu \in \mathbb{M}^H, \quad \sum_{v \in \mathbb{V}^{b_\mu}} X_{\mu,v} = 1 \tag{42}$$

$$\forall b \in \mathbb{B}, \forall \mu_i \neq \mu_j \in \mathbb{M}^b \cap \mathbb{M}^H, \quad X_{\mu_i,\mu_j}^{\text{before}} + X_{\mu_j,\mu_i}^{\text{before}} = 1 \tag{43}$$

$$\begin{aligned} &\forall b \in \mathbb{B}, \forall \mu_i \neq \mu_j \in \mathbb{M}^b \cap \mathbb{M}^H \quad && \beta_{\mu_i}^{\text{eff}} - \alpha_{\mu_j}^{\text{eff}} \\ &\quad \text{s.t. } d^{\mu_i} = d^{\mu_j}, \forall v \in \mathbb{V}^b, && +B \cdot \left(X_{\mu_i,\mu_j}^{\text{before}} + X_{\mu_i,v} + X_{\mu_j,v}\right) \leq 3 \cdot B - \Delta_f \end{aligned} \tag{44}$$

$$\begin{aligned} &\forall b \in \mathbb{B}, \forall \mu_i \neq \mu_j \in \mathbb{M}^b \cap \mathbb{M}^H, \quad && \beta_{\mu_i}^{\text{eff}} - \alpha_{\mu_j}^{\text{eff}} \\ &\quad \text{s.t. } d^{\mu_i} \neq d^{\mu_j}, \forall v \in \mathbb{V}^b, && +B \cdot \left(X_{\mu_i,\mu_j}^{\text{before}} + X_{\mu_i,v} + X_{\mu_j,v}\right) \leq 3 \cdot B - \Delta_m \end{aligned} \tag{45}$$

$$\forall t \in \mathbb{T}^H, \quad \beta_{\mu_{last}^t}^{\text{eff}} - B \cdot X_t^{\text{delay}} \leq \beta_{\mu_{last}^t}^{\text{ref}} \tag{46}$$

$$\forall t \in \mathbb{T}^H, \quad \beta_{\mu_{last}^t}^{\text{eff}} - B \cdot X_t^{\text{delay}} > \beta_{\mu_{last}^t}^{\text{ref}} - B \tag{47}$$

$$\forall \mu \in \mathbb{M}^H, \quad \beta_\mu^{\text{eff}} - B \cdot X_\mu^{\text{delay}} \leq \beta_\mu^{\text{ref}} \tag{48}$$

$$\forall \mu \in \mathbb{M}^H, \quad \beta_\mu^{\text{eff}} - B \cdot X_\mu^{\text{delay}} > \beta_\mu^{\text{ref}} - B \tag{49}$$

$$\forall \mu \in \mathbb{M}^H, \quad \alpha_\mu^{\text{eff}} + \delta_\mu^{\text{eff}} + \gamma_\mu^{\text{eff}} \leq T_d + H \tag{50}$$

$$\forall \mu \in \mathbb{M}^H, \quad \alpha_\mu^{\text{eff}} \in \mathbb{N} \tag{51}$$

$$\forall \mu \in \mathbb{M}^H, \quad \delta_\mu^{\text{eff}} \in \mathbb{N} \tag{52}$$

$$\forall \mu \in \mathbb{M}^H, \quad \gamma_\mu^{\text{eff}} \in \mathbb{N} \tag{53}$$

$$\forall b \in \mathbb{B}, \forall \mu_i \neq \mu_j \in \mathbb{M}^b \cap \mathbb{M}^H, \quad X_{\mu_i,\mu_j}^{\text{before}} \in \{0,1\} \tag{54}$$

$$\forall \mu \in \mathbb{M}^H, \forall v \in \mathbb{V}^{b_\mu}, \quad X_{\mu,v} \in \{0,1\} \tag{55}$$

$$\forall t \in \mathbb{T}^H, \quad X_t^{\text{delay}} \in \{0,1\} \tag{56}$$

$$\forall \mu \in \mathbb{M}^H, \quad X_\mu^{\text{delay}} \in \{0,1\} \tag{57}$$

Fig. 2. Mixed Integer Linear programming model.

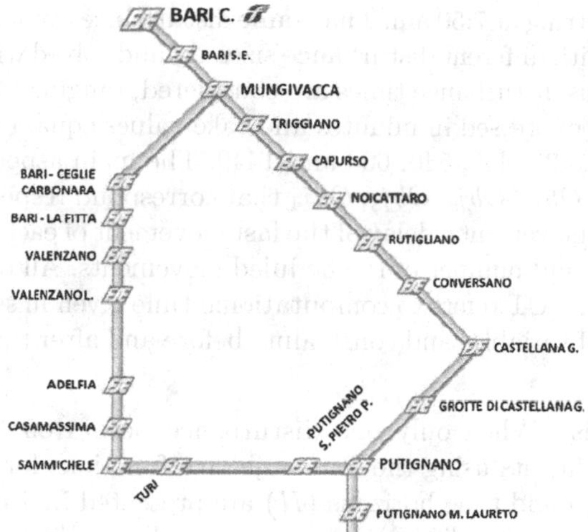

Fig. 3. Railway ring used for the numerical experiments.

Constraints (46) and (47) (resp. (48) and (49)) allow to assess if a train (resp. a movement) is delayed. Constraint (50) enforces each movement must end within time horizon. These constraints are active only when Obj_3 or Obj_4 is used.

Finally, Eqs. (51) to (57) give the domain of variables.

4 Numerical Experiments

The presented model is applied to Ferrovie del Sud Est (FSE), the largest public transport company operating in the Apulia region of Southern Italy. We analyze the railway ring connecting Mungivacca and Putignano stations (Fig. 3), where a CTC system is installed. The operation system is installed in Mungivacca station that is independent, not controlled by the CTC, as is Putignano, whereby these stations are not studied here. In particular, we refer to the line 1 of the railway ring – passing through Conversano – that is single-tracked except for the line connecting Noicattaro to Rutigliano, that is double-tracked. Moreover, Grotte di Castellana is a single track station, not chaired by an operator. 24 even trains and 22 odd trains run on the railway line during a day. We assume a safety time Δ_m of 3 min for two trains traveling in opposite directions and a time Δ_f of 1 min for trains in the same direction. In attempt to evaluate the optimality of the algorithm, we used IBM CPLEX 12.5 installed and run on an Intel Core 2 Duo 1.83 GHz CPU and 3 GB RAM, under Windows with the model formulated in AMPL.

4.1 One Disturbance on the Line

We consider a real data set referring to a train going from Putignano to Mungivacca that stops along the line that connects Castellana G. and Conversano due to a disturbance occurring at 7:50 am. That same disturbance event is used for all the experiments but with different disturbance sizes Δ_d and solved with different time horizons H. Various disturbance times are considered, ranging from 10 to 50 min. Time horizons are expressed in minutes and take values equal to 30, 60, 90, 120, 180, 240, 300, 360, 420, 480, 540, 600 and 1440. The main aspects considered to present results are Obj_1, Obj_2, Obj_3, Obj_4 that correspond respectively to sum of delays for all train movements, delay of the last movement of each train, number of rescheduled trains and number of rescheduled movements. All operational times are given in minutes. CT refers to computational time given in seconds. N and V refers to number of variables and constraints before and after the presolve phase.

Overall Analysis. When only one disturbance occurs on the railway line, results from experiments using the four objective functions for all different disturbance sizes (Δ_d) and time horizons (H) are presented in Table 5.

The methodology is applied to a real case study in which the occurence of short-term disturbances on the railway line is frequent, this is not a trivial problem. The model provides a proactive approach to solve, in real time, problems that occur on the railway line.

The computational time (CT) for all time horizons and disturbance sizes is of the order of a few seconds. Comparison between different time horizons is done in order to demonstrate that the model is able to quickly solve even considerables problems that take into account all trains on the railway line. In fact, the highest value of CT (equal to 69.77 s) is obtained using Obj_1 for $H = 1440$ min and $\Delta_d = 45$ min. This means that if a disturbance lasting 45 min occurs along the line, in just over a minute the train dispatcher can obtain the rescheduled timetable for the 24 hours following the occurrence of the fault. The speed of resolution is a very important factor for the presented problem, since the main objective of the real-time traffic management is to quickly establish a new timetable, in order to minimize the inconvenience for passengers.

Compared to the previous methodology used in [7] there is an improvement due to a reduction of total delay, number of rescheduled trains and computational time. The objective of this model is the minimization of delays of all movements. In Table 1 we compare results of the rescheduling process obtained with the application of the actual and the previous methodology (AM, PM) when $\Delta_d = 30$ min and $H = 180$ min. We noticed that Obj_1 is unchanged for all subsequent values of time horizon. This means that the delay caused by the disturbance is absorbed within the 180 min after its occurrence. For the previous methodology, we present values obtained for $H = 30$ min in addition to values obtained with the application of the heuristic algorithm after the time horizon. CT refers only to the optimization procedure.

By analyzing values we observe that with the actual methodology the new timetable is computed in 0.72 s. Three trains are involved in the rescheduling process for a total delay of 608 min. Previous methodology required 10.74 s to obtain the new timetable within a 30 min window after the occurrence of the fault. 10 trains are involved in the rescheduling process and total delay is equal

Table 1. Comparison between actual and previous methodology.

	Obj_1	Obj_3	CT
Our methodology	608	3	0.72
Dotoli et al.	665	10	10.74

to 665 min. Actual methodology allows to obtain the optimal solution with an exact approach, without the application of the heuristic algorithm which does not always provide optimal results. We should also take into account that solvers used by the two methodology are different. MATLAB with GLPK used by the previous methodology is replaced by IBM CPLEX in the actual one. Resolution methods used by the two solvers are different as well as their performance. The previous methodology has obtained an improvement compared to the current practice used by the train dispatcher; the actual methodology provides a further amelioration. This is in line with the objectives of the real-time traffic management.

Comparison Between Objective Functions. We compare values obtained using the four objective functions for $H = 1440$ and $\Delta_d = 50$ min, shown in Table 2. We analyze the time horizon of 1440 min because the complexity of the problem is high due to the presence, in the rescheduling process, of all trains movements on the railway line until the end of the day.

Table 2. Comparison between the four objective functions with $H = 1440$ and $\Delta_d = 50$.

	Obj_1	Obj_2	Obj_3	Obj_4
Obj_1	829	854	878	878
Obj_2	81	81	91	91
Obj_3	3	3	1	1
Obj_4	33	39	11	11
CT	30.07	16.86	6.56	26.18

Lowest values in terms of total delay are obtained using Obj_1. Comparing results obtained with the first and the second objective function we notice that although the value of Obj_2 is the same, Obj_1 changes. In particular, Obj_1 obtained is greater while Obj_2 is used. The reason is simple: in this case, minimizing the delay of the last movement of a train, the second objective function increases the number of its delayed movements. This means that although values of Obj_3 are unchanged using the first and the second objective function, values of Obj_4 varies. In general, Obj_2 increases the arrival time at intermediate stations, in order to minimize the delay at the last station of trains path. In this case, there are no differences between values of Obj_1 and Obj_2 obtained using Obj_3 and Obj_4. Extending the analysis to all time horizons, we notice that in some cases (e.g. $H = 180$ and $\Delta_d = 50$ min) there is a difference between the two values, due to the fact that these objective functions does not take into account the exact delay of trains. Thus, any solution showing the same number of delayed trains is optimal, whatever the value of Obj_1 and Obj_2 is. Multicriteria objective functions should be used to obtain an unique optimal solution. The minimization of the number of delayed trains (resp. movements) may imply an increase of the total delay of rescheduled trains.

Extending the analysis to values obtained in all time horizons (H) and for all disturbance sizes (Δ_d), we notice that minimizing Obj_1 provides better results in terms of Obj_1 and Obj_2. Minimizing Obj_4 provides better values in terms of Obj_3 and Obj_4.

Impact of Constraint (50). In order to prove the necessity of constraint (50) when using Obj_3 and Obj_4, we present a simple railway line made by 7 segments on which circulate 5 trains, represented in Fig. 4. Railway line is made by 4 single-tracked connection segments (b_1, b_3, b_5 and b_7) between 3 double-tracked stations (b_2, b_4 and b_6). Trains t_1 and t_3 are directed to the South, while t_2, t_4 and t_6 travel in the opposite direction.

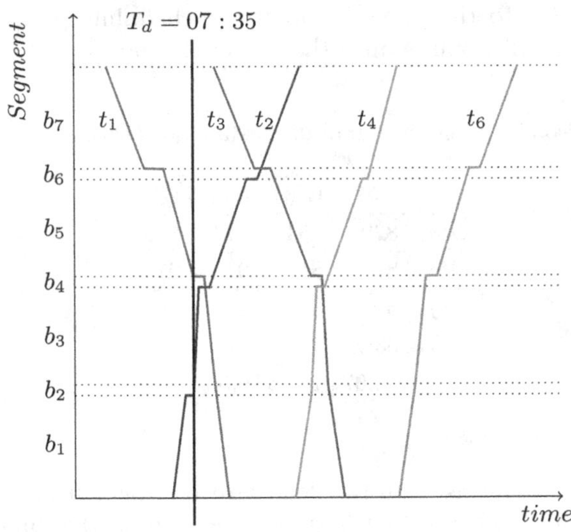

Fig. 4. Disturbance of t_1.

A disturbance occurs in the segment b_5 at 07:35 am and affects the train t_1. We analyze a problem with time horizon $H = 25$ min and a disturbance size $\Delta_d = 10$ min. The same scenarios have been solved without (WT) and with (W) the additional constraint using Obj_3. Results are presented in Table 3.

Table 3. Results from experiments using Obj_3 without and with additional constraint (50).

	Obj_1	Obj_2	Obj_3	Obj_4	CT
WT	10040	10000	1	5	0.01
W	50	10	1	5	0.01

By analyzing values of Obj_1 and Obj_2 we notice that despite the number of rescheduled trains is unchanged, without the additional constraints the departure or the duration of some movements is delayed for a long time. This means that minimizing Obj_3 tends to postpone movements of trains involved in the rescheduling process at the end of the time horizon, in order to affect the lowest number of trains on the line. In the example, t_1 is the only delayed train. By introducing the additional constraint, values of Obj_1 (and consequently Obj_2) are lower because the system is forced to reschedule movements within the time horizon. The number of rescheduled trains remains unchanged.

4.2 Two Disturbances on the Line

When two independent disturbances occur on the railway line, the rescheduling process of the first disturbance does not influence the rescheduling process of the second one. We consider the same railway line presented in Sect. 4 and we suppose that two independent disturbances occur at $T_d = 09{:}00$ a.m. respectively in the line that connects Rutigliano to Conversano stations (segment b_9) and Conversano to Castellana G. (segment b_{11}). We consider a time horizon $H = 1440$ min and a disturbance size $\Delta_d = 50$ min.

First, we solve the problem considering only the disturbance that occurs in the segment b_9 and that affects a train directed from Putignano to Mungivacca station. Then, we solve the problem considering only the disturbance that occurs in the segment b_{11} and that affects a train directed from Mungivacca to Putignano station. Finally, we solve the problem considering the two disturbances at the same time. We compare results with those given by the sum of values obtained solving the two problems separately. Table 4 presents values of Obj_1, Obj_2, Obj_3, Obj_4 and CT for the four scenarios.

Table 4. Two independent disturbances.

	FirstDisturbance					*SecondDisturbance*					*TwoDisturbances*					*Sum*				
	Obj_1	Obj_2	Obj_3	Obj_4	CT	Obj_1	Obj_2	Obj_3	Obj_4	CT	Obj_1	Obj_2	Obj_3	Obj_4	CT	Obj_1	Obj_2	Obj_3	Obj_4	CT
Obj_1	516	61	1	9	13.43	308	61	3	26	17.98	824	122	4	35	31.41	824	122	4	36	17.95
Obj_2	516	61	1	9	11.95	308	61	3	26	17.63	824	122	4	35	29.58	826	122	4	36	19.83
Obj_3	516	61	1	9	9.08	336	71	1	5	9.81	852	132	2	14	18.89	871	132	2	14	9.16
Obj_4	516	61	1	9	10.42	336	71	1	5	12.21	852	132	2	14	22.63	871	132	2	14	10.12

Table 5. Results of analysis using Obj_1, Obj_2, Obj_3 and Obj_4 for all H and Δ_d.

H	Tot.		PrSlv		Obj	$\Delta_d=10$					$\Delta_d=15$					$\Delta_d=20$					$\Delta_d=25$					$\Delta_d=30$					$\Delta_d=35$					$\Delta_d=40$					$\Delta_d=45$					$\Delta_d=50$				
	#V	#C	#V	#C		Obj_1	Obj_2	Obj_3	Obj_4	CT	Obj_1	Obj_2	Obj_3	Obj_4	CT	Obj_1	Obj_2	Obj_3	Obj_4	CT	Obj_1	Obj_2	Obj_3	Obj_4	CT	Obj_1	Obj_2	Obj_3	Obj_4	CT	Obj_1	Obj_2	Obj_3	Obj_4	CT	Obj_1	Obj_2	Obj_3	Obj_4	CT	Obj_1	Obj_2	Obj_3	Obj_4	CT	Obj_1	Obj_2	Obj_3	Obj_4	CT
30	6316	6555	1957	1897	1	70	10	1	7	0.04	105	15	1	7	0.04	140	20	1	7	0.03	175	25	1	7	0.06	210	30	1	7	0.03	245	35	1	7	0.04	280	40	1	7	0.03	315	45	1	7	0.03	350	50	1	7	0.10
30	6316	6555	1172	1897	2	70	10	1	7	0.05	105	15	1	7	0.05	140	20	1	10	0.08	175	25	1	10	0.01	210	30	1	10	0.05	245	35	1	10	0.03	280	40	1	10	0.16	315	45	1	10	0.03	350	50	1	10	0.06
30	6316	7047	1960	1897	3	70	10	1	7	0.14	105	15	1	7	0.47	140	20	1	7	0.05	175	25	1	7	0.05	210	30	1	7	0.03	245	35	1	7	0.05	280	40	1	7	0.05	315	45	1	7	0.05	350	50	1	7	0.05
30	6472	7359	1972	1921	4	70	10	1	7	0.06	105	15	1	7	0.03	140	20	1	7	0.05	175	25	1	7	0.03	210	30	1	7	0.05	245	35	1	7	0.05	280	40	1	7	0.03	315	45	1	7	0.05	350	50	1	7	0.20
60	7636	8190	2675	2818	1	110	10	1	11	0.07	211	23	3	25	0.09	296	31	2	19	0.09	367	36	2	21	0.09	428	43	2	13	0.08	433	43	2	13	0.07	450	45	2	13	0.09	515	55	2	13	0.06	580	65	2	13	0.06
60	7636	8190	2675	2818	2	110	10	1	11	0.06	211	23	3	19	0.08	300	31	2	21	0.09	367	38	3	21	0.12	428	43	2	21	0.18	437	43	2	23	0.09	450	45	2	23	0.09	470	55	2	26	0.09	610	56	1	26	0.08
60	7636	8758	2678	2814	3	110	10	1	11	0.06	304	38	2	19	0.09	300	31	2	19	0.08	531	61	2	13	0.08	667	75	2	19	0.08	595	56	1	11	0.06	600	56	1	11	0.06	605	56	1	11	0.08	610	56	1	11	0.09
60	7816	9118	2690	2831	4	110	10	1	11	0.07	304	38	2	19	0.12	502	69	2	13	0.09	567	79	2	13	0.08	428	43	2	13	0.11	595	56	1	11	0.08	600	56	1	11	0.08	605	56	1	11	0.09	610	56	1	11	0.08
90	8920	10074	3401	3995	1	110	10	1	11	0.12	266	31	2	21	0.30	310	31	2	21	0.20	431	44	6	44	0.23	512	51	5	34	0.21	517	55	5	34	0.20	554	58	4	32	0.30	697	76	3	28	0.28	751	69	1	11	0.45
90	8920	10074	3401	3995	2	110	10	1	11	0.12	289	31	2	21	0.12	310	31	2	23	0.05	431	44	6	23	0.90	540	55	4	23	0.20	517	55	5	34	0.16	572	58	5	34	0.22	752	69	1	35	0.23	751	69	1	35	0.20
90	8920	10712	3380	3937	3	110	10	1	11	0.06	913	83	1	11	0.17	891	83	1	11	0.18	913	83	1	11	0.18	891	83	1	11	0.18	892	83	1	11	0.17	913	83	1	11	0.12	378	83	1	11	0.18	913	83	1	11	0.20
90	9122	11916	3436	4848	4	110	10	1	11	0.08	913	83	1	11	0.20	891	83	1	11	0.22	913	83	1	11	0.20	913	83	1	11	0.23	892	83	1	11	0.22	891	83	1	11	0.22	751	69	1	11	0.18	913	83	1	11	0.23
120	10519	12275	4345	5762	1	110	10	1	11	0.20	266	31	2	21	0.42	310	31	2	21	0.28	477	46	7	67	0.51	586	68	5	39	0.36	591	68	5	39	0.36	616	69	5	39	0.56	753	82	5	45	0.73	814	80	2	18	0.88
120	10519	12275	4345	5762	2	110	10	1	11	0.17	289	31	2	21	0.50	310	31	2	23	0.37	477	46	7	23	0.40	604	59	6	23	0.45	609	59	6	35	0.40	672	65	6	35	0.55	839	80	2	36	1.18	838	80	2	39	0.62
120	10519	12996	4192	5290	3	110	10	1	11	0.14	864	82	2	18	0.45	406	47	2	18	0.40	864	82	2	18	0.40	857	82	2	18	0.40	864	83	2	18	0.28	847	80	2	18	0.28	878	82	2	18	0.30	830	81	2	18	0.26
120	10748	13454	4279	5467	4	110	10	1	11	0.18	855	82	2	18	0.44	864	83	2	18	0.72	856	82	2	18	0.50	833	80	2	18	0.50	863	82	2	18	0.36	861	82	2	18	0.33	863	82	2	18	0.30	830	81	2	18	0.31
180	13051	16644	5925	9048	1	110	10	1	11	0.3	266	31	2	21	0.95	310	31	2	21	0.50	523	48	8	90	0.83	608	76	3	29	0.72	613	76	3	29	0.62	638	77	3	29	1.17	829	81	3	33	0.51	829	81	3	33	1.40
180	13051	16644	5925	9048	2	110	10	1	11	0.31	267	31	2	21	1.12	310	31	2	23	0.60	523	48	8	23	0.73	714	65	8	23	0.68	719	65	8	35	0.80	825	73	8	35	1.30	853	81	3	36	1.68	853	81	3	39	0.99
180	13051	17485	5916	9008	3	110	10	1	11	0.17	875	91	1	11	0.64	875	91	1	11	0.65	875	91	1	11	0.64	875	91	1	11	0.61	875	91	1	11	0.61	875	91	1	11	0.67	875	91	1	11	0.64	948	91	1	11	0.48
180	13318	18019	6037	9249	4	110	10	1	11	0.20	875	91	1	11	0.70	875	91	1	11	0.72	875	91	1	11	0.75	875	91	1	11	0.81	875	91	1	11	0.69	875	91	1	11	0.80	875	91	1	11	0.75	875	91	1	11	0.72
240	18149	21612	7721	13225	1	110	10	1	11	0.31	266	31	2	21	3.01	310	31	2	21	0.98	523	48	8	90	1.87	608	76	3	29	1.99	613	76	3	29	1.64	638	77	3	29	2.20	829	81	3	33	8.16	829	81	3	33	8.35
240	15804	43004	5376	24173	2	110	10	1	11	0.30	267	31	2	21	2.24	314	31	2	23	1.46	523	48	8	23	1.31	716	65	8	23	1.56	721	65	8	35	1.50	820	73	8	35	2.11	855	81	3	36	2.84	853	81	3	39	1.98
240	15804	22574	7684	13117	3	110	10	1	11	0.31	875	91	1	11	0.70	875	91	1	11	0.78	875	91	1	11	0.15	875	91	1	11	0.84	875	91	1	11	0.78	875	91	1	11	0.90	875	91	1	11	0.81	875	91	1	11	0.96
240	16110	23186	7844	13436	4	110	10	1	11	0.36	875	91	1	11	0.95	875	91	1	11	0.90	875	91	1	11	0.95	875	91	1	11	0.90	875	91	1	11	1.01	875	91	1	11	0.99	875	91	1	11	0.90	875	91	1	11	1.01
300	17528	25234	8889	16141	1	110	10	1	11	0.57	266	31	2	21	3.91	310	31	2	21	2.31	523	48	8	90	2.88	608	76	3	29	2.77	613	76	3	29	2.56	638	77	3	29	2.96	829	81	3	33	7.05	829	81	3	33	5.04
300	17528	25234	8889	16141	2	110	10	1	11	0.37	289	31	2	21	2.04	318	31	2	26	1.23	523	48	8	23	1.90	716	65	8	23	2.42	721	65	8	35	1.95	820	73	8	35	2.43	854	81	3	36	3.62	853	81	3	39	2.01
300	17528	26266	8892	16115	3	110	10	1	11	0.40	875	91	1	11	1.01	875	91	1	11	0.89	875	91	1	11	0.98	875	91	1	11	0.94	875	91	1	11	1.01	875	91	1	11	1.07	875	91	1	11	0.93	875	91	1	11	1.06
300	17856	26922	9074	16478	4	110	10	1	11	0.44	875	91	1	11	1.25	875	91	1	11	1.17	875	91	1	11	1.20	875	91	1	11	1.34	875	91	1	11	1.25	875	91	1	11	1.23	875	91	1	11	1.12	875	91	1	11	1.22
360	20240	30791	10762	20933	1	110	10	1	11	0.78	266	31	2	21	3.04	310	31	2	21	1.77	523	48	8	90	3.44	608	76	3	29	3.33	613	76	3	29	2.85	638	77	3	29	4.29	829	81	3	33	12.48	829	81	3	33	5.55
360	20240	30791	10762	20933	2	110	10	1	11	0.51	319	31	2	21	3.20	324	31	2	23	2.07	523	48	8	23	2.48	716	65	8	23	2.92	721	65	8	35	2.84	820	73	8	36	4.47	855	81	3	36	4.37	854	81	3	39	4.70
360	20240	31929	10575	20491	3	110	10	1	11	0.87	875	91	1	11	1.34	875	91	1	11	1.22	875	91	1	11	1.30	875	91	1	11	1.18	875	91	1	11	1.38	875	91	1	11	1.26	875	91	1	11	1.34	875	91	1	11	1.32
360	20602	32653	10791	20922	4	110	10	1	11	1.12	875	91	1	11	1.48	875	91	1	11	1.57	875	91	1	11	1.51	875	91	1	11	1.48	875	91	1	11	1.64	875	91	1	11	1.46	875	91	1	11	1.40	875	91	1	11	1.64
420	23226	37541	12889	26619	1	110	10	1	11	0.98	266	31	2	21	3.52	310	31	2	21	2.29	523	48	8	90	5.66	608	76	3	29	4.76	613	76	3	29	3.66	638	77	3	29	6.50	829	81	3	33	19.40	829	81	3	33	10.28
420	23226	37541	12889	26619	2	110	10	1	11	0.68	289	31	2	21	3.80	314	31	2	23	3.20	523	48	8	23	3.43	716	65	2	23	3.57	721	65	8	35	3.40	840	73	8	35	5.27	853	81	3	36	6.32	853	81	3	39	4.15
420	23226	38787	12892	26585	3	110	10	1	11	0.69	875	91	1	11	1.73	875	91	1	11	1.62	875	91	1	11	1.68	875	91	1	11	1.64	875	91	1	11	1.73	875	91	1	11	1.80	875	91	1	11	1.65	875	91	1	11	1.81
420	23622	39579	13142	27084	4	110	10	1	11	0.76	875	91	1	11	2.03	875	91	1	11	2.06	875	91	1	11	2.07	875	91	1	11	2.12	875	91	1	11	2.07	875	91	1	11	1.60	875	91	1	11	1.88	875	91	1	11	2.12
480	26152	44235	14868	32237	1	110	10	1	11	1.74	266	31	2	21	5.52	310	31	2	21	2.71	523	48	8	90	5.21	608	76	3	29	4.66	613	76	3	29	4.62	638	77	3	29	8.70	829	81	3	33	31.04	829	81	3	33	9.70
480	26152	44235	14868	32237	2	110	10	1	11	1.36	267	31	2	21	5.63	324	31	2	23	4.08	523	48	8	23	4.93	716	65	8	23	4.85	721	65	8	35	4.77	820	73	8	35	7.63	853	81	3	36	8.05	853	81	3	39	6.47
480	26152	45601	14729	31903	3	110	10	1	11	1.35	876	91	1	11	2.54	876	91	1	11	2.21	876	91	1	11	2.18	876	91	1	11	2.18	876	91	1	11	2.14	876	91	1	11	2.18	876	91	1	11	2.26	876	91	1	11	2.21
480	26586	46469	15017	32478	4	110	10	1	11	1.57	876	91	1	11	2.40	876	91	1	11	2.37	876	91	1	11	2.43	876	91	1	11	2.37	876	91	1	11	2.40	876	91	1	11	2.30	876	91	1	11	2.42	876	91	1	11	2.42
540	29808	53122	17727	40361	1	110	10	1	11	1.66	266	31	2	21	12.53	310	31	2	21	4.30	523	48	8	90	7.72	608	76	3	29	6.75	613	76	3	29	8.83	638	77	3	29	12.96	829	81	3	33	30.15	829	81	3	33	18.48
540	29808	53122	17727	40361	2	110	10	1	11	1.72	267	31	2	21	7.58	314	31	2	24	5.55	523	48	8	24	2.14	716	65	8	24	5.60	721	65	8	36	6.16	820	73	8	36	9.36	853	81	3	36	10.98	853	81	3	39	7.59
540	29808	54588	17314	39441	3	110	10	1	11	1.74	875	91	1	11	2.71	875	91	1	11	2.80	875	91	1	11	2.73	875	91	1	11	2.93	875	91	1	11	2.70	875	91	1	11	2.79	875	91	1	11	3.85	875	91	1	11	2.87
540	30274	55520	17634	40080	4	110	10	1	11	1.87	875	91	1	11	2.87	875	91	1	11	2.99	875	91	1	11	2.92	875	91	1	11	3.00	875	91	1	11	2.94	875	91	1	11	3.32	875	91	1	11	3.07	875	91	1	11	3.20
600	33335	61819	20398	48293	1	110	10	1	11	2.07	266	31	2	21	12.87	310	31	2	21	4.60	523	48	8	90	10.12	608	76	3	29	8.52	613	76	3	29	8.13	638	77	3	29	10.15	829	81	3	33	38.12	829	81	3	33	15.33
600	33335	61819	20398	48293	2	110	10	1	11	2.15	267	31	2	21	8.92	321	31	2	24	7.84	523	48	8	24	6.86	716	65	8	24	6.58	721	65	8	36	7.06	834	73	8	36	10.75	867	81	3	36	12.45	853	81	3	39	9.41
600	33335	63394	20131	47675	3	110	10	1	11	2.23	877	91	1	11	3.26	877	91	1	11	3.66	878	91	1	11	3.54	877	91	1	11	3.43	877	91	1	11	3.60	877	91	1	11	3.45	877	91	1	11	3.54	877	91	1	11	3.62
600	33836	64396	20486	48384	4	110	10	1	11	2.12	877	91	1	11	3.46	877	91	1	11	3.54	878	91	1	11	3.66	877	91	1	11	3.60	877	91	1	11	3.62	877	91	1	11	3.70	877	91	1	11	3.52	877	91	1	11	3.73
1440	48247	101160	32059	84305	1	110	10	1	11	3.85	266	31	2	21	22.40	310	31	2	21	9.17	523	48	8	90	19.81	608	76	3	29	17.08	613	76	3	29	16.55	638	77	3	29	20.23	829	81	3	33	69.77	829	81	3	33	30.07
1440	48247	101160	32059	84305	2	110	10	1	11	3.96	266	31	2	21	21.60	332	31	2	23	17.94	523	48	8	24	14.43	716	65	8	24	14.24	721	65	8	36	14.22	820	73	8	34	22.37	853	81	3	36	24.27	854	81	3	39	16.86
1440	48247	103146	31766	83615	3	110	10	1	11	4.40	878	91	1	11	6.60	878	91	1	11	7.08	878	91	1	11	6.32	878	91	1	11	6.58	878	91	1	11	6.97	878	91	1	11	6.95	878	91	1	11	8.47	878	91	1	11	6.56
1440	48879	104428	32252	84586	4	110	10	1	11	2.99	878	91	1	11	5.63	878	91	1	11	5.27	878	91	1	11	6.64	878	91	1	11	9.64	878	91	1	11	14.21	878	91	1	11	14.28	878	91	1	11	11.23	878	91	1	11	26.18

By comparing values obtained from simultaneous resolution with those obtained from the sum of individual resolutions of disturbances, i.e. values presented in the third and the fourth block of Table 4, we notice that the simultaneous resolution provides a better result in terms of computational time for all objective functions. However, the sum of individual resolutions provides an improvement in terms of Obj_1 and Obj_4 using the four objective functions. In particular by applying Obj_1, there is a reduction of the number of rescheduled movements and by applying Obj_2, there is also a reduction of total delay. By using Obj_3 and Obj_4, there is a decrease of values of Obj_1. It is interesting to note that when multiple independent disturbances occur on the line, it is possible to decompose the problem in independent subproblems. In this way, the train dispatcher can give priority to the rescheduling of trains which paths include stations where a higher level of service is required or that have to comply connections with other trains. One could expect that two disturbances would be more difficult to solve but, according to the first experiments, this is not the case. More particularly, the time needed to solve the first disturbance is greater than the time needed to solve both, perhaps due to number of embedded variables.

5 Conclusions and Perspectives

In this paper, we propose a formalization of the rescheduling real-time problem for a regional single-tracked railway network in which a CTC control system is installed. We propose a mathematical model that operates as a decision support system for the train dispatcher. The main goal is to find a decision support system for the train dispatcher that is able to restore normal traffic conditions after the occurrence of a disturbance and to provide an adequate level of service to passengers. We analyze four alternative objective functions in order to find the optimal solution that is a good compromise between total delay, number of rescheduled trains and computational time.

There are many perspectives for this work:

- Increase the complexity of the analysis, considering a greater number of disturbances on the line that occur at different times and have different sizes.
- Introduce robustness in the rescheduling process. A robust rescheduled timetable is less subject to change if a new disturbance occurs on the railway line.
- Perform a structural analysis of the railway line in order to verify if there are independent sectors in which it is possible to predetermine an optimal solution to apply when a disturbance occours.
- Introduce indicators of the complexity of the problem in order to assess the sensibility of the computational time with these parameters.
- Include a resolution strategy that allow the cancellation of a train when the delay that it would accumulate along the line exceeds a certain threshold.
- Study other resolution methods most suitable to the complexity of the problem, such as constraints programming, able to produce a set of possible solutions.

References

1. D'Ariano, A.: Real-time train dispatching: models, algorithms and applications. Ph.D. thesis, Faculty of Civil Engineering and Geosciences, Delft University of Technology, Department of Transport and Planning (2008)
2. Assad, A.: Models for rail transportation. Transp. Res. Part A **14A**, 205–220 (1980)
3. Cordeau, J.F., Toth, P., Vigo, D.: A survey of optimization models for train routing and scheduling. Transp. Sci. **32**, 380–404 (1998)
4. Törnquist, J.: Computer-based decision support for railway traffic scheduling and dispatching: a review of models and algorithms. In: Proceedings of ATMOS 2005 (Algorithmic MeThods and Models for Optimization of RailwayS), Palma de Mallorca, Spain (2005)
5. Törnquist, J., Persson, J.A.: N-tracked railway traffic re-scheduling during disturbances. Transp. Res. Part B **41**, 342–362 (2007)
6. İsmail, S.: Railway traffic control and train scheduling based on inter-train conflict management. Transp. Res. Part B **33**, 511–534 (1999)
7. Dotoli, M., Epicoco, N., Falagario, M., Piconese, A., Sciancalepore, F., Turchiano, B.: A real time traffic management model for regional railway networks under disturbances. In: 9th annual IEEE Conference on Automation Science and Engineering, Madison, USA (2013)
8. FSE - Ferrovie del Sud Est: Fse - ferrovie del sud est e servizi automobilistici (2013). http://www.fseonline.it
9. Vicuna, G.: Organizzazione e tencica ferroviaria, vol. II. CIFI; Collegio Ingegneri Ferroviari Italiani, Roma (1989)

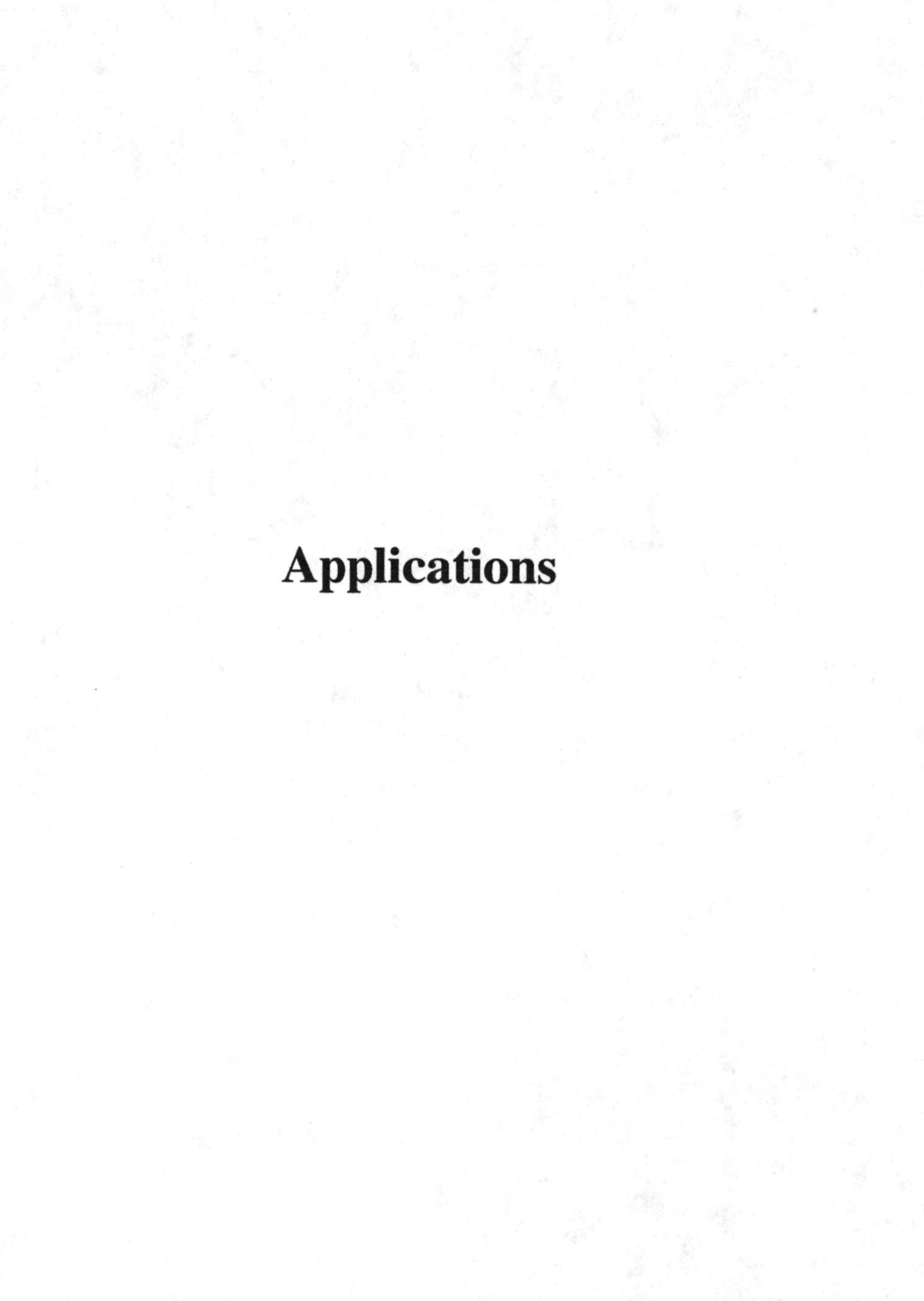

Applications

Solving to Optimality a Discrete Lot-Sizing Problem Thanks to Multi-product Multi-period Valid Inequalities

Céline Gicquel[1,2(✉)] and Michel Minoux[1,2]

[1] Laboratoire de Recherche en Informatique, Université Paris Sud, Campus d'Orsay, 91400 Orsay, France
celine.gicquel@lri.fr

[2] Laboratoire d'Informatique de Paris 6, Université Pierre et Maris Curie, Place Jussieu, 75005 Paris, France
michel.minoux@lip6.fr

Abstract. We consider a problem related to industrial production planning, namely the multi-product discrete lot-sizing and scheduling problem with sequence-dependent changeover costs. This combinatorial optimization is formulated as a mixed-integer linear program and solved to optimality by using a standard Branch & Bound procedure. However, the computational efficiency of such a solution approach relies heavily on the quality of the bounds used at each node of the Branch & Bound search tree. To improve the quality of these bounds, we propose a new family of multi-product multi-period valid inequalities and present both an exact and a heuristic separation algorithm which form the basis of a cutting-plane generation algorithm. We finally discuss preliminary computational results which confirm the practical usefulness of the proposed valid inequalities at strengthening the MILP formulation and at reducing the overall computation time.

Keywords: Production planning · Lot-sizing · Mixed-integer linear programming · Valid inequalities · Cutting-plane algorithm

1 Introduction

Capacitated lot-sizing arises in industrial production planning whenever changeover operations such as preheating, tool changing or cleaning are required between production runs of different products on a machine. The amount of the related changeover costs usually does not depend on the number of products processed after the changeover. Thus, to minimize changeover costs, production should be run using large lot sizes. However, this generates inventory holding costs as the production cannot be synchronized with the actual demand pattern: products must be held in inventory between the time they are produced and the time they are used to satisfy customer demand. The objective of lot-sizing is thus to reach the best possible trade-off between changeover and inventory holding costs while

E. Pinson et al. (Eds.): ICORES 2014, CCIS 509, pp. 237–250, 2015.
DOI: 10.1007/978-3-319-17509-6_16

taking into account both the customer demand satisfaction and the technical limitations of the production system.

An early attempt at modelling this trade-off can be found in [12] for the problem of planning production for a single product on a single resource with an unlimited production capacity. Since this seminal work, a large part of the research on lot-sizing problems has focused on modelling operational aspects in more detail to answer the growing industry need to solve more realistic and complex production planning problems. An overview of recent developments in the field of modelling industrial extensions of lot-sizing problems is provided in [7].

In the present paper, we focus on one of the variants of lot-sizing problems mentioned in [7], namely the multi-product single-resource discrete lot-sizing and scheduling problem or DLSP. As defined in [4], several key assumptions are used in the DLSP to model the production planning problem:

- A set of products is to be produced on a single capacitated production resource.
- A finite time horizon subdivided into discrete periods is used to plan production.
- Demand for products is time-varying (i.e. dynamic) and deterministically known.
- At most one product can be produced per period and the facility processes either one product at full capacity or is completely idle (discrete production policy).
- Costs to be minimized are the inventory holding costs and the changeover costs.

In the DLSP, it is assumed that a changeover between two production runs for different products results in a changeover cost. Changeover costs can depend either on the next product only (sequence-independent case) or on the sequence of products (sequence-dependent case). We consider in the present paper the DLSP with sequence-dependent changeover costs (denoted DLSPSD in what follows). Sequence-dependent changeover costs are mentioned in [7] as one of the relevant operational aspects to be incorporated into lot-sizing models. Moreover, a significant number of real-life lot-sizing problems involving sequence-dependent changeover costs have been recently reported in the academic literature: see among others [9] for a textile fibre industry or [3] for soft drink production.

A wide variety of solution techniques from the Operations Research field have been proposed to solve lot-sizing problems: the reader is referred to [2,6] for recent reviews on the corresponding literature. The present paper belongs to the line of research dealing with exact solution approaches aiming at providing guaranteed optimal solutions for the problem. A large amount of existing exact solution techniques consists in formulating the problem as a mixed-integer linear program (MILP) and in relying on a Branch & Bound type procedure to solve the obtained MILP. However the computational efficiency of such a procedure strongly depends on the quality of the lower bounds used to evaluate the nodes of the search tree. In the present paper, we seek to improve the quality of these lower bounds so as to decrease the total computation time needed to obtain guaranteed optimal solutions for medium-size instances of the problem.

Within the last thirty years, much research has been devoted to the polyhedral study of lot-sizing problems in order to obtain tight linear relaxations and improve the corresponding lower bounds: see e.g. [8] for a general overview of the related literature and [1,5,10] for contributions focusing specifically on the DLSP. However, these procedures mainly focus on the underlying single-product subproblems and thus fail at capturing the conflicts between multiple products sharing the same resource capacity. This leads in some cases to significant integrality gaps for multi-product instances of the DLSPSD. In what follows, we propose a new family of multi-product valid inequalities to partially remedy this difficulty and discuss both an exact and a heuristic algorithm to solve the corresponding separation problem. To the best of our knowledge, this is one of the first attempts focusing on improving the polyhedral description of multi-product lot-sizing problems.

The main contributions of the present paper are thus twofold. First we introduce a new family of valid inequalities representing conflicts on multi-period time intervals between several products simultaneously requiring production on the resource. Second we formulate the corresponding separation problem as a quadratic binary program and propose to solve it either exactly by relying on a quadratic programming solver or approximately through a Kernighan-Lin type heuristic algorithm. The results of the preliminary computational results carried out on medium-size instances show that the proposed valid inequalities are efficient at strengthening the linear relaxation of the problem and at decreasing the overall computation time needed to obtain guaranteed optimal solutions of the DLSPSD.

The remainder of the paper is organized as follows. In Sect. 2, we recall the initial MILP formulation of the multi-product DSLPSD and the previously published single-product valid inequalities. We then present in Sect. 3 the proposed new multi-product multi-period valid inequalities and discuss in Sect. 4 both an exact and a heuristic algorithm to solve the corresponding separation problem. Preliminary computational results are discussed in Sect. 5.

2 MILP Formulation

We first recall the initial MILP formulation of the DLSPSD. We use the network flow representation of changeovers between products, which was proposed among others by [1], as this leads to a tighter linear relaxation of the problem. We then discuss the valid inequalities first proposed by [10] to strengthen the underlying single-product subproblems.

2.1 Initial MILP Formulation

We wish to plan production for a set of products denoted $p = 1 \ldots P$ to be processed on a single production machine over a planning horizon involving $t = 1 \ldots T$ periods. Product $p = 0$ represents the idle state of the machine and period $t = 0$ is used to describe the initial state of the production system.

Production capacity is assumed to be constant throughout the planning horizon. We can thus w.l.o.g. normalize the production capacity to one unit per period and express the demands as binary numbers of production capacity units: see e.g. [4]. We denote d_{pt} the demand for product p in period t, h_p the inventory holding cost per unit per period for product p and S_{pq} the sequence-dependent changeover cost to be incurred whenever the resource setup state is changed from product p to product q.

Using this notation, the DLSPSD can be seen as the problem of assigning at most one product to each period of the planning horizon while ensuring demand satisfaction and minimizing both inventory and changeover costs. We thus introduce the following binary decision variables:

- y_{pt} where $y_{pt} = 1$ if product p is assigned to period t, 0 otherwise.
- w_{pqt} where $w_{pqt} = 1$ if there is a changeover from p to q at the beginning of t, 0 otherwise.

This leads to the following MILP formulation denoted DLSPSD0 for the problem.

$$Z_{LS0} = min \sum_{p=1}^{P} \sum_{t=1}^{T} h_p \sum_{\tau=1}^{t} (y_{p\tau} - d_{p\tau}) + \sum_{p,q=0}^{P} S_{p,q} \sum_{t=1}^{T-1} w_{p,q,t} \tag{1}$$

$$\sum_{\tau=1}^{t} y_{p\tau} \geq \sum_{\tau=1}^{t} d_{p\tau} \qquad \forall p, \forall t \tag{2}$$

$$\sum_{p=0}^{P} y_{pt} = 1, \qquad \forall t \tag{3}$$

$$y_{p,t} = \sum_{q=0}^{P} w_{q,p,t} \qquad \forall p, \forall t \tag{4}$$

$$y_{p,t} = \sum_{q=0}^{P} w_{p,q,t+1} \qquad \forall p, \forall t \tag{5}$$

$$y_{pt} \in \{0,1\} \qquad \forall p, \forall t \tag{6}$$

$$w_{p,q,t} \in \{0,1\} \qquad \forall p, \forall q, \forall t \tag{7}$$

The objective function (1) corresponds to the minimization of the inventory holding and changeover costs over the planning horizon. $\sum_{\tau=1}^{t}(y_{p\tau} - d_{p\tau})$ is the inventory level of product p at the end of period t. Constraints (2) impose that the cumulated demand over interval $[1, t]$ is satisfied by the cumulated production over the same time interval. Constraints (3) ensure that, in each period, the resource is either producing a single product or idle. Constraints (4)–(5) link setup variables y_{pt} with changeover variables w_{pqt} through equalities which can be seen as flow conservation constraints in a network. They ensure that in case product p is setup in period t, there is a changeover from another product q (possible $q = p$) to product p to at the beginning of period t and a changeover from product p to another product q (possible $q = p$) at the end of period t.

2.2 Single-Product Valid Inequalities

We now recall the expression of the valid inequalities proposed by [10] for the single product DLSP. We denote $d_{p,t,\tau}$ the cumulated demand for product p in the interval $\{t,\dots,\tau\}$ and $\Delta_{p,v}$ the vth positive demand period for product p. $\Delta_{p,d_{p,1,t}+v}$ is thus the period with the vth positive unit demand for product p after period t occurs.

$$\sum_{\tau=1}^{t}(y_{p\tau}-d_{p\tau})+\sum_{v=1}^{w}\Big[y_{p,t+v}+\sum_{\tau=t+v+1}^{\Delta_{p,d_{p,1,t}+v}}\sum_{q\neq p}w_{q,p,\tau}\Big]\geq w$$
$$\forall p, \forall t, \forall w\in[1,d_{p,t+1,T}] \quad (8)$$

The idea underlying valid inequalities (8) is to compute a lower bound on the inventory level of a product p at the end of a period t, $\sum_{\tau=1}^{t}(y_{p\tau}-d_{p\tau})$, by considering both the demands and the resource setup states for this product in the forthcoming periods $\tau=t+1...\Delta_{p,d_{p,1,t}+v}$. The reader is refered to [10] for a full proof of validity for these inequalities. In the computation experiments to be presented in Sect. 5, we use a standard cutting-plane generation algorithm to strengthen the formulation DLSPSD0 by adding violated valid inequalities of family (8). The resulting improved formulation is denoted DLSPSD1.

Constraints (8) can be understood as a way to strengthen the demand satisfaction constraints (2) by expressing in a more detailed way the need for each individual product to access the resource in order to satisfy its own demand on a given subinterval of the planning horizon. However, in the resulting DLSPSD1 formulation, the conflicts between different products simultaneously requiring production on the resource will only be handled by the single-period capacity constraints (3). In what follows, we propose to improve this representation of the conflicts between different products by considering multi-period multi-product valid inequalities.

3 New Multi-product Valid Inequalities

We now present the multi-period multi-product valid inequalities we propose to strengthen the linear relaxation of the multi-product DLSPSD.

Proposition 1
Let $SP\subset\{0...P\}$ be a subset of products.
Let $t\in[1,T]$ be a period within the planning horizon. Let $(\theta_1,...,\theta_p,...,\theta_P)\in[0,T]^P$ be a set of periods such that $\theta_p<t$ if $p\in SP$. For each period $\tau\in[1,T]$, we denote $SD_\tau=\{p=1...P|\theta_p>\tau\}$.
The following inequalities are valid for the multi-product DLSPSD.

$$\Big[\sum_{q=1}^{P}d_{q,1,\theta_q}\Big]\Big[\sum_{p\in SP}y_{pt}\Big]\leq\sum_{\tau=1}^{T}\tilde{C}_\tau \quad (9)$$

where $\tilde{C}_\tau$ *is defined by:*

$$\begin{aligned}
\tilde{C}_\tau &= min\Big(\sum_{q\in SD_\tau} y_{q,\tau}, \sum_{p\in SP} y_{p,t}\Big) \text{ if } \tau \notin [t-1;t+1] \\
\tilde{C}_{t-1} &= \sum_{q\in SD_{t-1}, p\in SP} w_{qpt} \\
\tilde{C}_t &= 0 \\
\tilde{C}_{t+1} &= \sum_{p\in SP, q\in SD_{t+1}} w_{pq,t+1}
\end{aligned}$$

Before providing the proof for Proposition 1, we briefly explain the idea underlying valid inequalities (9). We choose a subset SP of products. If none of these products is assigned for production in period t (i.e. $\sum_{p\in SP} y_{pt} = 0$), all corresponding valid inequalities are trivially respected. But if one of these products is produced in period t (i.e. $\sum_{p\in SP} y_{pt} = 1$), then we have to make sure that we are able to satisfy the total cumulated demand $\sum_{q=1}^{P} d_{q,1,\theta_q}$ on the remaining periods $1...t-1, t+1...T$. In this case, the right hand side of inequalities (9) computes a tight upper bound $(\sum_{\tau=1}^{T} \tilde{C}_\tau)$ of the total production capacity remaining to satisfy this cumulated demand.

Proof. Let (y, w) be a feasible solution of the DLSPSD. We arbitrarily choose a subset of products SP, a period t and a vector of periods $(\theta_1, ..., \theta_p, ..., \theta_P)$ such that $\theta_p < t$ if $p \in SP$ and show that all proposed inequalities (9) are valid for the considered feasible solution.

We distinguish two main cases:

- Case 1: $\sum_{p\in SP} y_{pt} = 0$
  In this case, the left hand side of the inequalities is equal to 0 whereas the right hand side is nonnegative. All inequalities (9) are thus trivially true.
- Case 2: $\sum_{p\in SP} y_{pt} = 1$
  In this case, the left hand side of inequalities (9) is equal to the total cumulated demand over intervals $[1, \theta_q]$ for products $q = 1..P$, i.e. to $\sum_{q=1}^{P} d_{q,1,\theta_q}$.

$\sum_{p\in SP} y_{pt} = 1$ means that period t is devoted to the production of one of the products in SP. As we have $\theta_p < t$ for each product $p \in SP$, period t cannot be used to satisfy the cumulated demand $d_{p,1,\theta_p}$ of any product in SP. Hence (y, w) can be a feasible solution of the DLSPSD if and only if the remaining total cumulated production capacity over the periods $1...t-1, t+1...T$ is sufficient to satisfy the cumulated demand $\sum_{q=1}^{P} d_{q,1,\theta_q}$.

We now seek to compute a tight upper bound for the production capacity C_τ available in each period $\tau \in [1, t-1] \cap [t+1, T]$ to satisfy the cumulated demand $\sum_{q=1}^{P} d_{q,1,\theta_q}$:

- By capacity constraints (3), we have $C_\tau \leq 1$, i.e. $C_\tau \leq \sum_{p\in SP} y_{pt}$.
- Moreover, the cumulated demand $d_{q,1,\theta_q}$ for a product q can only be satisfied by a production for q in period τ if $\tau \leq \theta_q$ as demand backlogging

is not allowed here. Hence period τ can be used to satisfy part of demand $\sum_{q=1}^{P} d_{q,1,\theta_q}$ only if the resource is setup for one of products $q = 1..P$ such that $\tau \leq \theta_q$. This gives $C_\tau \leq \sum_{q \in SD_\tau} y_{q,\tau}$.

We thus obtain $C_\tau \leq min(\sum_{q \in SD_\tau} y_{q,\tau}, \sum_{p \in SP} y_{pt})\ \forall \tau \in [1, t-1] \cap [t+1, \theta]$.

Now, we can exploit our knowledge of the setup state of the resource in period t to further strengthen these inequalities. Namely, we know that a product p belonging to SP is produced in period t. A changeover to (resp. from) this product p thus has to take place at the beginning (resp. at the end) of period t. This means that:

- If period $t-1$ is to be used to satisfy the demand of one of the products belonging to SD_{t-1}, there must be a changeover from this product $q \in SD_{t-1}$ to the product $p \in SP$ at the beginning of period t. The production capacity available in period $\tau = t-1$ for the products in SD_{t-1} is thus limited by $C_{t-1} \leq \sum_{p \in SP, q \in SD_{t-1}} w_{q,p,t}$.
- Similarly, if period $t+1$ is to be used to satisfy the demand of one of the products belonging to SD_{t+1}, there must be a changeover from the product $p \in SP$ to this product at the end of period t. The production capacity available in period $\tau = t+1$ for the products in SD_{t+1} is thus limited by $C_{t+1} \leq \sum_{p \in SP, q \in SD_{t+1}} w_{p,q,t+1}$.

We can thus strengthen the upper bound of C_{t-1} (resp C_{t+1}) by replacing the term $min(\sum_{q \in SD_\tau} y_{q,\tau}, \sum_{p \in SP} y_{pt})$ by $\sum_{p \in SP, q \in SD_{t-1}} w_{q,p,t}$ (resp. $\sum_{p \in SP, q \in SD_{t+1}} w_{p,q,t+1}$)
and obtain the inequalities (9) discussed in Proposition 1.

4 Separation Problem

The number of valid inequalities (9) grows very fast with the problem size. It therefore not possible to include them a priori in the MILP formulation of the problem. This is why we use a cutting-plane generation strategy to add to the MILP formulation only the most violated valid inequalities of the family. This requires solving the corresponding separation algorithm which, given a fractional solution $(\overline{y}, \overline{w})$ of the DLSPSD, will either identify a violated valid inequality or prove that no such inequality exists.

4.1 Exact Separation Algorithm

We first discuss an exact separation algorithm, i.e. an algorithm which is guaranteed to find an inequality violated by a fractional solution $(\overline{y}, \overline{w})$ of the DLSPSD if one exists. We consider each period t and seek to identify the subset SP and the vector $(\theta_1, ..., \theta_p, ..., \theta_P)$ which provide the largest violation of inequalities (9). To achieve this, we formulate the separation problem for a given t as follows.

We introduce the following decision variables:

- $\alpha_p = 1$ if product $p \in SP$, 0 otherwise.
- $\beta_{q,\theta} = 1$ if $\theta_q = \theta$, 0 otherwise.

- $\gamma_\tau = 1$ if capacity C_τ is limited by $\sum_{p=0}^{P} \overline{y_{pt}}\alpha_p$, 0 if C_τ is limited by $\sum_{q=0}^{P}\sum_{\theta=\tau}^{T} \overline{y_{q,\tau}}\beta_{q,\theta}$.

With this notation, the separation problem QBP_t for a given t and a solution $(\overline{y}, \overline{w})$ is formulated as:

$$max \sum_{p=0}^{P}\sum_{q=1}^{P}\sum_{\theta=1}^{T} d_{q,1,\theta}\overline{y_{pt}}\alpha_p\beta_{q,\theta} - \sum_{p=0}^{P}\sum_{q=1}^{P}\sum_{\theta=t-1}^{T} \overline{w_{q,p,t}}\alpha_p\beta_{q,\theta}$$

$$-\sum_{p=0}^{P}\sum_{q=1}^{P}\sum_{\theta=t+1}^{T} \overline{w_{p,q,t+1}}\alpha_p\beta_{q,\theta}$$

$$- \sum_{\tau=1...t-2t+2...\theta} \left[\sum_{p=0}^{P} \overline{y_{pt}}\alpha_p\gamma_\tau + \sum_{q=1}^{P}\sum_{\theta=\tau}^{T} \overline{y_{q,\tau}}\beta_{q,\theta}(1-\gamma_\tau)\right] \quad (10)$$

$$\alpha_p + \sum_{\theta=t}^{T} \beta_{p,\theta} \leq 1 \quad \forall p \quad (11)$$

$$\sum_{\theta=0}^{T} \beta_{p,\theta} = 1 \quad \forall p \quad (12)$$

$$\alpha_p \in \{0,1\} \quad \forall p \quad (13)$$

$$\beta_{p,\theta} \in \{0,1\} \quad \forall p, \forall \theta \quad (14)$$

$$\gamma_\tau \in \{0,1\} \quad \forall \tau \quad (15)$$

The objective function (10) corresponds to the maximimization of the violation of the inequalities, i.e. we seek to identify SP and$(\theta_1, ..., \theta_p, ..., \theta_P)$ so as to maximize the difference between the left and the right hand side of the inequality. If this value is strictly positive, we obtain a violated valid inequality. In case this value is less than or equal to 0, it means that all valid inequalities for period t are satisfied by the fractional solution $(\overline{y}, \overline{w})$. Constraints (11) state that for a given product p, we cannot simultaneously include it in SP and choose a period θ_p such that $\theta_p \geq t$. Constraints (12) guarantee that for each product p, exactly one value of θ_p is chosen .

Problem QBP_t is a binary program with a quadratic objective function and a series of linear constraints. It can be solved to optimality by a quadratic binary programming solver such as the one embedded in CPLEX 12.5.

4.2 Heuristic Separation Algorithm

As can be seen from the computational experiments to be presented in Sect. 5, solving to optimality a sequence of quadratic binary programs QBP_t leads to prohibitively long computation times for the cutting-plane generation algorithm, even for small-size instances. We are thus currently investigating the development of a heuristic separation algorithm capable of identifying violated valid inequalities more quickly.

We discuss here a first version of this separation algorithm which focuses on a special case of the proposed multi-product valid inequalities. This special case consists in choosing a period θ such that $\theta \geq t$, in restricting the possible values for periods θ_1,..., θ_p,...,θ_P to the set $\{0, \theta\}$ and in imposing $\theta_p = 0$ if $p \in SP$.

In this case, for a given pair of periods (t, θ), the separation problem amounts to finding a tripartition of the set of products $\{0...P\}$ into 3 subsets: SP, $SDem_\theta = \{q = 1..P|\theta_q = \theta\}$ and $SDem_0 = \{q = 1..P|\theta_q = 0\}$ such that the quadratic expression (10) is maximized. This problem shares some common features with graph partitioning problems. We therefore propose to solve it using the following Kernighan-Lin type heuristic as this type of algorithm is known to be rather efficient at solving graph partitioning problems.

Choose a tripartition of $\{0...P\}$, Π_{ref}, and compute its violation V_{ref}.
While (test =0):
  Let $test = 1$, $PossMove = P + 1$ and $\Pi_{cur} = \Pi_{ref}$.
  Allow all possible moves to explore the neighbourhood of Π_{cur}.
  While ($PossMove > 0$):
    Evaluate all partitions obtained by carrying out each of the allowed moves in the neighbourhood of Π_{cur}
    Select the best partition obtained in this neighbourhood of Π_{cur}, Π_{best}, forbid the move used to obtain Π_{best} from Π_{cur}, decrease $PossMove$ by 1 and set $\Pi_{cur} = \Pi_{best}$
    If $V_{best} > V_{ref}$, $test = 0$ and $\Pi_{ref} = \Pi_{best}$

The neighbourhood of a tripartition Π of $\{0...P\}$ is defined as the set of tripartitions obtained by moving a single product from its current subset in Π to one of the two other subsets. Moreover, in the computational experiments to be presented in Sect. 5, five different types of partitions are used to initialize the heuristic.

4.3 Cutting-Plane Generation Algorithm

We now briefly describe the cutting-plane generation used to strengthen formulation DLSPSD1 by adding to it some multi-product valid inequalities (9).

Compute the initial LP relaxation of the DLSPSD using formulation DLSPSD1.
While ($test = 0$):
  Denote $(\overline{y}, \overline{w})$ the solution of the current linear relaxation.
  For t=1...T such that $\exists p$ such that $0.0001 < \overline{y_{pt}} < 0.9999$;
    Let $\theta = t$ and $found$ =0.
    While ($\theta \leq T$) and ($found == 0$),
      Solve the separation problem for periods (t, θ) using either the exact or the heuristic algorithm.
      If a violated valid inequality has been found, let $found = 1$.
      $\theta = \theta + 1$
  If at least one violated valid inequality is found, add all the found violated valid inequalities to the current formulation and compute its LP relaxation.
  Else set $test = 1$ to stop the cutting-plane generation.

5 Computational Results

We now discuss the results of some preliminary computational experiments carried out to evaluate the effectiveness of the proposed multi-product valid inequalities at strengthening the formulation of the multi-product DLSPSD and to assess their impact on the total computation time.

We randomly generated instances of the problem using a procedure similar to the one described in [11] for the DLSP with sequence-dependent change-over costs and times. More precisely, the various instances tested have the following characteristics:

- *Problem dimension.* The problem dimension is represented by the number of products P and the number of periods T: we solved medium-size instances involving 4–10 products and 15–75 periods.
- *Inventory holding costs.* For each product, inventory holding costs have been randomly generated from a discrete uniform $DU(5, 10)$ distribution.
- *Changeover costs.* We used two different types of structure for the changeover cost matrix S. Instances of sets A1–A7 have a general cost structure: the cost of a changeover from product p to product q, S_{pq}, was randomly generated from a discrete uniform $DU(100, 200)$ distribution. Instances of sets B1–B7 correspond to the frequently encountered case where products can be grouped into product families: there is a high changeover cost between products of different families and a smaller changeover cost between products belonging to the same family. In this case, for products p and q belonging to different product families, S_{pq} was randomly generated from a discrete uniform $DU(100, 200)$ distribution; for products p and q belonging to the same product family, S_{pq} was randomly generated from a discrete uniform $DU(0, 100)$ distribution.
- *Production capacity utilization.* Production capacity utilization ρ is defined as the ratio between the total cumulated demand ($\sum_{p=1}^{P} \sum_{t=1}^{T} d_{pt}$) and the total cumulated available capacity (T). We set $\rho = 0.95$ for all instances.
- *Demand pattern.* Binary demands $d_{pt} \in \{0, 1\}$ for each product have been randomly generated according to the a procedure similare to the used by [11].

For each considered problem dimension, we generated 10 instances, leading to a total of 140 instances.

All tests were run on an Intel Core i5 (2.7 GHz) with 4 GB of RAM, running under Windows 7. We used a standard MILP software (CPLEX 12.5) with the solver default settings to solve the problems with one of the following formulations:

- DLPSD1: initial MILP formulation DLSPSD0, i.e. formulation (1)–(7), strengthened by single-product valid inequalities (8). We used a standard cutting-plane generation strategy based on a complete enumeration of all possible valid inequalities to add them into the formulation.

- DLSPSD2e: formulation DLSPSD1 strengthened by multi-product valid inequalities (9). We used the cutting-plane generation algorithm presented in Sect. 4.3 to add only the most violated valid inequalities and relied on the exact separation algorithm discussed in Sect. 4.1.
- DLSPSD2h: formulation DLSPSD1 strengthened by multi-product valid inequalities (9). We used the cutting-plane generation algorithm presented in Sect. 4.3 to add only the most violated valid inequalities and relied on the heuristic separation algorithm discussed in Sect. 4.2.

Tables 1 and 2 display the computational results. We provide for each set of 10 instances:

- P and T: the number of products and planning periods involved in the production planning problem.
- V and Cst: the number of variables and constraints in the initial formulation DLSPSD0.
- SP: the number of single-product violated valid inequalities (8) added in the three formulations.
- MPe and MPh: the number of multi-product violated valid inequalities added in formulation DLSPSD2e by the exact separation algorithm and in formulation DLSPSD2h by the heuristic separation algorithm.
- Gap_{LP1} (resp. Gap_{LP2e}, Gap_{LP2h}): the average percentage gap between the linear relaxation of formulation DLSPSD1 (resp. DLSPSD2e, DLSPSD2h) and the value of an optimal integer solution.
- N_{IP1} (resp. N_{IP2e}, N_{IP2h}): the average number of nodes explored by the Branch & Bound procedure before a guaranteed optimal integer solution is found or the computation time limit of 2700 s is reached.
- T_{IP1} (resp. T_{IP2e}, T_{IP2h}): the total computation time (cutting-plane generation and Branch & Bound search) needed to find a guaranteed optimal integer solution (we used the value of 2700s in case a guaranteed optimal integer solution could not be found within the computation time limit).

Results from Table 1 show that the proposed valid inequalities (9) are efficient at strengthening formulation DLSPSD1. Namely, the integrality gap is

Table 1. Preliminary computational results: exact separation algorithm.

						DLSPSD1			DLSPSD2e			
	P	T	V	Cst	SP	Gap_{LP1}	N_{IP1}	T_{IP1}	MPe	Gap_{LP2e}	N_{IP2e}	T_{IP2e}
A1	4	15	425	250	106	2.6 %	2	0.3s	9	0.0 %	0	38.5s
A2	6	15	840	315	108	0.9 %	0	0.3s	3	0.1 %	0	50.2s
A3	4	20	600	300	193	2.6 %	5	0.4s	13	0.1 %	0	2386.0s
B1	4	15	425	250	105	11.5 %	6	0.3s	12	0.02 %	0	51.2s
B2	6	15	840	315	107	5.3 %	1	0.3s	17	1.3 %	0	273.0s
B3	4	20	600	300	192	8.3 %	9	0.5s	20	0.3 %	2	3609.9s

Table 2. Preliminary computational results: heuristic separation algorithm.

						DLSPSD1			DLSPSD2h			
	P	T	V	Cst	SP	Gap_{LP1}	N_{IP1}	T_{IP1}	MPh	Gap_{LP2h}	N_{IP2h}	T_{IP2h}
A1	4	15	425	250	106	2.6 %	2	0.3*s*	9	0.0 %	0	0.1s
A2	6	15	840	315	108	0.9 %	0	0.3*s*	3	0.2 %	0	0.2s
A3	4	20	600	300	193	2.6 %	5	0.4*s*	15	0.2 %	0	0.3s
A4	6	25	1400	625	315	4.3 %	9	1.0*s*	27	0.7 %	4	1.0s
A5	6	50	2800	1050	1153	1.6 %	32	6.7*s*	20	0.9 %	11	4.7s
A6	10	50	6600	1650	1949	2.1 %	99	21.0*s*	51	1.1 %	30	22.7s
A7	8	75	6750	2025	2776	2.7 %	856	151.9*s*	23	2.5 %	660	147.5s
B1	4	15	425	250	105	11.5 %	6	0.3*s*	16	0.1 %	0	0.1s
B2	6	15	840	315	107	5.3 %	1	0.3*s*	10	2.1 %	1	0.3s
B3	4	20	600	300	192	8.3 %	9	0.5*s*	21	0.4 %	0	0.4s
B4	6	25	1400	625	307	9.2 %	13	1.2*s*	30	0.8 %	1	0.7s
B5	6	50	2800	1050	1248	12.2 %	1753	47.7*s*	48	9.5 %	983	37.6s
B6	10	50	6600	1650	1274	15.7 %	25937	901.0*s*	97	11.9 %	11284	496.0s
B7	8	75	6750	2015	2681	15.3 %	25015	1961.9*s*	53	10.7 %	22323	1904.7.0s

reduced from an average of 5.3 % with formulation DLSPSD1 (see Gap_{LP1}) to an average of 0.3 % with formulation DLSPSD2e (see Gap_{LP2e}). We note that this reduction is particularly significant for instances B1–B3 featuring a product family changeover cost structure. Moreover this formulation strengthening is obtained thanks to a relatively small number of multi-product inequalities as can be seen from the average value of MPe (12). However, even if the number of nodes needed by the Branch & Bound procedure to find a guaranteed optimal solution is slightly reduced when using formulation DLSPSD2e, it does not lead to an overall reduction of the computation time. This is mainly explained by the fact that the cutting-plane generation algorithm based on an exact separation algorithm requires prohibitively long computation times to identify the violated multi-product valid inequalities to be added to the formulation. It is thus necessary to resort to a heuristic separation algorithm such as the one proposed in Sect. 4.2.

Comparison of the results obtained with the exact and the heuristic separation algorithm for the instances A1–A3 and B1–B3 (Tables 1 and 2) shows that the proposed heuristic is efficient at finding violated valid inequalities for small size instances. Namely, the average integrality gap for these 60 instances when using the heuristic algorithm is the $Gap_{LP2h} = 0.5\,\%$ which is close to the one obtained when using the exact algorithm ($Gap_{LP2e} = 0.3\,\%$). Moreover, the number of violated valid inequalities found by the heuristic algorithm is nearly the same as the number of violated valid inequalities found by the exact algorithm.

Results from Table 2 also confirm that the proposed heuristic is rather efficient at finding violated valid inequalities for larger instances. This can be seen by looking at the results for instances A4–A7 and B4–B7. We first note that, for these instances, the integrality gap is reduced from an average of 7.9 % while using formulation DLSPSD1 to an average of 4.7 % while using formulation DLSPSD2h. Moreover a significant decrease in the overall computation time is obtained for instances B4–B7 when using formulation DLSPSD2h.

6 Conclusions

We considered the multi-product discrete lot-sizing and scheduling problem with sequence-dependent changeover costs and proposed a new family of multi-product valid inequalities for this problem. This enabled us to better take into account in the MILP formulation the conflicts between different products simultaneously requiring production on the resource. We then presented both an exact and a heuristic separation algorithm in order to identify the most violated valid inequalities to be added in the initial MILP formulation within a cutting-plane generation algorithm. Our preliminary results show that the proposed valid inequalities are efficient at strengthening the MILP formulation and that their use leads to a significant reduction of the overall computation time for instances featuring a product family changeover cost structure. Research work is currently ongoing in order to extend the proposed heuristic separation algorithm so as to identify violated valid inequalities from the whole family.

Acknowledgements. This work was funded by the French National Research Agency (ANR) through its program for young researchers (project ANR JCJC LotRelax).

References

1. Belvaux, G., Wolsey, L.A.: Modelling practical lot-sizing problems as mixed-integer programs. Manage. Sci. **47**(7), 993–1007 (2001)
2. Buschkühl, L., Sahling, F., Helber, S., Tempelmeier, H.: Dynamic capacitated lot-sizing problems: classification and review of solution approaches. OR Spectrum **32**, 231–261 (2010)
3. Ferreira, D., Clark, A.R., Almada-Lobo, B., Morabito, R.: Single-stage formulations for synchronized two-stage lot-sizing and scheduling in soft drink production. Int. J. Prod. Econ. **136**, 255–265 (2012)
4. Fleischmann, B.: The discrete lot sizing and scheduling problem. Eur. J. Oper. Res. **44**, 337–348 (1990)
5. Gicquel, C., Minoux, M., Dallery, Y.: On the discrete lot-sizing and scheduling problem with sequence-dependent changeover times. Oper. Res. Lett. **37**, 32–36 (2009)
6. Jans, R., Degraeve, Z.: Meta-heuristics for dynamic lot sizing: a review and comparison of solution approaches. Eur. J. Oper. Res. **177**, 1855–1875 (2007)
7. Jans, R., Degraeve, Z.: Modelling industrial lot sizing problems: a review. Ind. J. Prod. Res. **46**(6), 1619–1643 (2008)
8. Pochet, Y., Wolsey, L.A.: Production Planning by Mixed Integer Programming. Springer Science, New York (2006)
9. Silva, C., Magalhaes, J.M.: Heuristic lot size scheduling on unrelated parallel machines with applications in the textile industry. Comput. Ind. Eng. **50**, 79–89 (2006)

10. van Eijl, C.A., van Hoesel, C.P.M.: On the discrete lot-sizing and scheduling problem with Wagner-Whitin costs. Oper. Res. Lett. **20**, 7–13 (1997)
11. Salomon, M., Solomon, M., van Wassenhove, L., Dumas, Y., Dauzère-Pérès, S.: Solving the discrete lotsizing and scheduling problem with sequence dependant set-up costs and set-up times using the travelling salesman problem with time windows. Eur. J. Oper. Res. **100**, 494–513 (1997)
12. Wagner, H.M., Whitin, T.M.: Dynamic version of the economic lot size model. Manage. Sci. **5**(1), 89–96 (1958)

A Quadratic Knapsack Model for Optimizing the Media Mix of a Promotional Campaign

Ulrich Pferschy[1(✉)], Joachim Schauer[1], and Gerhild Maier[2]

[1] Department of Statistics and Operations Research, University of Graz, Universitaetsstr. 15, 8010 Graz, Austria
{pferschy,joachim.schauer}@uni-graz.at
[2] UPPER Network GmbH, Seering 7/2, 8141 Unterpremstaetten, Austria
g.maier@uppernetwork.com

Abstract. We consider the decision problem of a marketing manager who has to decide on the best selection of advertising media to be used in a promotional campaign. In this paper an optimization model is developed as part of the marketing management software solution MARMIND. It estimates the effect of each single medium and each pair of media from the evaluation data recorded for past campaigns. These evaluations are weighted by similarity measures which represent the distance between campaigns based on their attributes and goals. Furthermore, a memory effect is introduced to give lower weight to campaigns of the more distant past and higher weight to more recent campaign valuations. The resulting discrete optimization model is a Quadratic Knapsack Problem (QKP) which can be solved almost to optimality by a genetic algorithm. Then the given campaign budget is allocated to all selected advertising media based again on estimations from previous campaigns.

Keywords: Advertising media optimization · Quadratic Knapsack Problem · Genetic algorithm

1 Introduction

Every company striving to sell its products has to find the right way for their marketing, whatever industry or market it is concerned with. In contrary to classical production planning where a certain set of inputs is transformed into a known and fixed set of outputs (disregarding aspects of unknown yields in certain technologies), the task of marketing management is made much more difficult by the lack of quantifiable *cause and effect* relationships between the chosen advertising media and the obtained results, such as increase in sales etc. Under this uncertainty, the central task of marketing management, namely how to use the budget of a promotional campaign, becomes very challenging.

This research was supported by the Austrian Research Promotion Agency (FFG) under project "MARMIND media mix optimization". Ulrich Pferschy and Joachim Schauer were supported by the Austrian Science Fund (FWF): [P 23829-N13].

E. Pinson et al. (Eds.): ICORES 2014, CCIS 509, pp. 251–264, 2015.
DOI: 10.1007/978-3-319-17509-6_17

The increase of available marketing options during the last decade with new possibilities such as targeted social media advertising and context sensitive web banners have further added to this difficulty. On the other hand, some electronically based advertising media also allow for a better monitoring of their reached effect (e.g. click-rates). In this environment of increasing diversity in the marketing decision space and partial but limited feedback information the suitable selection of advertising media for a promotional campaign, i.e. deciding on the *media mix*, definitely asks for an automated decision support system.

The software platform *MARMIND*[1] produced and offered by *UPPER Network* [2] provides a wide range of tools to support the daily tasks of a marketing department from planning to realization. In collaboration with the University of Graz, Austria, an optimization tool was developed and added to the solution which computes a suggestion for the media mix of a planned promotional campaign. This tool is now an integral part of MARMIND and recently started being used by marketing managers.

In the literature contributions to finding the best media mix were given for particular industry sectors, e.g. in [1,2], and from an optimization point of view in several papers going back to [3] and more recently e.g. by [4–6].

A central question of marketing planning concerns the effect and efficiency of advertising media (see e.g. the survey paper [7,8] on internet advertisements). While many statistical methods have been employed to find partial answers to this questions, these require survey data or other means of market research, which is usually not available for the full range of marketing options available to the decision maker in a typical planning scenario. Therefore, we aim to gain information from past campaigns.

Let us give a brief overview of our optimization tool: MARMIND keeps a data base of all past promotional campaigns with ratings of their overall success and the degree of attainment for the different goals of the campaign. These valuations can also incorporate empirical data from electronic advertising media. Based on these observations of past campaigns, we estimate the effect of every advertising medium for the currently planned campaign. To this end we take the "similarity" between planned and past campaigns into account. Moreover, we derive estimations for the pairwise effect of advertising media, since many media influence each other or are dependent on each other and thus cannot be separated into unconnected decisions. Both kinds of effect estimations may also be subject to a memory effect, which attributes less influence to campaigns of the more distant past and more influence to quite recent campaigns. This allows to incorporate changes in the general marketing landscape into the planning of the current campaign.

Based on these effect estimations we draw up an optimization model which turns out to be a classical *Quadratic Knapsack Problem* (QKP). After solving this model by an improved genetic algorithm almost to optimality, we assign the

[1] www.marmind.com/en.

[2] www.uppernetwork.com.

available budget to the selected advertising media by considering the proportional budget allocation of past campaigns.

In-house tests indicate that the media mix selected by the optimization tool gets highly positive appraisals from experts in the field. The various possibilities of para-metrization allow a flexible adaptation for every domain.

2 Formal Problem Formulation

In this paper a promotional campaign is described by a number of attributes, some of them represented by nominal values such as target groups, product classes and general strategic goals, others expressed by numerical values such as desired market share, increase in revenue, etc.

Formally, a promotional campaign t is defined by a k-dimensional vector of parameters $t(1), \ldots, t(k)$, where for some fixed k' with $0 \leq k' \leq k$ there are nominal values $t(1), \ldots, t(k')$ and positive cardinal values $t(k'+1), \ldots, t(k)$. A campaign may also consist of only a subset of these parameters and leave the remaining entries of the vector empty.

To express and measure the goals of promotional campaigns there is set of operative goals $g_1, \ldots, g_l$ defined such as number of new customers, awareness level, number of repeat customers, etc. Each promotional campaign t is assigned a subset G_t of these operative goals with $l_t := |G_t|$. For convenience we impose an upper bound $l_t leq L$ on the number of selected goals, which is of moderate size in practice (think of single digit numbers), i.e. $L \ll l$. Furthermore, the chosen goals in G_t are ranked in a total ordering to indicate their relative importance. This preference relation between goals is represented by a rank number $r_t(g_j)$ for each goal $g_j \in G_t$, where $r_t = l_t$ signifies the most important, i.e. highest ranked, goal and $r_t = 1$ the least important. Clearly, each number in $1, \ldots, l_t$ is assigned to exactly one goal as a rank r_t.

Finally, there is a total budget B_t given for the promotional campaign t.

After completion of the promotional campaign t the responsible manager should be able to state the degree of achievement of each operative goal $g_j \in G_t$ of the campaign by assigning a numerical value representing the achieved percentage of the goal. For simplicity we will assume that this value is scaled into an achievement level $a_t(g_j) \in [0, 1]$ with $a_t(g_j) = 1$ indicating perfect achievement of goal g_j. In addition, the marketing manager will be asked to evaluate the overall success of a completed promotional campaign by assigning a discrete value $s_t \in \{1, \ldots, S\}$, where S indicates the best outcome and 1 the worst. Usually, S is a single digit number.

Of course, it would be desirable to extract more information on the impact of the applied advertising media. However, one should keep in mind that an overly complicated feedback system will often be ignored or filled with data of low quality. Practical experience suggests to keep the evaluation system as simple as possible.

To reach the goals of a promotional campaign there are n different advertising media $m_1, \ldots, m_n$, available ($n \approx 200$), e.g. TV spots for different stations, newspaper ads in various publications, flyers, catalogs, social media ads, promotional events, etc., each with different characteristics.

After choosing the parameters and operative goals of a promotional campaign the central task of the marketing manager as a decision maker consists of the selection of a subset of advertising media and the allocation of a budget b_i to each selected medium m_i, such that the defined goals are met to a high degree while the available budget B_t is not exceeded. The decision on this so-called media mix is crucial for the success of any campaign.

Unfortunately, the effect of each advertising medium on the defined goals in connection with the selected parameters of the promotional campaign are mostly impossible to be quantified. Moreover, the effects of different media can not be separated but are highly interdependent, e.g., a promotional event with a celebrity will hardly have any effect without appropriate news coverage, and an evening TV spot will be better remembered if its tune is repeated by a morning radio spot. Under these circumstances, only educated guesses and general rules of thumb gained from experience can be used by the decision maker to allocate the promotional budget.

The existing software solution MARMIND can keep track of all tasks involved with the realization of a promotional campaign including accounting, managing orders with advertisement companies, etc. In this contribution we describe an optimization system developed to give the decision maker an automatically generated suggestion for the media mix.

There are two core features of our system: (1) an estimation of the direct effect and the interdependencies between advertising media based on the evaluation of past promotional campaigns by the managers, (2) the incorporation of these values into a discrete optimization model, which is basically a Quadratic Knapsack Problem (QKP), possibly with additional constraints.

3 Quadratic Knapsack Model

Given the parameters and operative goals of a promotional campaign t we will derive in Sects. 4 and 5 an estimation of the following three values for all advertising media. For simplicity of notation we omit the reference to the current campaign t.

1. Direct effect p_i on the promotional campaign caused by selecting medium m_i.
2. Joint effect q_{ij} on the promotional campaign caused by selecting both media m_i and m_j.
3. Estimated budget b_i allocated to medium m_i, if it is selected in the promotional campaign.

With these estimations we can set up the following mathematical optimization model with binary variables $x_i \in \{0, 1\}$ representing the selection of advertising medium m_i. The objective function consists of a convex combination of a linear (direct effect) and a quadratic (joint effect) term with a parameter $\lambda \in (0,1)$ to be chosen appropriately. As a starting value we set $\lambda = 0.5$.

$$\max \ \lambda \sum_{i=1}^{n} p_i x_i + (1-\lambda) \sum_{i=1}^{n}\sum_{j=1}^{n} q_{ij}\, x_i x_j \tag{1}$$

$$\text{s.t.} \sum_{i=1}^{n} b_i x_i \leq B_t \tag{2}$$

$$x_i \in \{0,1\} \tag{3}$$

The model (1)–(3) is the well-known Quadratic Knapsack Problem (QKP), see e.g. [9, Chap. 12] or [10].

It may seem reasonable to restrict the number of different advertising media selected for one promotional campaign by adding a cardinality constraint

$$\sum_{i=1}^{n} x_i \leq K. \tag{4}$$

However, it will turn out that the estimation of budget allocations b_i produces values of a certain proportion w.r.t. B_t which implicitly restricts the number of chosen advertising media and thus makes (4) redundant.

Practical considerations also suggest that certain advertising media (e.g. TV spots) are more costly and require a minimum budget to make sense. Thus, we will eliminate in a preprocessing step all advertising media whose minimum budget requirement would consume most of the available budget B_t.

The final suggestion of the media mix presented to the user of the system follows directly from the solution of (1)–(3). Exactly those advertising media m_i should be used whose decision variables have value $x_i = 1$ in the solution. Allocating the final budget $\bar{b}_i$ to each selected medium m_i requires a bit more care and will be treated in Sect. 5.2.

4 Linear and Quadratic Effect Estimation

It should be pointed out that all our estimations are based on the evaluation of past promotional campaigns and are not founded on some strict stochastic model. They were developed in several rounds of interaction with practitioners and validated with real-world case data. The fact that the convex combination of several terms allows the setting of a number of weighting parameters should be seen as an advantage since it permits the adaptation of the optimization system to the special customs and practices of the particular domain the system is applied in. By no means we can expect to deliver a "plug-and-play" system ready for use in any domain for every type of company.

Let $T(i)$ be the set of all past promotional campaigns containing advertising medium m_i. The linear profit value p_i will be expressed by a convex combination of the general success attributed to medium m_i in the past and the level of goal achievement reached by similar campaigns if they included m_i, i.e.

$$p_i := \lambda_p\, ps_i + (1 - \lambda_p) pg_i \tag{5}$$

with $\lambda_p \in (0,1)$. The first term ps_i represents the average scaled success of all past promotional campaigns containing medium m_i. The underlying argument says that every medium contributed in some way to the overall success of past campaigns. Formally, we have:

$$ps_i := \frac{1}{|T(i)|} \sum_{t \in T(i)} \frac{s_t}{S} \tag{6}$$

Clearly, ps_i is in $[0,1]$.

The second term pg_i considers achievement of operative goals and similarity of parameters in more detail and will be described in the following subsection.

4.1 Considering Similarity of Campaigns

The value pg_i should reflect the principle that it is a good idea to repeat strategies that worked well in the past for campaigns with similar parameters. To formalize this principle we will express "working well" by the degree of goal achievement and "similar parameters" by introducing a similarity measure between campaigns.

Let $\tilde{T}(j)$ be the set of all past promotional campaigns containing operative goal g_j. Then the overall goal achievement a_t of a promotional campaign t will be defined as follows:

$$a_t := \frac{1}{\sum_{j \in G_t} r_t(g_j)} \left(\sum_{j \in G_t} r_t(g_j) \cdot \left(\frac{1}{2} \left(a_t(g_j) - \frac{1}{|\tilde{T}(j)|} \sum_{\tau \in \tilde{T}(j)} a_\tau(g_j) \right) + \frac{1}{2} \right) \right) \tag{7}$$

The term in the inner capital brackets computes the difference of the goal achievement for goal g_j from the average goal achievement over all promotional campaigns τ containing goal g_j. This number lies in $(-1,1)$ and is transformed to lie in $(0,1)$. Finally, the terms are weighted by their rank number and scaled by the sum of rank numbers.

Now we introduce a measure to express the similarity between two promotional campaigns t and t'. Formally, we will define a function $sim(t,t') \rightarrow [0,1]$, such that higher values of sim indicate closer similarity of two campaigns. Measures of distance and similarity are used in many fields of applied mathematics and statistics, in particular in cluster analysis (see e.g. [11,12]). Our similarity function will deal separately with a linear combination of nominal and cardinal parameters of campaigns expressed by *sim_par* and with the similarity of the ordinally ranked operative goals *sim_goal*.

$$sim\_par(t,t') := +\frac{1}{\sum_{i=1}^{k} c_i}\left(\sum_{i=1}^{k'} c_i \cdot sim\_nom(t(i),t'(i)) + \sum_{i=k'+1}^{k} c_i \cdot sim\_card(t(i),t'(i))\right) \tag{8}$$

The weighting parameters $c_i \in (0,1)$ can be used to indicate the importance of different parameters.

Comparing nominal parameters is done simply by an inverted Hamming distance, i.e. assigning $sim\_nom(t(i),t'(i)) = 1$ if $t(i) = t'(i)$ and 0 otherwise, for $i = 1,\ldots,k'$. Clearly, also more complicated measures such as the Jaccard index, the Sørensen coefficient or the Tanimoto distance might be used, see e.g. [13].

For cardinal parameters $i = k'+1,\ldots,k$ the similarity is computed from the relative deviation by

$$sim\_card(t(i),t'(i)) = 1 - \frac{|t(i) - t'(i)|}{\max\{t(i),t'(i)\}}, \tag{9}$$

which is clearly in $[0,1]$. Basically, any Minkowski metric could be used and scaled into the corresponding similarity measure.

For comparing the ordered selection of goals between two campaigns in a similarity measure $sim\_goal(t,t')$, classical distance measures of orderings such as Kendall tau rank distance (similar to Kemeny distance) could be used (see [14,15] for recent contributions). In our case, out of the available set of l goals each campaign is assigned only subset of goals of small, but varying size. Hence, we use the following rather unorthodox approach.

Define a decreasing sequence of positive bonus points $\beta_1 > \beta_2 > \ldots > \beta_L$ and translate rank numbers into bonus points by assigning the goal g of a promotional campaign t with rank $r_t(g)$ exactly $\beta_{l_t - r_t(g)+1}$ points, i.e. the best ranked goal receives β_1 points and the lowest ranked goal with $r_t(g) = 1$ gets β_{l_t} points. The remaining points $\beta_{l_t+1},\ldots,\beta_L$ are not assigned at all.

For any pair (t,t') of campaigns we determine the intersection of selected goals and add the bonus points accrued by every such goal in both campaigns. I.e. if some goal g' is ranked on first position in t and on third position in t', then g' contributes $\beta_1 + \beta_3$ to the total sum, while goals appearing in only one of the two campaigns do not contribute at all. This sum is scaled by the maximum possible number of points $\sum_{j=1}^{\min\{l_t,l_{t'}\}} 2\,\beta_j$ which guarantees a final value $sim\_goal(t,t')$ in $[0,1]$, with the desired property that identical orderings of goals yield a similarity of 1 while disjunctive sets of goals have similarity 0.

Finally, we put together the two similarity measures with a weighting parameter λ_g.

$$\begin{aligned} sim(t,t') := (1-\lambda_g)\, sim\_par(t,t') \\ + \lambda_g \cdot sim\_goal(t,t') \end{aligned} \tag{10}$$

A drawback of the above definitions can be found in the "averaging effect" which means that taking a linear combination over many different factors may dilute the effect of strong similarity or deviance in some components and tends to produce moderate values for almost any pair of promotional campaigns.

Thus, we aim at strengthening the influence of strong or weak similarities by increasing values closer to 1 and decreasing values closer to 0. This will be done by applying the following *sigmoid function* $F(x)$ on every partial similarity measure $sim\_nom\ (t(i), t'(i))$, $sim\_card(t(i), t'(i))$ and $sim\_goal(t, t')$. $F(x)$ is depicted in the following figure. It contains a tuning parameter k which we set to $k = 10$ in our implementation.

$$F(x) = \frac{1}{1+e^{(\frac{k}{2}-kx)}} + \frac{1}{1+e^{\frac{k}{2}}} \cdot (2x-1) \tag{11}$$

It remains to put together the expressions of goal achievement and similarity. This is done by simply summing up achievement values of past campaigns weighted by their similarity to the current campaign t^c. Formally, we have

$$pg_i := \frac{1}{|T(i)|} \sum_{t \in T(i)} sim(t, t^c) \cdot a_t. \tag{12}$$

Again, pg_i is in $[0, 1]$.

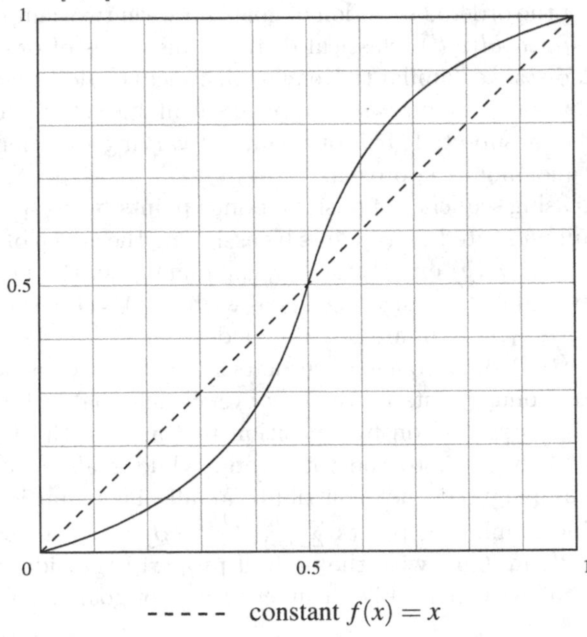

4.2 Estimation of Media Interaction

We proceed to estimate the effect of having two advertising media m_i and m_j together in a promotional campaign. This is done by separating from the set of all past campaigns a subset of particularly effective campaigns which stood out among the remaining campaigns. Then we will simply count the occurrence of every pair of advertising media in the effective campaigns relative to all its occurrences. Thereby, we aim to detect a systematic effect of successful pairs that happened to be chosen together in conspicuous frequency among the more effective campaigns. Note that our existing sample of campaigns is too small to allow statistical tests on this hypothesis.

Formally, we sort the set of past promotional campaigns in decreasing order of their goal achievement a_t and determine a threshold a^T such that only a prescribed percentage of campaigns exceeds this achievement value, e.g. 25 %. Then we set:

$$q_{ij} := \frac{|T(i) \cap T(j) \text{with}\, a_t \geq a^T|}{|T(i) \cap T(j)|} \tag{13}$$

It turned out that there are certain pairs of media that marketing managers generally want to use together and which appear in pairs in almost all campaigns (if they appear at all), no matter whether the campaigns worked well or not. This effect is not captured by (13) which was hence extended to include the presence of pairs of media in past campaigns with strong similarities to the current campaign t^c. For some similarity threshold δ let

$$T^c := \{t \mid sim(t, t^c) \geq \delta\}.$$

Then we define the final quadratic effect as:

$$q'_{ij} := \lambda_q \, q_{ij} + (1 - \lambda_q) \cdot \frac{|(T(i) \cap T(j)) \cap T^c|}{|T^c|} \tag{14}$$

5 Budget Allocation

5.1 Estimation of Budget Values

While it may seem quite reasonable that one can learn from past promotional campaigns which advertising media, resp. which combination of media, worked well to reach certain goals for campaigns with a certain set of parameters, it is less clear how to assign a budget value to an advertising medium after deciding to use it. However, one can not separate media selection from budget allocation since one may end up with a collection of advertising media that can not be realized within the given budget B_t considering the natural lower bounds on the budget for each medium.

To allow a plausible estimation of the budget values b_i in the optimization model, we consider a subset of past campaigns T^B with a budget in similar range as the current campaign t^c, i.e.

$$T^B := \{t \mid k_1 \, B_{t^c} \leq B_t \leq k_2 \, B_{t^c}\} \tag{15}$$

with suitably chosen parameters $k_1 < 1$, $k_2 > 1$. Then we determine for each advertising medium the relative proportion of budget allocated in the past (depending on its assigned budget b_i^t) and take the mean over these values as an estimation of b_i. Formally,

$$b_i := \frac{B_{t^c}}{|T(i) \cap T^B|} \sum_{t \in T(i) \cap T^B} \frac{b_i^t}{B_t}. \tag{16}$$

Note that different from Sect. 4 we do not take similarity of campaigns into account in this estimation. Discussions with marketing managers and analysis of available data exhibit that the choice of advertising media is very much tailored to the particular goals and parameters of a campaign. But once a medium is selected the invested budget is mostly dependent on technical constraints and the "size", i.e. budget, of the overall campaign. But clearly, it would be straightforward to restrict the summation in (16) to campaigns in T^c with a certain similarity to t^c.

5.2 Actual Budget Allocation

After solving the optimization model (1)–(3) we obtain a solution set $S := \{i \mid x_i = 1\}$ of all selected advertising media. Assigning the actual budget values $\bar{b}_i$ to all media $m_i \in S$ could be done by simply resorting to the estimations b_i from (16).

We suggest a more refined procedure taking into account two aspects: First, the discrete solution of optimization model will most likely leave a certain amount of budget $B_t - \sum_{i \in S} b_i$ unused and thus miss chances for a better utilization of the available budget. Secondly, and more important, it should make sense to consider the particular combination of media in S, which we already targeted specifically by the quadratic coefficients q_{ij}.

To do so, we give the relative budget proportions in a promotional campaign t, i.e. $\frac{b_i^t}{B_t}$, more weight if t shares more advertising media with the solution for the current campaign t^c. This is achieved by the following formula for every medium m_i, $i \in S$:

$$\bar{b}_i := \frac{B_{t^c}}{|S| - 1} \sum_{j \in S, j \neq i} \frac{1}{|T(i) \cap T(j) \cap T^B|} \sum_{t \in T(i) \cap T(j) \cap T^B} \frac{b_i^t}{B_t} \tag{17}$$

Allocating budgets according to (17) may result in infeasible solutions or (as before) in leftover budget. We propose the following allocation process to overcome this issue.

The budget estimation b_i in (16) can be seen as an estimator in the strict statistical sense. Hence, we can also compute the associated empirical standard deviation σ_i based on the sum of squared distances from the mean and defined as follows:

$$\sigma_i := \sqrt{\frac{B_{t^c}}{|T(i) \cap T^B| - 1} \sum_{t \in T(i) \cap T^B} \left(\frac{b_i^t}{B_t} - \frac{b_i}{B_{t^c}} \right)^2} \tag{18}$$

Now we start the budget allocation procedure by assigning each advertising medium $m_i \in S$ in decreasing order of profit values p_i a conservative budget value of $b_i - \sigma_i$, i.e. the estimated value reduced by one standard deviation. Then we enter into a second round and increase the budget to b_i as long as the budget B_t permits, again in decreasing order of p_i. Finally, if there is still budget left, we take a third round and increase the allocated budget to $b_i + \sigma_i$ until B_t is completely used up. Clearly, the last advertising medium considered by this procedure may obtain a budget allocation in between the three prescribed values by consuming all the remaining budget.

An analogous procedure is done for the more sophisticated budget values $\bar{b}_i$ (with the corresponding empirical standard deviation $\bar{\sigma}_i$) where it can be expected to be more relevant, since there is a larger difference from the budget values used in the optimization model. Note that in this case it may happen that we run out of budget already in the first round of allocations, since the values b_i used in the weight constraint of the optimization model may deviate considerably from $\bar{b}_i$.

6 Memory Perspective

The presented optimization system is clearly meant for long-term use within a marketing department. In fact, since the estimations of input values for the optimization model rely on data of past campaigns, the system can be expected to deliver its full benefits only after a certain number of campaigns were entered into the data base. Also for the first time use of the system, a sufficiently large data set of past campaigns should be available.

On the other hand, the effects of advertising media can not be expected to remain constant over a long time period but will be subject to change. Obviously, technological possibilities and consumer behavior are constantly evolving and thus the valuation of promotional campaigns completed years ago should not have the same impact on the current decision as a very recent campaign.

To account for this time perspective we introduce a *memory factor* in our model. Therefore, we introduce a time line starting at some initial point $D = 0$ and assign a point in time $D^t > 0$ to every promotional campaign t. Then a memory factor $\mu^t \in [0, 1]$ will be defined for every past campaign t indicating the influence level of that campaign. μ^t depends only on the current time D^c and on time D^t, where $\mu^t = 0$ implies total oblivion of campaign t while $\mu^t = 1$ represents full consideration of t.

In the literature similar discount factors are used e.g. in the smoothing of trend functions. Frequently, μ^t is defined by some exponential function depending on an exponent $\delta \in [0, 1]$, where $\delta = 1$ means low rate of oblivion while $\delta = 0$ indicates rapid oblivion. An obvious choice would be the following:

$$\mu^t = \frac{(D^c - D^t)^\delta}{D^c - D^t} \tag{19}$$

A second possibility with a similar tendency is given by:

$$\mu^t = e^{-\frac{(D^c - D^t)}{\delta}} \tag{20}$$

Incorporating this memory factor into the profit estimations of p_i and q_{ij} can be done in a straightforward way taking scaling into account since the memory factors of a certain profit estimation will usually not add up to 1. As an example, the direct success ps_i attributed to medium m_i as defined in (6) will be transformed into:

$$\tilde{ps}_i := \frac{1}{\sum_{t \in T(i)} \mu^t} \sum_{t \in T(i)} \frac{\mu^t \cdot s_t}{S} \tag{21}$$

7 Solution of the Quadratic Knapsack Problem

The model introduced in Sect. 3 is a standard Quadratic Knapsack Problem (QKP) with no additional side-constraints. This is somewhat rare, since practical applications usually require additional constraints and do not fit into the mould of standard models.

Important exact solution methods for QKP were given by [16,17]. The former approach uses Lagrangian relaxation and is able to solve instances containing up to 200 variables. It is especially well suited for dense instances. [17] uses Lagrangian decomposition and is able to solve instances of roughly the same size, however it outperforms the previous approach on instances of medium and low density.

The currently best working strategy was given by [18]. It succeeds in reducing the size of many instances dramatically by fixing items that will or will not occur in an optimal solution. The reduced problem can then be solved by any algorithm for QKP. Combining this approach with an exact solution algorithm [18] were able to solve instances with up to 1500 items. Unfortunately, this code is not available, therefore we used the implementation described in [16] for solving benchmark problems of MARMIND and managed to solve instances to optimality with up to $n = 200$ advertising media in less than 10 min on a simple standard PC with 2.2 GHz and 2 GB Ram.

For ensuring a good user experience *UPPER Network* however requested that the optimized marketing campaign of MARMIND has to be computed in less than 3 s. Moreover, we recall that all data of our QKP instances is based on estimates and does not represent assured values. Thus, we can easily settle for a good approximate solution. Note that a theoretical analysis of the approximability of QKP was recently given in [19].

For our optimization tool we implemented a genetic algorithm and imposed a time limit of 3 s. It turned out that this gave solutions for all instances of the required size (≥ 200 items) with an average deviation of less than 1 % from optimality.

Our algorithm is a modified version of [20] which worked well for the random test instances generated according to the same method used in [16]. Reference [20] reports test data for ten instances of 100 items and ten instances of 200 items. Every instance was solved 50 times and the algorithm was able to find the optimal solution value in about 90 percent of the runs, although the running time sometimes exceeds 1 min. Note that our implementation was especially tuned for getting high quality results in a very short time but often succeeded to yield results similar or better than [20].

Recently [21] published a well performing metaheuristic that combined GRASP with tabu search. On 100 randomly generated benchmark instances that follow the same scheme as in [16] the metaheuristic was able to find the optimal solution 99 times in less than 0.8 s. In the remaining case the gap to the optimal solution was negligibly small. Moreover, they were able to get good solutions for instances of up to 2000 variables (the solution quality was justified by comparison to known upper bounds) in less than 300 s.

Currently, we are working on a project to systematically test our genetic algorithm, compare it to the other existing methods listed above and to introduce harder benchmark instances for QKP. The results of this comprehensive computational study will be published as they become available.

8 Conclusions

We developed an optimization system to offer marketing managers an evidence-based suggestion for the media mix to be used for a given promotional campaign. It relies on a comparison of the current campaign to past campaigns based on their parameters and goals. The direct and pairwise effect of advertising media is computed in a fairly complicated estimation scheme by evaluating the performance of these media in past campaigns. A special memory factor takes into the account the time elapsed since these performance values were recorded.

Building an optimization model with these effect estimations gives rise to a Quadratic Knapsack Problem which can be solved almost to optimality in all real-world scenarios within a time limit of 3 s. The optimization tool is currently used within the industrial software solution MARMIND.

Future developments include a revision of some of the effect estimations by stochastic models as soon as a suitable set of test data derived from real world applications is available. Furthermore, based on classical tools of statistical analysis it should be possible to detect certain trends of advertising media increasing or decreasing in importance, or in their effect for certain goals or target groups.

References

1. Färe, R., Grosskopf, S., Seldon, B., Tremblay, V.: Advertising efficiency and the choice of media mix: a case of beer. Int. J. Ind. Organ. **22**, 503–522 (2004)
2. Reynar, A., Phillips, J., Heumann, S.: New technologies drive CPG media mix optimization. J. Advertising Res. **50**, 416–427 (2010)

3. Balachandran, V., Gensch, D.: Solving the "marketing mix" problem using geometric programming. Manage. Sci. **21**, 160–171 (1974)
4. Sorato, A., Viscolani, B.: Using several advertising media in a homogeneous market. Optim. Lett. **5**, 557–573 (2011)
5. Nobibon, F., Leus, R., Spieksma, F.: Optimization models for targeted offers in direct marketing: exact and heuristic algorithms. Eur. J. Oper. Res. **210**, 670–683 (2011)
6. Sönke, A.: Optimizable and implementable aggregate response modeling for marketing decision support. Int. J. Res. Mark. **29**, 111–122 (2012)
7. Vakratsas, D., Ambler, T.: How advertising works: what do we really know? J. Mark. **63**, 26–43 (1999)
8. Pergelova, A., Prior, D., Rialp, J.: Assessing advertising efficiency: does the internet play a role? J. Advertising **39**, 39–54 (2010)
9. Kellerer, H., Pferschy, U., Pisinger, D.: Knapsack Problems. Springer, Berlin (2004)
10. Pisinger, D.: The quadratic knapsack problem - a survey. Discrete Appl. Math. **155**, 623–648 (2007)
11. Everitt, B., Landau, S., Leese, M., Stahl, D.: Cluster Analysis, 5th edn. Wiley, New York (2011)
12. Guldemir, H., Sengur, A.: Comparison of clustering algorithms for analog modulation classification. Expert Syst. Appl. **30**, 642–649 (2006)
13. Tan, P.N., Steinbach, M., Kumar, V.: Introduction to Data Mining. Addison-Wesley, Boston (2006)
14. Sculley, D.: Rank aggregation for similar items. In: Proceedings of the 7th SIAM International Conference on Data Mining, pp. 587–592. SIAM (2007)
15. Kumar, R., Vassilvitskii, S.: Generalized distances between rankings. In: Proceedings of the 19th International World Wide Web Conference, pp. 571–580. ACM (2010)
16. Caprara, A., Pisinger, D., Toth, P.: Exact solution of the quadratic knapsack problem. INFORMS J. Comput. **11**, 125–137 (1999)
17. Billionnet, A., Soutif, É.: An exact method based on lagrangian decomposition for the 0–1 quadratic knapsack problem. Eur. J. Oper. Res. **157**, 565–575 (2004)
18. Pisinger, D., Rasmussen, A., Sandvik, R.: Solution of large quadratic knapsack problems through aggressive reduction. INFORMS J. Comput. **19**, 280–290 (2007)
19. Pferschy, U., Schauer, J.: Approximating the quadratic knapsack problem on special graph classes. In: Kaklamanis, C., Pruhs, K. (eds.) WAOA 2013. LNCS, vol. 8447, pp. 61–72. Springer, Heidelberg (2014)
20. Julstrom, B.: Greedy, genetic, and greedy genetic algorithms for the quadratic knapsack problem. In: GECCO 2005: Proceedings of the 2005 Conference on Genetic and Evolutionary Computation, pp. 607–614. ACM (2005)
21. Yang, Z., Wang, G., Chu, F.: An effective GRASP and tabu search for the 0–1 quadratic knapsack problem. Comput. Oper. Res. **40**, 1176–1185 (2013)

A Decomposition Approach to Solve Large-Scale Network Design Problems in Cylinder Gas Distribution

Tejinder Pal Singh[1(✉)], Nicoleta Neagu[2], Michele Quattrone[2], and Philippe Briet[3]

[1] Air Liquide, 12800 W. Little York Rd., Houston 77041, USA
tejinder.singh@airliquide.com
[2] Air Liquide, 1 chemin de la Porte des Loges, 78350 Les Loges-en-Josas, France
{nicoleta.neagu,michele.quattrone}@airliquide.com
[3] Air Liquide, 75 Quai d'Orsay, 75321 Paris, France
philippe.briet@airliquide.com

Abstract. The logistics network has to be optimally designed for an effective supply chain. The focus of this research is to solve network design problem occurring in packaged gases (e.g., cylinder) supply chain. The integrated logistics network design problem for packaged gases is defined as follows: given a set of potential locations for filling plants and hubs, and customers with deterministic demands, determine the configuration of the production-distribution system i.e., optimal facility locations, the filling plant production capacities, the inventory at plants and hubs, and the number of packages to be routed in primary and secondary transportation. The problem is modeled as a deterministic mixed integer program and a decomposition approach is developed which allows a natural split of the production and distribution decisions. The proposed framework is illustrated with numerical examples from real-life packaged gases supply chain. The results show that the decomposition approach is effective in solving a broad range of problem sizes. The results from the decomposition approach are benchmarked by solving optimally the whole packaged gases network design model for smaller test cases. In the end, we perform sensitivity analysis for parameters that are likely to change in the future for better understanding of their impact.

Keywords: Network design · Optimization · Location-routing · Packaged gases · Inventory management · Decomposition approach · Heuristics · Mixed integer linear programming model · Cylinder gas distribution

1 Introduction

Supply chain networks are essential within the world wide economic activities. They are fundamental to stay competitive in today's markets through efficient delivery of products (e.g., energy, food, pharmaceutics, and clothing). The optimal supply chain network design is the basis for its efficiency. Moreover, the network design is a complex topic as it needs to take into account and integrate many aspects of real life problems.

E. Pinson et al. (Eds.): ICORES 2014, CCIS 509, pp. 265–284, 2015.
DOI: 10.1007/978-3-319-17509-6_18

In this paper we consider the packaged gases supply chain with its specific characteristics. Network configuration in packaged gases (also referred as cylinder distribution) is a strategic decision that impacts the tactical delivery planning and daily scheduling and transportation operations. A typical cylinder supply chain network consists of filling plants, hubs/distribution centers, and customers. Filling plants supply cylinders to hubs which distribute them to customers. It is also possible for filling plants to directly supply the customers. Filling plants and hubs manage cylinder stocks in order to enable the supply chain to maintain an adequate service level. The agility of the supply chain and the operational efficiency are constrained by the structure of the network determining the flow of material.

Optimizing the network design problem for cylinder distribution consists of determining the locations for filling plants/hubs, the production tools to be installed at the plants, the primary and the secondary flows, as well as the inventory at plants and customer locations. The framework based on a mixed-integer linear programming (MILP) model is developed to capture a real-life packaged gases business model. The mathematical model contains constraints on network structure, primary transportation, i.e., flow of cylinders among different supplier locations, secondary transportation, i.e., flow of cylinders from supplier locations to customers, stock management and assets management. The proposed framework has been developed by leveraging the best practices and knowledge of logistics experts within packaged gases supply chain. Therefore, the framework can be used to determine a new cylinder supply chain/logistics network for a new market or to study the impact of change in different elements of the supply chain, e.g., when new customer accounts are opened or old accounts are closed, change in customer demand, impact of new filling centers/hubs and assets like filling tools and manpower. We show the efficiency of the proposed framework for real-life test cases provided by the packaged gases supply chain managers.

The paper is organized as follows. Section 2 outline the state of the art related to integrated supply chain decision models. In Sect. 3, we provide the problem description and in Sect. 4 we represent the mathematical model with the objective function and the business constraints. In Sect. 5, we discuss the solution approaches used to solve the integrated model. Section 6 presents the obtained results and Sect. 7 concludes the paper with possible future research directions.

2 Literature Review

The network design problem in packaged gases consists of three main sub-problems:

- Facility location: It involves the improvement of the existing network and the determination of the best configuration.
- Inventory management: It consists of determining the best inventory levels at hubs/plants.
- Routing: Optimization of flows determining the optimal flows of goods through the network.

The network design problems are complex as they involve strategic decisions which influence tactical and operational decisions [3]. The strategic decisions are mainly related to facility locations, their capacities and what products need to be produced at each plant, etc. The tactical decisions are related to inventory management and manpower, and depend on the strategic decisions whereas operational decisions like routing are directly related to tactical and strategic decisions made earlier. In other words, it means that if facility location decisions are sub-optimal, even if production, inventory and distribution plans are fully optimized, the supply chain may still be operating inefficiently. Therefore, for determining the best network configuration, all the costs at the three levels need to be taken into account to optimize the system-wide production, inventory and distribution costs. One of the challenges in the network configuration is that customer demands and cost parameters may change over time and it is very hard to change the facility location decisions once a supply chain network is configured. Thus, it is critical to design a supply chain network that is optimal and is not sensitive to changes in the operational parameters. The integrated network design problem has been usually solved by considering the integration of two sub-problems while approximating the third one. We provide following few approaches in the literature for solving the integrated network design problem.

The facility location problem integrated with routing is proved to be NP-complete by [8]. The objective function and the constraints of the models they propose are linear. The reader is referred to the reviews provided by [7, 10]. The facility location problem and its variants have been widely researched on theoretical models but the problem is rarely approached from a supply chain management and real-life perspective [9].

Most of the papers in the literature study the integration of two of the above three important decisions: location-routing models (LR), inventory-routing models, and location-inventory (LI) models. For reviews on location-routing models, readers can refer to [2]. In LR models, both the location problem and the vehicle routing problem (VRP) are typically NP-hard, which makes the integrated model even more complex. In this paper, the VRP problem is solved by approximating the routes based on either a heuristic approach or historical data. The resulting routing costs are then fed as an input to the location model. For inventory-routing models, please refer to [1, 6]. LI models also study the location, inventory and distribution coordination issues by either ignoring the inventory costs or approximating the non-linear costs with linear functions. In this paper, inventory costs are considered but assumed to be linear similar to some papers that consider inventory costs. Refer to the papers [4, 5, 11], for a better understanding of LI models.

The case that motivated this research deals with the network design for packaged gases distribution. The problem addressed in the current paper combines some elements of LR and LI models to determine an optimal network design by minimizing the sum of the production costs, the transportation costs and the inventory costs. Our mathematical model can be classified as a deterministic single-period MILP model with multiple products applied to a three-level network. The main contribution of our work is that it integrates supply chain network design decisions without fixing the fillings plant locations with inventory and resource allocation decisions required at the plants. We also consider the transportation costs for the entire supply chain including the transshipment costs among different facilities by deciding the replenishment frequency.

3 Problem Description

We address the network design problem specific to packaged gases supply chain occurring in real-life. The problem consists of determining the number and the location of the production plants and the distribution centers, the allocation of customer demands to distribution centers, and the allocation of distribution centers to production plants. The main goal is to identify the optimal configuration for producing and delivering packaged gas products to customers at the lowest cost while satisfying the network constraints.

More specifically, this network design problem aims in helping the decision making on locations for building plants, the production tools to be installed at the filling plants, the primary and the secondary flows, as well as the inventory at plants and hubs. A diagram of the packaged gases distribution network is shown in the Fig. 1. The nodes of the network are classified in four categories: filling plants, hubs, agent distributors, and end users (or reseller). Each location has a certain inventory capacity to satisfy customer demand. Customers manage their own inventories by placing orders at the right time. Therefore, in the current problem we consider the inventory decisions only at filling plants and hub locations.

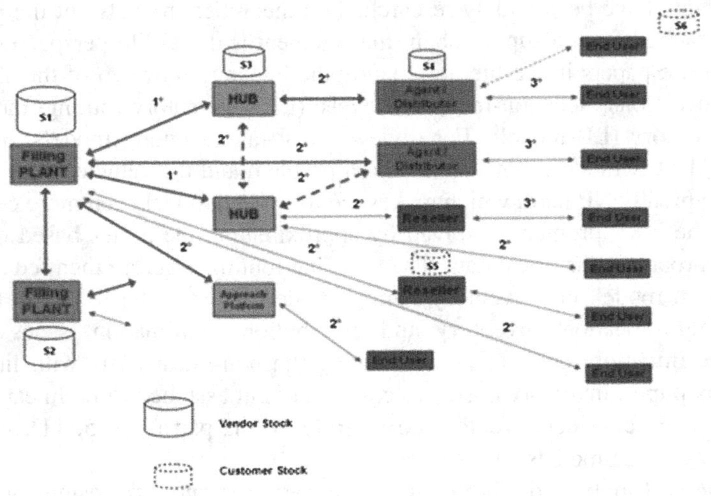

Fig. 1. Diagram of the distribution network.

The arrows in Fig. 1 represent the transport of packaged gases which is classified as:

- Primary transport which occurs between filling plants and hub locations.
- Secondary transport which represents the transport between hub/filling plants and client/agent/reseller locations.
- Tertiary transport which happens between agent/reseller and client locations.

This paper will not handle the whole distribution network but rather will focus on the primary and secondary transport. Agents/distributors, resellers and end users will all be called customers without distinction in the rest of the paper. Since tertiary flows happen between customers of different types, they are not considered in this problem. The word "plant" by itself is referring to both hubs and filling plants. In this paper we also assume that the vendor who supplies cylinders to the customers owns the whole packaged gases supply chain network. Therefore, we do not consider any ordering costs between different plants. We do consider the transportation cost of transshipments among different plants which is a function of replenishment periods for the primary flows. The main goals of the proposed methodology are related to the location decisions of plants, production, and to the hubs/filling plants transport and inventory. The primary goals of the network design problem for packaged gases are as follows, see also Fig. 2:

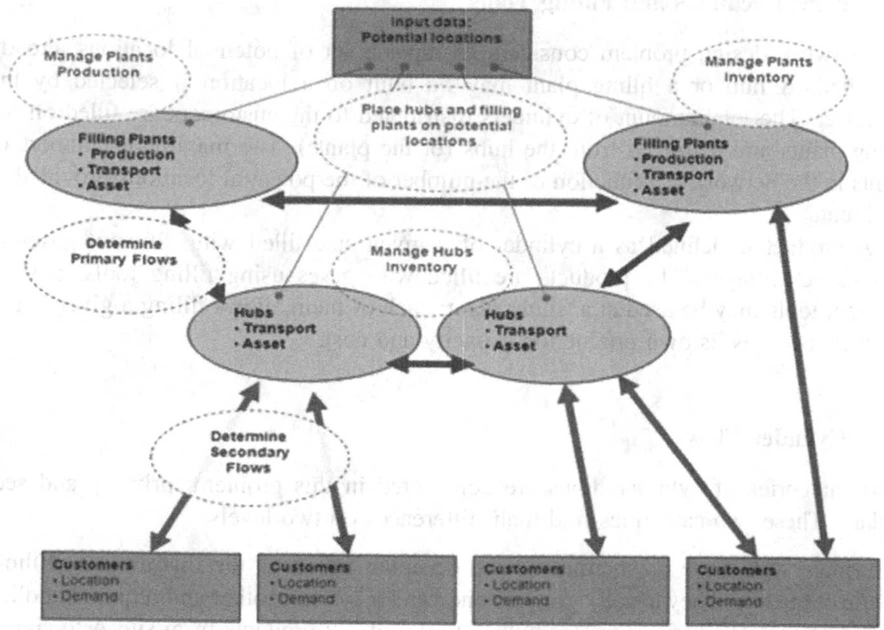

Fig. 2. Network design problem.

- Determine the number and the locations of the hubs and the filling plants.
- Determine the production of different products at the filling plants.
- Determine the primary and the secondary transportation cylinder flows, i.e. the customer-plant allocation decisions.
- Determine the inventory levels at plants consisting of working stock & the safety stocks at the plants.

These four issues are fundamental in the structure of a supply chain. Nevertheless, these issues are interrelated by the cylinders flows and it is clear that it would be a source of improvement to treat them all at the same time. A general description of all the key aspects of the problem as well as the hypotheses assumed at this stage is presented in the following sub-sections.

3.1 Multi-products Network

We assume that different products may be considered while designing the distribution network. This means that different products are filled at a plant, and transported to others plants and customers. Therefore, each product has to be characterized depending on its package, its composition and the filling tools that are compatible with it.

3.2 Plant Locations and Filling Tools

The network design problem considers as input a set of potential locations already identified. A hub or a filling plant may be built on a location if selected by the optimizer. The total amount of cylinders distributed to the customers are filled on the filling plants and delivered from the hubs (or the plants). The maximum numbers of plants in the network is a function of the number of the potential locations provided in the input.

A product is defined as a cylinder of a given size filled with different gases in various percentages. The products are filled with gases using filling tools. Several different tools may be used at a filling plant; each of them allows filling a given set of products and has its own production capacity and cost.

3.3 Cylinder Flows

Two categories of cylinder flows are considered in this problem: primary and secondary. These two categories find their differences on two levels:

- Primary flows are an internal choice of optimization of the distribution without direct income. They usually go from one vendor site to another and require handling during the round trip: a tractor pulls a trailer of full products from site A to site B, leaves the trailer on site B and takes back a trailer of empty products from site B to site A.
- Secondary flows are a direct source of income for the vendor as customers have to pay for the delivered cylinders. A secondary round trip is usually composed with several drops on different customers' sites where full products have to be unloaded before empty products are loaded on the trailer.

The transport cost is usually composed of a fixed cost and a variable cost. The fixed cost consists of truck costs, driver costs, and extra fees and the variable cost is dependent on the distance and the duration of the round trips. However, we model the cost of a round trip with an average variable cost per driver distance taking every cost into consideration. The handling cost of the cylinders on plants is taken into consideration independently. In this paper, we are approximating the routing costs to serve each customer and therefore, do not consider the routing decisions in the model.

3.4 Primary Transport Cost

The primary transport cost is quite straightforward to estimate. As primary trips are defined as full trailer load deliveries in the model, the cost of primary round trips between two identified plants can be known before solving the network design.

3.5 Secondary Transport Cost

Contrary to the primary transport, secondary transport cost is difficult to estimate precisely. In the network design model, we do not consider day-to-day demand data which implies that it is impossible to build actual secondary round trips. In the model, each customer has a global demand over a year and it is not possible to know which customers will order on the same day. Also, secondary unit transport costs are provided as an input to the problem and therefore, the actual costs cannot be calculated until the plant-customer allocation decisions are made. Therefore, we determine the cost to deliver a unit cylinder to a customer from each potential location by modeling the average round trips during which the customer will be delivered. The round trip does not consist of a single delivery but multiple deliveries and this makes the secondary transport cost approximation realistic. For a given customer, an average round trip starting from a given plant is modeled by:

- The driven route split in one dispersion ring which represents the zone where the delivered customers are located and an approach distance to go from the plant to the dispersion ring.
- The average number of customers visited during such a round trip.
- The average number of cylinders delivered during such a round trip.

The radius of the dispersion ring (see Fig. 3) is set for every customer to a value determined experimentally from real round trips or from round trips generated from a heuristic approach used during the pre-processing of data. The heuristic approach used is not discussed in this paper. The dispersion ring has its centre on the customer under consideration. The approach distance is the shortest distance from the plant to the dispersion ring. Another value found experimentally determines the percentage of customers included in the same dispersion ring which can be delivered in the same

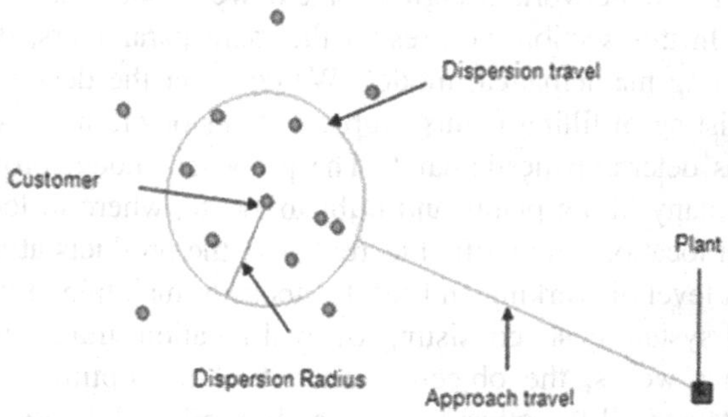

Fig. 3. Secondary round trip model.

round trip. This coefficient aims at correcting the fact that one dispersion ring could withhold several round trips. The secondary round trips have to respect the following constraints:

- The average number of cylinders delivered during the round trip cannot exceed the capacity of the trailer used.
- The duration of the round trip cannot overcome the maximum driver work time. The round trip duration is mainly a function of the number of customers visited during the trip.

3.6 Trucks

Only two standard trucks are considered in the problem. One truck type is dedicated to primary transport and the other one to the secondary transport. Each type of truck is characterized by its capacity, speed and cost per distance travelled. We assume no limit on the number of trucks of each type that are available for distribution in the model.

3.7 Inventory Management

It is important for the cost evaluation of a plant to determine the investment cost necessary for the stocks on its site. The required stock at a plant is composed of:

- Replenishment stock which includes the products filled on the plant everyday and the products delivered from other plants at each primary round trip.
- Delivery stock which represents all the products which are being delivered to customers and other plants every day. When calculating the size of this stock, we assume that the same number of products is delivered every day for this plant.

The stocks take into consideration the variation of demands over a year through a variance of the cylinder flows. The variance of the flows is supposed to be directly proportional to the average volume delivered per day.

4 Mathematical Model

To solve the integrated network design problem we propose a mixed integer programming model. In this section we present the main parameters, decision variables and the corresponding mathematical model. We consider the design of a three-tiered supply chain consisting of filling plants, hubs, and customers as described in Sect. 3. Each customer has deterministic demand. The proposed model provides the needed decisions on how many filling plants and hubs to locate, where to locate them among the list of potential locations, how often to replenish the products at the hubs from the filling plants, what level of working and safety stocks to maintain at the plants, so as to minimize the total system costs consisting of total location, transportation and inventory costs. In other words, the objective is to find the optimal trade-off between transportation costs and all the other costs, mainly the location costs. Inventory costs are a function of the replenishment periods and the demand allocation to the plants.

To simplify the mathematical model, we define two units of measure. We define Equivalent Cylinder (EqCyl) as the unit of area occupied by a 50-litre water capacity compressed gas cylinder for transportation on a truck. As the model deals with more than one product, and a truck is allowed to transport many products together, to quantify the capacity of the trucks and also to define the demands of different customers, EqCyl would be used. We also introduce a measure of time called Work Unit (WU). A WU is a unit of time to a physical activity for which time is the main factor to represent work e.g., filling and handling of cylinders. All parameters and variables that denote time are expressed in terms of WU.

Inputs and Parameters

I: Set of customers
J: Set of potential locations
P: Set of products
T: Set of filling tools
R: Set of replenishment periods between plants
f_j: Fixed cost (yearly) of locating a filling plant at location j, for each $j \in J$
g_j: Fixed cost (yearly) of locating a hub at location j, for each $j \in J$
h_j: Fixed inventory holding cost per EqCyl per year at location j, for each $j \in J$
χ: Fixed cost of a full time employee per year
α: Primary handling productivity at any plant (WU/year/employee)
β: Secondary handling productivity at any plant (WU/year/employee)
π_p: Work time (in WU) needed to handle one package of product p at a plant for primary transport, for each $p \in P$
θ_p: Work time (in WU) needed to handle one package of product p at a plant for secondary transport, for each $p \in P$
w_{pt}: Work time (in WU) necessary to fill one package of product p using tool t, for each $p \in P$ and $t \in T$
a_p: Area (in EqCyl) occupied by one package of product p, for each $p \in P$
m_t: Filling productivity (WU/employee/year) of a filling tool t, for each $t \in T$
b_t: Fixed cost of using a tool t per year, for each tool $t \in T$
z_t: Maximum time (in WU) available to fill packages with tool t per year, for each tool $t \in T$
s_{pt}: Binary parameter, 1 if a filling tool t can fill a package of product p, for each $p \in P$ and $t \in T$, 0 otherwise
μ_{ip}: Average number of packages consumed (yearly) at customer i for product p, for each $i \in I$ and $p \in P$
σ_p: Variance of demand (yearly) for product p, for each $p \in P$
η: Constant representing number of working days per year (e.g. 250)
τ: Truck capacity for primary transportation
c_r: Average cost per distance travelled during primary transport for a replenishment period r, for each $r \in R$
λ_{ji}: Average cost per EqCyl from location j to serve customer i, for each $j \in J$ and $i \in I$
M: Maximum number of tools at any filling plant

Decision Variables

p_j: Binary variable, 1 if a filling plant is build on location j, for each $j \in J$, 0 otherwise
q_j: Binary variable, 1 if a hub is build on location j, for each $j \in J$, 0 otherwise
e_j: Total number of employees working on location j, for each $j \in J$
x_{jpt}: Number of packages of product p filled per year at location j by tool t, for each $j \in J$, $p \in P$, and $t \in T$, a discrete variable
Ψ_{jt}: Number of filling tools of type t required at the location j, for each $j \in J$, and $t \in T$, a discrete variable
u_{jkr}: Binary variable, 1 if primary trips are used between locations j & k after replenish period r such that $j \neq k$, for each $j \in J$, $k \in J$ and $r \in R$, 0 otherwise
v_{jkpr}: Number of EqCyl of product p delivered from location j to location k during primary trips undergone every replenish period r such that $j \neq k$, for each $j \in J$, $k \in J$, $p \in P$, and $r \in R$
Φ_{ji}: Binary variable, 1 if customer i can be delivered products from location j, for each $j \in J$, $i \in I$, 0 otherwise
Ω_{jip}: Number of EqCyl of product p delivered from location j to customer i during secondary trips, for each $j \in J$, $i \in I$, $p \in P$
ω_{jp}: Stock at location j of packages of product p, for each $j \in J$, $p \in P$

The objective function is composed of five main parts as shown below:

- Fixed costs of hubs and filling plants.
- Fixed costs of filling tools.
- Manpower cost dedicated to filling and handling packages.
- Total inventory cost.
- Transport cost, excluding the handling cost at the filling plants and hubs.

Location costs are strategic costs that are incurred when configuring the network. The first two terms in the objective function ensure that fixed costs for either a hub or a filling plant are applied to each selected location. Filling tool costs, manpower costs, and inventory costs are the costs associated with the tactical decisions whereas transport costs are the operational costs. The mathematical formulation of the objective function is given below in Eq. (1).

$$\text{Minimize} \sum_{j\in J}\left((q_j - p_j)g_j + p_j f_j + \chi e_j + \sum_{\substack{k \in J\\ r \in R}} \frac{\eta}{r} u_{jkr} c_r + \sum_{\substack{i \in I\\ p \in P}} \lambda_{ji} \Phi_{jip} + \sum_{p\in P} a_p \omega_{jp} h_j + \sum_{t\in T} b_t \Psi_{jt}\right) \tag{1}$$

The business constraints which are related to the network structure and flow, primary and secondary transport, and inventory management are given below, (2)–(11):

$$p_j \le q_j, \forall j \in J \tag{2}$$

$$e_j \leq \sum_{\substack{p \in P \\ t \in T}} \frac{w_{pt} x_{jpt}}{m_t} + \sum_{k \in J} \sum_{\substack{p \in P \\ r \in R}} \frac{\alpha \pi_p}{a_p} \left(v_{jkpr} + v_{kjpr} \right) + \sum_{\substack{i \in I \\ p \in P}} \frac{\beta \theta_p}{a_p} \Omega_{jip}, \forall j \in J \tag{3}$$

$$\sum_{p \in P} r \frac{\left(v_{jkpr} + v_{kjpr} \right)}{\eta} \leq \tau \, \mathrm{u_{jkr}}, \forall j \in J, k \in J, r \in R \tag{4}$$

$$\sum_{r \in R} u_{jjr} \leq 0, \forall j \in J \tag{5}$$

$$\sum_{rR} u_{jkr} \leq 1, \forall j \in J, k \in J \tag{6}$$

$$\sum_{p \in P} \Omega_{jip} \leq \Phi_{ji} \sum_{p \in P} a_p \mu_{ip}, \forall j \in J, i \in I \tag{7}$$

$$\sum_{t \in T} \Psi_{jt} \leq p_j \mathrm{M}, \forall j \in J \tag{8}$$

$$\sum_{j \in J} w_{pt} x_{jpt} \leq z_t s_{pt} \sum_{j \in J} \Psi_{jet}, \forall p \in P, t \in T \tag{9}$$

$$\sum_{p \in P} w_{pt} x_{jpt} \leq z_t \Psi_{jt}, \forall j \in J, t \in T \tag{10}$$

$$\omega_{jp} = \sum_{t \in T} \frac{x_{jpt}}{\eta} (1 + \sigma_p) + \frac{1 + \sigma_p}{\eta \, a_p} \left(\sum_{\substack{k \in J \\ r \in R}} \frac{v_{kjpr}}{\eta \, a_p} \max(1, r) + \sum_{\substack{k \in J \\ r \in R}} v_{jkpr} + \sum_{i \in I} \Omega_{jip} \right), \forall j \in J, p \in P \tag{11}$$

5 Solution Approach

The mathematical formulation of network design is a MILP problem. As the traditional facility location problem is NP-complete (Krarup & Pruzan, 1983), we simplify the model by approximating the routing costs. Moreover, in this paper we are dealing with a real-life large-scale problem occurring in packaged gases supply chain. Therefore, we analyzed various solving techniques: from near-optimal methods up to approximate ones. The near-optimal approach can be used for small problem instances whereas approximate methods can be applied in the context of large-scale problems. Moreover, we can compare the near-optimal solutions to the approximate ones to benchmark the approximate solutions. In this paper, we provide details about the approximate approaches in order to achieve a reasonable computation time.

5.1 Mono-Product Approximation

As the number of products occurring in the packaged gases network design problem implies high complexity, the first approximate approach considered consists of grouping the multi-products into a single product which we call a mono-product problem. To that aim, each product is treated relatively to its volume of equivalent cylinder (EqCyl) and its type is ignored. Converting multiple product constraints into single product constraints may cause solution infeasibility; the constraints are modified carefully to minimize the likelihood of such infeasibility. As the modified model becomes a single-product model, variables are no longer depending on the number of products available. For example, consider the variable xjpt representing the number of packages of product p filled per year at location j by tool t, for each $j \in J$, $p \in P$, and $t \in T$. In the mono-product approximation, xjpt is changed to xjt defining the number of EqCyl of the single product filled per year by tool t at location j. Similarly, constraints (3, 4, 7, 9, 10), and (11) are modified along with the objective function to represent a single product network design problem.

The resulting problem is also a MILP problem but we do not show the modified model in this paper. The network design problem becomes a unique flow problem and thus, it is easier to solve. We compare the results of this approach with the results obtained by solving the complete model in Sect. 6. It is shown that this approach gives good solutions especially for placing the hubs' locations and satisfactory results for secondary transport decisions. As this approach does not treat different types of products, the number of filling tools installed at the filling plants is underestimated compared to the optimal solutions. This approach can be used when the problem size is very large and the main interest is to find the network configuration i.e. location of hubs and allocation of customers to hubs whereas resource/inventory optimization can be done separately. Figure 4 shows the physical representation of the mono-product approach.

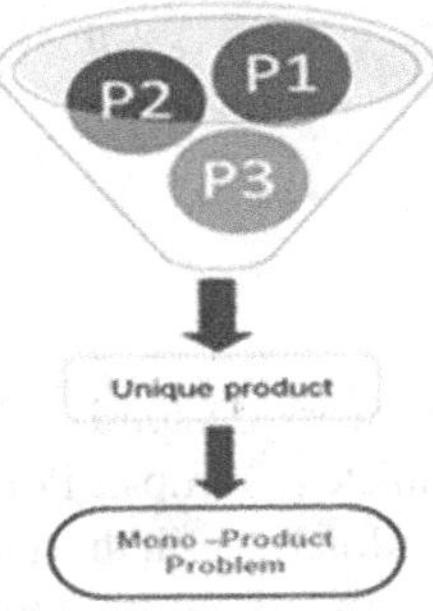

Fig. 4. Scheme of mono-product approximation.

5.2 Two-Steps Decomposition

In order to reduce the computation time, a typical approach for large-scale problems is based on problem decomposition. We consider a two-step decomposition approach to generate an approximate solution. In the first step of the decomposition approach, the hubs' locations and the hub-customer allocation decisions are determined by solving the mono-product flow problem with minimization of the secondary transportation costs and the hub costs. Secondary transportation cost is more a function of number of cylinders transported between hubs and customers and independent of different products. Therefore it is a safe approximation to determine hubs through optimization of mono-product flow problem. In the second-step, the residual problem is solved based on the multi-product model. The second step optimization determines if the hub built on a given location is a filling plant or not and decides the tools associated with this given filling plant by minimizing the production and the primary transportation costs (tools, sourcing, manpower). Moreover, it optimizes the inventory management by defining the frequency of trips between plants and the flow quantities for the primary transportation. Figure 5 shows the two-step decomposition decisions graphically.

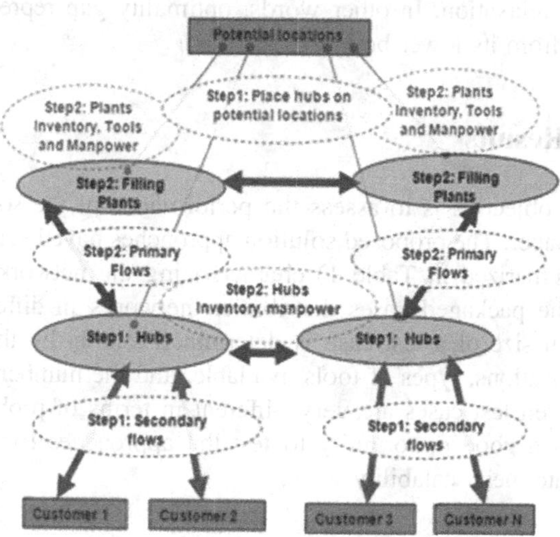

Fig. 5. Two-step decomposition decisions.

The size of the residual problem in the second step can be further reduced by grouping the products into families of products. The product families are created by selecting the products among the most requested customer's products. Thus, the whole set of products is aggregated into families of products. A family essentially is a set of products that can be produced by the same tools. This further reduces the problem size and helps to obtain good results in a reasonable time compared to the complete problem. The grouping does not change the model as it is done in the input data. The second-step model is also an MILP problem and is still hard to be solved to optimality for large-scale network problems. One of the reasons of the complexity to solve the second-step model optimally is that a significant number of binary variables still remain to be optimized for primary transportation.

The MILP mathematical models in our testing are solved on a 2.66 GHz, 16 GB RAM server using CPLEX®. By tuning the CPLEX parameters, the performance of the CPLEX has been improved on the test cases used. The upper limit on the number of cutting plane passes CPLEX performs when solving the root node of a MILP model is set to 1. The number of rows in the problem with cuts added is set to 30 times the original number of rows. Relaxation Induced Neighborhood Search (RINS) heuristic explores the neighborhood of the current incumbent solution to try to find a new, improved incumbent after every 70 nodes are visited. It is also important to manage the memory problems that occur on a server when solving a large-scale problem. Therefore, the number of stored solutions kept in the solution pool on the server is set to 10. If the node file parameter in CPLEX is set to 0, when the tree memory limit is reached, optimization is terminated. By setting the node file parameter to 3, the node files are transferred to disk in compressed form and CPLEX actively manages which nodes remain in memory for processing. An optimality gap has been used for test cases as it is hard to solve the test cases for full optimization. The optimality gap represents the maximum ratio between the optimal solution of the MILP program and its Linear Programming (LP) relaxation. In other words, optimality gap represents how far the current solution is from its lower bound.

6 Numerical Results

In this section our objective is to assess the performance of the solution approaches considered in this paper. The proposed solution approaches have been applied to 3 real-life test cases (summarized in Table 1) characterizing the network design problems occurring within the packaged gases distribution networks in different geographical zones. The problem size of a test case is determined mainly by the number of customers, potential locations, types of tools available, and the number of products to be distributed. The given test cases are very different in terms of problem structure and size. This provides a good opportunity to test the approaches for different problem settings and evaluate their scalability.

Table 1. Real-life network design test cases

Instance	Customers	Potential locations	Tools	Products
1	520	4	3	5
2	1,964	14	6	4
3	12,036	26	22	43

Each test case in Table 1 has been solved by different approaches providing near-optimal and approximate solutions. It has been observed for the small test cases (e.g., containing up to 2000 customers), a near-optimal solution can be reached in a reasonable computation time (e.g., 60 min for test case 2).

For test cases 1 and 2, the facility location decisions i.e. the number and the set of locations to be opened as filling plants/hubs from the mono-product and the two-steps approaches are exactly the same as from the near-optimal solution. This shows that both the approximate approaches are successful in determining optimal strategic decisions. In the test results, mono-product approach underestimates the total network cost which is expected due to the simplification of the model. Therefore, we do not consider mono-approach for tactical and operational decisions as it solves an approximate model. For test case 1, two-steps approach provides the same network cost as from the near-optimal method with the same optimality gap. For test case 2, the two-steps approach provides a solution with 1.19 % higher network cost than the near-optimal solution. For test case 3, the near-optimal solution was not generated as we could not solve the complete problem within an acceptable optimality gap. For real-life network design, we believe that the computation time in a few hours is acceptable due to the fact that the opportunities to setup a new supply chain network or completing overhauling an existing one are not very frequent. The computation time for test cases 1 and 2 with the two-steps is relatively small but test case 3 takes more than 15 hours to obtain a solution within the optimality gap of 0.43 %. It is also possible to achieve a solution in few hours by increasing the optimality gap to 5 % as shown in the Table 2.

Table 2. Performance test results.

Case	Gap (%)	Solution approach	Solver time (min)	Total cost (€)	Filling plants	Hubs
1	0.5	Mono-product	2	561,891	2	3
		Two-steps	6	782,070	2	3
		Near-optimal	4	782,070	2	3
2	0.5	Mono-product	11	2,147,515	6	11
		Two-steps	593	2,276,267	6	12
		Near-optimal	63	2,303,703	6	12
3	0.5	Mono-product	121	19,289,480	11	18
	5.0	Two-steps	332	24,125,572	11	18
	0.43	Two-steps	948	23,160,386	11	18

For test case 3, the benchmarking of two-steps decomposition solution is done by comparing the facility location decisions with a manual solution based on the network designer's experience. The gap analysis with the manual solution shows that the two-steps solution for test case 3 provides a solution with total network cost which is 6.4 % less than the manual solution.

Figure 6 shows the different cost components as percentages of the total network design cost for test case 3 with optimality gap of 0.43 %. It also shows that the facility location costs and the secondary transportation costs are the highest cost components of the total network cost and therefore, have more influence on the network design decisions. Since the network design model studied in this paper is deterministic, we also perform sensitivity analysis to check the impact of different parameters on the facility location and other network decisions. The parameters which are more likely to change over time are demand, unit transportation cost, and manpower cost.

Figure 7 shows the location decisions of scenarios obtained by changing the demand at each customer equally for test case 2 which is solved with near-optimal approach. The results show that the facility location decisions i.e. the number and the set of opened facilities do not change even with more than 5 % increase or decrease in

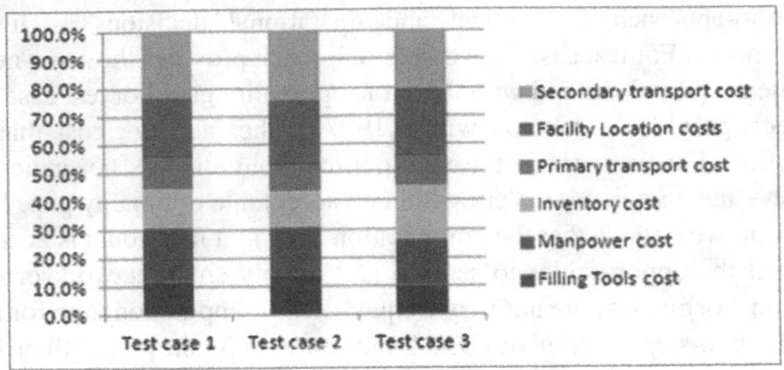

Fig. 6. Network design costs for 3 test cases.

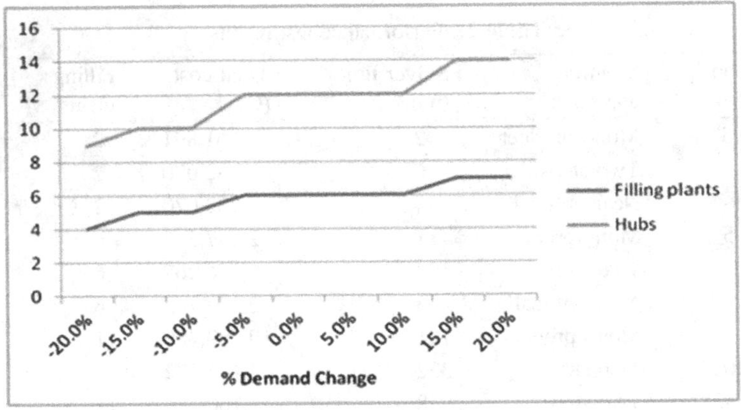

Fig. 7. Demand sensitivity analysis.

demand at each customer location. The main reason for such stability is that we increased the product demand equally for all the customers to perform sensitivity analysis. In reality, the demand of different customers does not homogenously increase or decrease over time. Also, the impact of demand change on inventory and transportation costs (both primary and secondary) is more compared to the other costs. When facility location decisions don't change with modified demand, the change in inventory and transportation costs is nearly linear with demand change. Table 3 provides similar sensitivity analysis results for unit transportation cost and manpower cost for test case 2. Manpower is mainly a function of demand and therefore, does not influence facility location decisions significantly as evident from the results.

For test case 2, manpower costs have to increase or decrease at least 20 % to make a change in the facility location decisions. Since the primary and the secondary transportation costs represent a significant portion of network design costs, the facility location decisions are sensitive to the unit transportation cost. An increase in the unit transportation cost causes more hubs to open along with filling plants which is expected to minimize the secondary transportation costs. It is also quite easy to introduce

Table 3. Cost sensitivity analysis.

	Standard	Potential Value
Primary truck capacity	260 EqCyl	240 EqCyl
Cylinder cost	30	25
Primary transport cost	1.2	1.38
Primary handling Productivity	25,000	22,000
Payback period of type-F tools	3 years	2 years

constraints to fix strategic decisions and just optimize for tactical decisions. We perform such an analysis for test case 3 considering the decision to install new filling tools called Type-F tools at already open filling plants. We also perform sensitivity analysis to check whether the decision to install Type-F tools changes when some parameters are modified. Table 4 shows 5 parameters with standard values and possible potential values.

Table 4. Parameter values.

	Unit transportation cost		Manpower cost	
% Change	Filling plants	Hubs	Filling plants	Hubs
–20.0 %	4	10	7	13
–15.0 %	4	11	6	12
–10.0 %	5	11	6	12
–5.0 %	6	12	6	12
0.0 %	6	12	6	12
5.0 %	6	13	6	12
10.0 %	7	14	6	12
15.0 %	7	14	6	12
20.0 %	7	14	5	11

In the original test case 3, we had 22 filling tools and we introduce 6 new Type-F tools in the problem and then re-solve using Two-Steps approach with standard parameters. Figure 8 shows the allocation of 6 Type-F tools in the new optimal solution which has a lower cost compared to the solution with no Type-F tools. Note that we had 11 fillings plants in the optimal solution of original test case 3.

To validate the robustness of the above solution, we perform sensitivity analysis for 5 parameters in Table 4 with their potential values. When payback period is changed to 2 years, we get a completely different solution as shown in Fig. 9. The changed solution is due to the reason that with the decrease in payback period, the annual savings have to be more to justify the investment in the new tools. The results show that the decision to invest into Type-F tools is sensitive to all 5 parameters which are summarized in Table 5. In the first scenario, the reduction in the truck capacity causes a loss in the primary transport efficiency: in order to deliver the same quantity of products more transport is needed. In that way, primary transport cost increases and to compensate for this, the solution favors the local production. When cylinder cost decreases, the overall

TOOLS	Location 1	Location 2	Location 3	Location 4	Location 5	Location 6	Location 7	Location 8	Location 9	Location 10	Location 11
Tool 1						X					
Tool 2											
Tool 3		X									
Tool 4											
Tool 5											
Tool 6											

Fig. 8. Type-F filling tools allocation.

TOOLS	Location 1	Location 2	Location 3	Location 4	Location 5	Location 6	Location 7	Location 8	Location 9	Location 10	Location 11
Tool 1						X			X	X	X
Tool 2						X					
Tool 3		X									
Tool 4						X					
Tool 5											
Tool 6											

Fig. 9. Solution with 2 years of payback period.

Table 5. Type-F filling tools sensitivity analysis.

	Loc. 1	Loc. 2	Loc. 6	Loc. 9	Loc. 10	Loc. 11	Total cost
Standard parameters	X	X	X	X	X	X	23,070,000
Primary truck Capacity = 240		X	X	X	X	X	23,230,000
Cylinder cost = 25		X					22,980,340
Primary transport cost = 1.38	X	X	X	X	X	X	23,560,000
Primary WU = 22,000		X	X	X	X		23,380,000
Payback= 2 years		X	X				23,100,000

cost including production, stock, handling and transport are reduced. Therefore, the solver is comparing the profitability of installing tools with the depreciation cost of tools. According to the results provided, the primary transport is favored instead of production, hence the small number of tools added. In the third scenario, the production is reduced and thus, the primary handling productivity is decreasing and more work has to be done to satisfy the same production demand. The result recommends adding several filling tools in order to reduce the primary transport and thus, the primary handling activities. When fuel price increases transport cost rises as well so in order to avoid transport cost, the solver takes advantage of installing many tools and develops local production. By comparing the results on all scenarios, and analyzing where and which production tools are frequently installed by solver, a final recommendation on new assets investment is made. It consists of recommending at which locations more filling tools should be added in order to reduce the total costs.

7 Conclusions

In practice supply chain network configuration typically involves optimizing strategic decisions without considering their impact on all the tactical delivery planning and daily scheduling decisions. In this paper we optimize not only strategic decisions but also consider all tactical and operational decisions in the mathematical model for the network configuration. We specifically consider the integrated network design problem dedicated to the packaged gases distribution. The main goals for solving the integrated network design problem include determining the locations of the hubs and the filling plants, the production capacity of the filling plants, the primary and the secondary cylinders flows and the inventory of both the filling plants and the hubs. To solve it, we propose a mathematical model which combines both the location-routing and the location-inventory integrated models and approximates the routing cost used in both the integrated models. In order to solve real large-scale problems, we propose approximate decomposition based approach. We applied near-optimal and approximate approaches on 3 real-life test cases from packaged gases cylinder distribution. The obtained solutions are within an acceptable optimality gap from the optimal solutions. The results indicate that mono-approach and two-steps approaches are capable to generate good facility location solutions in a reasonable time and are comparable to near-optimal solutions on smaller test cases. The difference between mono-product and two-steps is that two-steps method provides a better estimate of tactical and operational costs. For large-scale test cases, it is hard to obtain near-optimal solutions whereas two-steps approximation can generate good solutions in an acceptable time. Therefore, near-optimal approach is suitable for smaller test cases and approximation approaches for large-scale test cases.

In the future, further studies on improving the computation time to solve the complete model without using decomposition approach can be envisioned. Also, further analysis is needed to better benchmark the approximate approaches considered in this paper for large-scale test cases. Even though we performed sensitivity analysis for few input parameters, future work also needs to be focused on developing and solving a robust model in designing a resilient packaged gases network.

References

1. Adelman, D.: A Price-Directed Approach to Stochastic Inventory/Routing. Working Paper. University of Chicago, Chicago (2003)
2. Balakrishnan, A., Ward, J.E., Wong, R.T.: Integrated facility location and vehicle routing models: recent work and future prospects. Am. J. Math. Manage. Sci. **7**, 35–61 (1987)
3. Crainic, T.G., Laporte, G.: Planning models for freight transportation. Eur. J. Oper. Res. **97**, 409–438 (1997)
4. Daskin, M.S., Owen, S.H.: Location models in transportation. In: Hall, R. (ed.) Handbook of Transportation Science, pp. 311–360. (1999)
5. Erlebacher, S.J., Meller, R.D.: The interaction of location and inventory in designing distribution systems. IIE Trans. **32**, 155–166 (2000)
6. Kleywegt, A., Nori, V.S., Savelsbergh, M.W.P.: The stochastic inventory routing problem with direct deliveries. Transp. Sci. **36**, 94–118 (2002)
7. Klose, A., Drexl, A.: Facility location models for distribution system design. Eur. J. Oper. Res. **162**, 4–29 (2005)
8. Krarup, J., Pruzan, P.M.: The simple plant location problem-survey and synthesis. Eur. J. Oper. Res. **2**, 36–81 (1983)
9. Melo, M., Nickel, S., Saldanha-da-Gama, F.: Facility location and supply chain management—a review. Eur. J. Oper. Res. **196**(2), 401–412 (2009)
10. ReVelle, C., Eiselt, H.: Location analysis: a synthesis and survey. Eur. J. Oper. Res. **165**, 1–19 (2005)
11. Shen, Z.M.: Approximation Algorithms for Various Supply Chain Problems, Ph.D. thesis. Department of Industrial Engineering and Management Sciences, Northwestern University, Northwestern (2000)

Elaboration of General Lower Bounds for the Total Completion Time in Flowshop Scheduling Problems through MaxPlus Approach

Nhat Vinh Vo(✉), Pauline Fouillet, and Christophe Lenté

Université François-Rabelais de Tours, CNRS, LI EA 6300, OC ERL CNRS 6305, Tours, France
{nhat.vo,christophe.lente}@univ-tours.fr, pauline.fouillet@etu.univ-tours.fr
http://li.univ-tours.fr

Abstract. As a type of scheduling problem, the flowshop problem has been largely studied for 60 years. The total completion time is a very interesting criterion because it reflects "the total manufacturing waiting time experienced by all customers" (Emmons and Vairaktarakis). There have been many studies in the past but they focused on a limited number of machines and/or on specific constraints. Therefore, this study presents a new approach to tackle a general permutation flowshop problem, with various additional constraints, to elaborate on lower bounds for the total completion time. These lower bounds can take into account several constraints, like delays, blocking or setup times, but they imply solving a Traveling Salesman Problem. The theory is developed first, based on a MaxPlus modeling of flowshop problems and experimental results of a branch-and-bound procedure with a lower bound selection strategy are then presented.

Keywords: Scheduling · Flowshop · Total completion time · Lower bound

1 Introduction

The m-machine flowshop scheduling problem has been largely studied for 60 years. The makespan is the most studied criterion, especially for permutation flowshop problems. However, the total completion time criterion also receives a great amount of attention. It reflects "the total manufacturing waiting time experienced by all customers" [10]. Even with only two machines, problem $F_2||\sum C_i$ is $NP - hard$ in the strong sense and so are problems with more machines. Therefore, results that help to solve these problems are interesting.

Total completion time criterion has also been greatly studied. A branch-and-bound algorithm, incorporating a lower bound, dominance relation and an upper bound is presented by Allahverdi and Al-Anzi in [1]. That study solves total completion time minimization problem $F_3|perm, S_{nsd}|\sum C_i$ where separate setup times are taken into account. The number of visited nodes and the percentage

E. Pinson et al. (Eds.): ICORES 2014, CCIS 509, pp. 285–299, 2015.
DOI: 10.1007/978-3-319-17509-6_19

between this number and that of possible nodes are considered. This percentage shows us that their lower bound is effective as the number of visited nodes is quite small. Separate setup times are also investigated by Su and Lee [23] in a two-machine flowshop no-wait scheduling problem with a single server in order to minimize total completion time. In another research, eleven heuristics based on the Shortest Processing Time (SPT) rule are implemented by Aydilek and Allahverdi [3]. Their study is about minimizing total completion time of a two-machine flowshop scheduling problem, in which processing times are bounded. In [9], a lower bound based on the first machine of problem $F_2||\sum C_i$ is presented as the sum of a previously existing lower bound and the optimum of an asymmetric traveling salesman problem (ATSP). These aforementioned studies only deal with limited number of machines and few constraints. It is not easy to generalize them to any number of machines or any constraint.

In this study, the proposed approach is based on MaxPlus algebra (see Sect. 2.1). It has been widely used in control systems, especially in relation with Petri Nets but rarely in the scheduling theory. Nevertheless, some articles can be cited on project scheduling problems [13], on cyclic parallel machine problems [16], on cyclic flowshop scheduling problems [8,11] and on cyclic jobshop scheduling problems [12]. The MaxPlus algebra is applied to modeling and scheduling flowshop problems with minimal delays, setup and removal times [5,19]. It is also applied to flowshop problems with minimal-maximal delays for two-machines [4] or for any number of machines [2]. In these studies, jobs are associated to MaxPlus square matrices and lower bounds, upper bounds and/or dominance conditions are derived by applying transformations to those matrices. Moreover, this approach is used effectively to model flowshop problems with minimal-maximal delays, setup and removal times and to highlight a central problem [25].

The objective of this study is to address a general permutation flowshop problem in terms of constraints taken into account. We elaborate on lower bounds for the total completion time that are based on the resolution of two sub-problems: one problem similar to the one machine total completion time minimization problem and the other similar to a traveling salesman problem. These lower bounds are incorporated in a branch-and-bound procedure to be tested. Additionally, a selection strategy of adequate lower bounds is also implemented in order to accelerate the branch-and-bound procedure.

The next section presents the background of the study: MaxPlus algebra and flowshop scheduling problem. We recall in Sect. 3 how MaxPlus algebra can be used to model a general flowshop problem. The lower bound construction is then explained in Sect. 4. Finally, a branch-and-bound algorithm with a lower bound selection strategy is described and some tests concerning problem $F_3|perm; S_{nsd}|\sum C_i$ (previously studied in [1]) and problem $F_m|perm|\sum C_i$ are presented as experimental results.

2 Context and Definitions

2.1 MaxPlus Algebra

MaxPlus algebra is briefly described as follows; a more detailed presentation can be found in [15].

In MaxPlus algebra, the maximum is denoted by $\oplus$ and the addition by $\otimes$. The former, $\oplus$, is idempotent, commutative, associative and has a neutral element $(-\infty)$ denoted by $\mathbb{0}$. The latter, $\otimes$, is associative, distributive on $\oplus$ and has a neutral element (0) denoted by $\mathbb{1}$. The null element, $\mathbb{0}$, is an absorbing element for $\otimes$. These properties can be summarized by stating that $\mathbb{R}_{max} = (\mathbb{R} \cup \{-\infty\}, \oplus, \otimes)$ is a dioid. It is important to note that in MaxPlus algebra in particular, and in dioids in general, the first operator $\oplus$ can not be simplified, that is $a \oplus b = a \oplus c \nRightarrow b = c$. Furthermore, in $\mathbb{R}_{max}$, the second operator $\otimes$ is commutative, and except $\mathbb{0}$, every element is invertible: the inverse of x is denoted by x^{-1} or $\mathbb{1}/x$. For simplicity, we denote the ordinary subtractions by x/y instead of $x \otimes y^{-1}$ and by xy the product $x \otimes y$.

It is possible to extend these two operators to $m \times m$ matrices of elements of $\mathbb{R}_{max}$. Let A and B be two matrices of size $m \times m$, operators $\oplus$ and $\otimes$ are defined by

$$\forall (i,j) \in \{1, \ldots, m\}^2, [A \oplus B]_{ij} = [A]_{ij} \oplus [B]_{ij}$$

$$\forall (i,j) \in \{1, \ldots, m\}^2, [A \otimes B]_{ij} = \bigoplus_{k=1}^{m} [A]_{ik} \otimes [B]_{kj}$$

where $[.]_{ij}$ is the element at the ith row and j^{th} column of the corresponding matrix. It is not difficult to show that the set of $m \times m$ matrices in $\mathbb{R}_{max}$ is a dioid. However, $\otimes$ is not commutative and not every matrix is invertible.

The two following lemmas can be derived from the previous definitions. They will be useful for the development of the lower bound.

Lemma 1. $\forall j \in \{1, \ldots, m\}$:

$$[A \otimes B]_{1j} = \bigoplus_{k=1}^{m} [A]_{1k} \otimes [B]_{kj} \geq [A]_{11} \otimes [B]_{1j} \tag{1}$$

Lemma 2. $\forall \ell, j \in \{1, \ldots, m\}$:

$$[A \otimes B]_{1j} \geq [A]_{1\ell} \otimes [B]_{\ell j} \tag{2}$$

$$[A \otimes B]_{\ell j} \geq [A]_{\ell\ell} \otimes [B]_{\ell j} \tag{3}$$

2.2 Flowshop Scheduling Problem

Since the paper of Johnson [18], flowshop problems have been studied largely [10]. Basically, a flowshop scheduling problem consists of a set of n-jobs $\mathcal{J} = \{J_1, \ldots, J_n\}$ and another set of m-machines $\{M_1, \ldots, M_m\}$. Each job must go through all machines in the same predefined order, let us say from M_1 to M_m and each machine can load only one job at a time [6]. If all jobs must be executed in the same order over all machines, the problem is called a permutation flowshop problem. In this case, there exists an ordered list of jobs (or a sequence) σ that is identically scheduled on all machines. We limit our current study to permutation flowshop problems.

Each job J_i is composed of m operations O_{ik} $(1 \leq k \leq m)$: one per machine. An operation is at least described by a processing time p_{ik}: the processing time of job J_i on machine M_k (or equivalently, the processing time of the kth operation of job J_i). The completion time of job J_i on machine M_k (C_{ik}) and the completion time of job J_i (C_i) are related by $C_i = C_{im}$.

Over the years, several additional constraints have been taken into consideration [10]. Some of them can be modeled using MaxPlus algebra [25]. One of the most common constraints is the permutation constraint ($perm$) which has just been mentioned above. A constraint of $no-wait$ appears in problems where there is no delay allowed between two successive operations of a job. On the contrary, constraints of $min-delay$, $max-delay$, $min-max\ delay$ indicate a flowshop problem with delays between two successive operations of a job. Depending on the case, these delays may have to meet a lower bound, an upper bound or both. It may also exist separate non-sequence dependent setup times (S_{nsd}) and/or removal times (R_{nsd}) before and after each operation. Finally, some authors have considered blocking constraints, due to the non-existence of intermediate storage between consecutive machines or to specific interactions between machines. These constraints are referred to as RSb, RCb and RCb^* in [24].

The most studied criterion is the makespan, or the maximal completion time (C_{max}). It is defined by the completion time of the last operation scheduled on the last machine (M_m). In this article we focus on the total completion time ($\sum C_i$) which is the sum of the completion times of the different jobs in a given schedule.

3 MaxPlus Modeling of Flowshop Scheduling Problems

Our problem can be noted $F_m|perm\ \beta|\gamma$ using notations proposed by Graham et al. [14]. It is a m machine permutation flowshop problem with a set of constraints β that is a subset of $\{min-max\ delay,\ no-wait,\ S_{nsd},\ R_{nsd},\ RSb, RCb, RCb^*\}$. Criterion γ can be whatever we desire since it does not interfere in the modeling process. The total completion time criterion is investigated in the following of this article.

Basically, the modeling process follows this scheme:

- Given the kth operation O_{ik} of a job J_i, four dates are considered: date δ_k of availability of machine M_k (before execution of operation O_{ik}), starting time St_{ik} of operation O_{ik}, its completion time C_{ik} and date of liberation D_{ik} of machine M_k (after execution of operation O_{ik}), that is the date when job J_i leaves machine M_k to be placed in a stock or on the following machine. In most flowshop problems, dates C_{ik} and D_{ik} are equal; however, they can be different in case of blocking constraints or removal times. Date of liberation of the last machine (D_{im}) is equal to the completion time C_i of the job, except if there exist removal times. In this case D_{im} is equal to C_i plus the removal time of operation O_{im}. Figure 1 shows an example of flowshop problem $F_m|perm, min-max\ delay, S_{nsd}, R_{nsd}|\sum C_i$ where triangles illustrate setup and removal times and rectangles illustrate processing times.

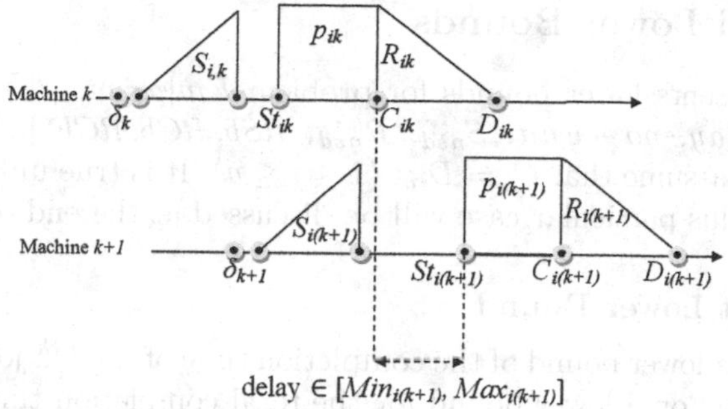

Fig. 1. Example of flowshop problem: $F_m|perm, min - max\ delay, S_{nsd}, R_{nsd}|\sum C_i$.

- Formulate the system (S) of inequalities that link these different variables.
- Calculate the smallest (D_{ik}) $(1 \leq k \leq m), (1 \leq i \leq n)$ solutions of the system (S).

Whatever the set of constraints β is, these calculations lead to a MaxPlus linear relation between dates of liberation D_{ik} and dates of availability δ_k [20,25]. More precisely, we can state the following proposition:

Proposition 1 (Matrix Associated to a Job). *Let $\boldsymbol{\delta}$ (resp. $\boldsymbol{D}_i$) be the line vector of the m dates δ_k (resp. D_{ik}): it exists a $m \times m$ MaxPlus matrix T_i computed from data of job J_i such that*

$$\boldsymbol{D}_i = \boldsymbol{\delta} \otimes T_i \tag{4}$$

Matrix T_i is called the associated matrix of job J_i: it entirely characterizes job J_i.

Various elements of matrix T_i will be denoted $t^i_{\ell c}$, in other words, $t^i_{\ell c} = [T_i]_{\ell c}$. This matrix sums up the job data (processing times, setup times, delays and so on) and the flowshop constraints.

$$T_i = \begin{pmatrix} t^i_{11} & t^i_{12} & \dots t^i_{1m} \\ t^i_{21} & t^i_{22} & \dots t^i_{2m} \\ \dots & \dots & \dots\dots \\ t^i_{m1} & t^i_{m2} & \dots t^i_{mm} \end{pmatrix} \tag{5}$$

These results can be generalized to a sequence of jobs [5,20].

Definition 1 (Matrix Associated to a Sequence). *Let σ be a sequence of ν jobs: its associated matrix is matrix T_σ defined by*

$$T_\sigma = \bigotimes_{i=1}^{\nu} T_{\sigma(i)} \tag{6}$$

Proposition 2. *If $\boldsymbol{\delta}$ is the vector of dates of availability of machines and $\boldsymbol{D}_\sigma$ the vector of dates of liberation of machines, after the execution of sequence σ, we have the relation*

$$\boldsymbol{D}_\sigma = \boldsymbol{\delta} \otimes T_\sigma \tag{7}$$

4 Proposed Lower Bounds

This section presents lower bounds for problem $Fm|perm,\ \beta|\sum C_i$, with $\beta \subset \{min - max\ delay,\ no-wait,\ S_{nsd},\ R_{nsd},\ RSb, RCb, RCb^*\}$. To develop the calculations, we assume that $C_i = D_{im}$ $(1 \leq i \leq n)$. It is true unless there exists removal times: this particular case will be discussed at the end of this section.

4.1 The First Lower Bound

We first present a lower bound of the completion time of the kth job in a sequence before elaborating on a lower bound for the total completion time.

Lower Bound of Completion Time of a Job

Proposition 3. *Let σ a sequence of jobs and $\boldsymbol{\delta}$ the line vector of dates of availability of the machines ($\boldsymbol{\delta} = (\delta_1, \delta_2, \ldots, \delta_m)$). The completion time of the job in k^{th} position in the sequence verifies relation:*

$$\begin{aligned} &if\ k = 1 : C_{\sigma(1)} \geq \delta_1 \left[T_{\sigma(1)}\right]_{1m} \\ &if\ k = 2 : C_{\sigma(2)} \geq \delta_1 \left[T_{\sigma(1)} T_{\sigma(2)}\right]_{1m} \\ &if\ k > 2 : C_{\sigma(k)} \geq \delta_1 \bigotimes_{j=1}^{k-2} (t_{11}^{\sigma(j)}) \left[T_{\sigma(k-1)} T_{\sigma(k)}\right]_{1m} \end{aligned}$$

Proof. Let τ be the sub-sequence composed of the first k jobs of sequence σ. Proposition 2 and definition 1 result in:

$$\boldsymbol{D}_\tau = \boldsymbol{\delta} \otimes T_\tau = \boldsymbol{\delta} \otimes \bigotimes_{i=1}^{k} T_{\tau(i)} = \boldsymbol{\delta} \otimes \bigotimes_{i=1}^{k} T_{\sigma(i)} \tag{8}$$

Moreover $\boldsymbol{D}_{\sigma(k)} = \boldsymbol{D}_\tau$ and by assumption $C_{\sigma(k)} = D_{\sigma(k)m}$, which is the last element of vector $\boldsymbol{D}_{\sigma(k)}$, we have,

$$C_{\sigma(k)} = [\boldsymbol{D}_\tau]_{1m} = \left[\boldsymbol{\delta} \otimes \bigotimes_{i=1}^{k} T_{\sigma(i)}\right]_{1m} \tag{9}$$

If $k = 1$, the application of Lemma 1 results in:

$$C_{\sigma(1)} \geq [\boldsymbol{\delta}]_{11} \left[T_{\sigma(1)}\right]_{1m} = \delta_1 \left[T_{\sigma(1)}\right]_{1m} \tag{10}$$

If $k \geq 2$, by iteratively applying this lemma into Eq. (9), we obtain:

$$C_{\sigma(k)} \geq [\boldsymbol{\delta}]_{11} \otimes \bigotimes_{j=1}^{k-2} \left[T_{\sigma(j)}\right]_{11} \left[T_{\sigma(k-1)} T_{\sigma(k)}\right]_{1m} \tag{11}$$

Inequality (11) can be rewritten as

$$C_{\sigma(k)} \geq \delta_1 \bigotimes_{j=1}^{k-2} (t_{11}^{\sigma(j)}) \left[T_{\sigma(k-1)} T_{\sigma(k)}\right]_{1m} \tag{12}$$

Lower Bound of the Total Completion Time

Definition 2 (Lower Bound LB^1_{VFL}). *Given a sequence σ of n-jobs, we define:*

$$A_1(\sigma) = \bigotimes_{j=1}^{n-1} (t_{11}^{\sigma(j)})^{n-j}$$

$$B_1(\sigma) = \left[T_{\sigma(1)}\right]_{1m} \left(\bigotimes_{j=2}^{n} \frac{\left[T_{\sigma(j-1)}T_{\sigma(j)}\right]_{1m}}{t_{11}^{\sigma(j-1)}} \right)$$

Proposition 4. $\forall \sigma$ *sequence* : $\bigotimes_{i=1}^{n} C_{\sigma(i)} \geq \delta_1^n \otimes A_1(\sigma) \otimes B_1(\sigma)$

Proof. Considering proposition 3, we have:

$$\bigotimes_{i=1}^{n} C_{\sigma(i)} \geq \delta_1 \left[T_{\sigma(1)}\right]_{1m} \otimes \bigotimes_{i=2}^{n} \left(\delta_1 \bigotimes_{j=1}^{i-2} (t_{11}^{\sigma(j)}) \left[T_{\sigma(i-1)}T_{\sigma(i)}\right]_{1m} \right) \tag{13}$$

Rearranging the factors on the right side of inequality (13):

$$\bigotimes_{i=1}^{n} C_{\sigma(i)} \geq (\delta_1)^n \otimes \bigotimes_{i=1}^{n-2} \left(t_{11}^{\sigma(i)}\right)^{n-i-1} \otimes \left[T_{\sigma(1)}\right]_{1m} \bigotimes_{i=2}^{n} \left[T_{\sigma(i-1)}T_{\sigma(i)}\right]_{1m} \tag{14}$$

and then multiplying the inequality (14) by $\frac{\bigotimes_{i=1}^{n-1} t_{11}^{i}}{\bigotimes_{i=1}^{n-1} t_{11}^{i}}$, we complete the proof.

At this point, we can obtain a lower bound of the total completion time by computing the optimal values of factors A_1 and B_1. The two following propositions explain how to do.

Proposition 5 (Minimisation of A_1). *Let σ^1_{SPT} the sequence obtained by sorting jobs in non-decreasing order of the coefficient t_{11}. This sequence minimizes criterion A_1.*

Proof.

$$A_1(\sigma) = \bigotimes_{j=1}^{n-1} (t_{11}^{\sigma(j)})^{n-j} = \bigotimes_{j=1}^{n} (t_{11}^{\sigma(j)})^{n-j+1} . \frac{1}{\bigotimes_{j=1}^{n} (t_{11}^{\sigma(j)})} \tag{15}$$

The second factor is a constant, so we have to minimize $\bigotimes_{j=1}^{n} (t_{11}^{\sigma(j)})^{n-j+1}$. It is similar to the total completion time criterion in a one-machine problem

$(1||\sum C_i)$ where processing times are the t_{11}^is. This criterion is minimized by using Smith's rule [22].

Proposition 6 (Minimisation of B_1). *Let us consider an Asymmetric Traveling Salesman Problem (ATSP) defined by the following distances between $n+1$ towns, numbered from 0 to n:*

$$\begin{cases} \forall i \in \{1,\dots,n\}: & d(0,i) = [T_i]_{1m} \\ \forall i \in \{1,\dots,n\}: & d(i,0) = \mathbb{1}\ (=0) \\ \forall (i,j) \in \{1,\dots,n\}^2: & d(i,j) = \dfrac{[T_iT_j]_{1m}}{t_{11}^i} \end{cases} \tag{16}$$

Let sequence σ_{ATSP}^1 be an optimal cycle of this ATSP: $B_1(\sigma_{ATSP}^1)$ is the optimal value of criterion B_1.

Proof. With these notations, $B_1(\sigma)$ can be rewritten as the length of a cycle:

$$B_1(\sigma) = d(0,\sigma(1)) \left(\bigotimes_{i=1}^{n-1} d(\sigma(i),\sigma(i+1)) \right) d(\sigma(n),0) \tag{17}$$

All these results lead to the next proposition.

Proposition 7 (Lower Bound LB_{VFL}^1). *Let $LB_{VFL}^1 = (\delta_1)^n \otimes A_1(\sigma_{SPT}^1) \otimes B_1(\sigma_{ATSP}^1)$: LB_{VFL}^1 is a lower bound of the total completion time. In usual notations, this lower bound is defined by:*

$$LB_{VFL}^1 = n\delta_1 + A_1(\sigma_{SPT}^1) + B_1(\sigma_{ATSP}^1) \tag{18}$$

It is needed to solve a traveling salesman problem to compute this lower bound; however, the procedures for solving that problem are rather effective on medium size instances.

This lower bound is similar to the one presented by Della Croce et al. [9] for two machines, but it works with m machines and more constraints.

Existence of Removal Times. If there are removal times, the date of liberation of machine M_m by job J_i (D_{im}) is equal to the sum of completion time C_i of job J_i and removal time of the last operation of O_{im} of J_i. Thus, the total sum of D_{im} ($1 \leq i \leq n$) is equal to the total completion time plus a constant term which is equal to the sum of removal times of all last operations. Therefore, to obtain a lower bound of the total completion time we only have to subtract this constant from LB_{VFL}^1.

4.2 Additional Similar Lower Bounds

A similar approach to the construction of the first lower bound can be developed to achieve the ℓth lower bound ($\forall \ell \in \{1,\dots,m\}$). Using iteratively Lemma 2, we obtain:

$$C_{\sigma(1)} \geq \delta_\ell \left[T_{\sigma(1)}\right]_{\ell m}$$
$$C_{\sigma(2)} \geq \delta_\ell \left[T_{\sigma(1)}T_{\sigma(2)}\right]_{\ell m}$$
$$\forall i > 2 \ : C_{\sigma(i)} \geq \delta_\ell \left(\bigotimes_{j=1}^{i-2} t_{\ell\ell}^{\sigma(j)}\right) \left[T_{\sigma(i-1)}T_{\sigma(i)}\right]_{\ell m} \tag{19}$$

Defining $A_\ell(\sigma)$ and $B_\ell(\sigma)$:

$$A_\ell(\sigma) = \bigotimes_{j=1}^{n-1} (t_{\ell\ell}^{\sigma(j)})^{n-j} \tag{20}$$

$$B_\ell(\sigma) = \left[T_{\sigma(1)}\right]_{\ell m} \left(\bigotimes_{j=2}^{n} \frac{\left[T_{\sigma(j-1)}T_{\sigma(j)}\right]_{\ell m}}{t_{\ell\ell}^{\sigma(j-1)}}\right) \tag{21}$$

we have

$$\bigotimes_{i=1}^{n} C_{\sigma(i)} \geq \delta_\ell^n A_\ell(\sigma) B_\ell(\sigma) \tag{22}$$

Similarly to propositions 5 and 6, we can find σ_{SPT}^ℓ to minimize $A_\ell(\sigma)$ and σ_{ATSP}^ℓ to minimize $B_\ell(\sigma)$.

The ℓth lower bound of the total completion time of the initial flowshop problem is then:

$$LB_{VFL}^\ell = (\delta_\ell)^n A_\ell(\sigma_{SPT}^\ell) B_\ell(\sigma_{ATSP}^\ell) \tag{23}$$

5 Branch-and-Bound Algorithm

5.1 Proposed Branch-and-Bound Procedure

In order to validate the lower bounds we proposed, we have incorporated them in a branch-and-bound procedure. A branch-and-bound procedure is an enumeration method that builds dynamically a search tree. Lower bounds or other criteria like dominance relations are used to cut some useless branches. We have used the separation scheme introduced by Ignall and Schrage [17]: a partial sequence is progressively built as we go deeper in the search tree. A node corresponds to a partial sequence and a set of free jobs. The separation of a node consists in adding a free job at the end of the sequence. A node has as many children as its free jobs. The branching strategy is Depth-First-Search (DFS). An upper bound is computed at the root node and updated at each node. For this purpose, we have used heuristic $PR4(15)$ presented by Pan and Ruiz [21].

The branch-and-bound procedure is detailed in Algorithm 1 and numerical results are presented in Sect. 6. In this algorithm, L is the list of nodes that have not yet been separated and LC the list of child nodes built after separation of a node.

Algorithm 1. Branch-and-Bound.

```
procedure BRANCH-AND-BOUND
    BestUB ← ∞
    Generate Root tree NRoot                                  // an empty node
    Compute LB(NRoot) and UB(NRoot)
    Add NRoot to list L
    while L is not empty do
        N ← top(L)                                  // move the first node of list L
        BestUB ← min{BestUB, UB(N)}
        if LB(N) < BestUB then
            Generate children list LC of N
            for each NChild in LC do
                Compute LB(NChild)
                Compute UB(NChild)
            end for
            Sort LC                        // in non-increasing order of LB(NChild)
            for each NChild in LC do
                if LB(NChild) < BestUB then
                    top(L) ← NChild
                else
                    Delete NChild
                end if
            end for
        else
            Delete N
        end if
    end while
end procedure
```

5.2 Lower Bound Selection Strategy

It may take time to solve traveling salesman problems. It means that it is not reasonable to compute all lower bounds LB_{VFL}s at each node, especially when the number of jobs is increasing. Therefore, we are looking for a strategy so that we could avoid useless calculations and accelerate the branch-and-bound procedure. It may be an improvement in the algorithm as well as a way to combine lower bounds LB_{VFL} according to each node.

Firstly, we decide to compute the upper bound only once at the root node. This step seems to reduce the computation time for calculating upper bounds although it increases the number of visited nodes.

Secondly, we propose a lower bound selection strategy ($LBSS$). For each of the first ten thousand visited nodes, we compute all lower bounds LB_{VFL}s. For these whole ten thousand visited nodes, we compute the number of times α_i that lower bound LB^i_{VFL} $(1 \leq i \leq m)$ is dominant (i.e. the greatest). For the next nodes, lower bound LB^k_{VFL} $(1 \leq k \leq m)$ is computed if and only if its relative reference value (see (24)) exceeds a given threshold α^*:

$$\frac{\alpha_k}{\max\limits_{1 \le i \le m} (\alpha_i)} \ge \alpha^* \tag{24}$$

A good choice of this threshold may help to avoid computing bad lower bounds. The effectiveness of $LBSS$ is shown in Table 1.

6 Experimental Results

There are few studies on exact resolution of flowshop scheduling problems with criterion of total completion times. We decided to compare our branch-and-bound procedure to the one developed by Allahverdi and Al-Anzi [1] for problem $F_3|perm; S_{nsd}|\sum C_i$. According to the approach proposed by Allahverdi and Al-Anzi, the processing and setup times values were randomly generated respectively from the uniform distribution on the interval $[1, 100]$ and on the interval $[0, 100k]$. We considered problems of n-jobs (n=7,8,9,10,11,12,13,14,15,16, 17,18,20). A class of thirty instances was generated for each number of jobs and each k value. The k value for each data set was assigned to 0.3, 0.5 and 0.8. It was assumed that all machines were available from the time zero ($\delta_k = 0$, $1 \le k \le m$). To compute lower bounds LB_{VFL}s, we used the ATSP solving procedure developed by G. Carpaneto, M. Dell'amico and P. Toth [7]. The used machine is based on an Intel Duocore 2.6 GHz 4 GB RAM.

We have reported in Table 1 the mean computation times (in seconds) for each class (n, k) of instances of three versions of the branch-and-bound procedure: the basic one, in which all lower bounds LB_{VFL}s are computed at each nod, the $LBSS$ one which uses the lower bound selection strategy presented in 5.2 and the SLB one which uses a simple lower bound (SLB). Lower bound SLB of a node is equal to the total completion time of its corresponding partial sequence. This SLB version allows us to evaluate the effectiveness of lower bounds LB_{VFL}s. When optimal solutions have never been found over the thirty instances within the time limit, we indicate "> 1500" if 1500 s is the time limit.

Table 1 shows that the basic version of the branch-and-bound procedure is more effective than the SLB version, despite the fact that LB_{VFL}s need to solve traveling salesman problems. It also confirms that lower bound selection strategy $LBSS$ reduces strongly computation times. In other words, LB_{VFL}s with the current lower bound selection strategy are effective to eliminate unworthy branches. However, we still need to improve this strategy as we are now limited to twenty jobs.

In their study, Allahverdi and Al-Anzi did not indicate computation times, they prefer computing the percentage of visited nodes to solve an instance relatively to the total number of nodes of the whole search tree. Therefore, we did the same in order to perform a comparison. By default, the result is obtained from thirty instances. However, in following tables, when optimum could not be achieved within the time limit, the number of solved instances is indicated between parentheses. Percentage of visited nodes and computation times are then computed over these solved instances. We limited computation times to 1500 (3000 and 9000, respectively) seconds in case of $n \in \{7, 8, 9, 10, 11, 12, 13, 14, 15\}$

Table 1. The mean computation time for each class.

Jobs	$k = 0.3$	$k = 0.5$	$k = 0.8$	LB
7	0.010	0.010	0.010	$LBSS$
	0.048	0.049	0.058	$LB^1_{VFL}, LB^2_{VFL}, LB^3_{VFL}$
	0.030	0.039	0.041	SLB
8	0.010	0.010	0.010	$LBSS$
	0.100	0.098	0.176	$LB^1_{VFL}, LB^2_{VFL}, LB^3_{VFL}$
	0.099	0.102	0.121	SLB
9	0.011	0.012	0.012	$LBSS$
	0.212	0.165	0.193	$LB^1_{VFL}, LB^2_{VFL}, LB^3_{VFL}$
	0.683	0.602	0.709	SLB
10	0.020	0.021	0.024	$LBSS$
	0.291	0.285	0.353	$LB^1_{VFL}, LB^2_{VFL}, LB^3_{VFL}$
	6.654	5.629	6.748	SLB
11	0.041	0.043	0.054	$LBSS$
	0.730	0.838	0.932	$LB^1_{VFL}, LB^2_{VFL}, LB^3_{VFL}$
	73.674	66.924	74.569	SLB
12	0.073	0.093	0.125	$LBSS$
	1.313	1.779	2.330	$LB^1_{VFL}, LB^2_{VFL}, LB^3_{VFL}$
	906.030	1173.178	905.417	SLB
13	0.160	0.241	0.277	$LBSS$
	4.589	7.593	7.995	$LB^1_{VFL}, LB^2_{VFL}, LB^3_{VFL}$
	> 1500	> 1500	> 1500	SLB
14	0.624	0.653	1.420	$LBSS$
	16.317	15.306	34.569	$LB^1_{VFL}, LB^2_{VFL}, LB^3_{VFL}$
	> 1500	> 1500	> 1500	SLB
15	2.981	2.877	4.385	$LBSS$
	68.559	50.609	68.467	$LB^1_{VFL}, LB^2_{VFL}, LB^3_{VFL}$
	> 1500	> 1500	> 1500	SLB
16	6.367	6.537	24.310	$LBSS$
	152.996	155.756	404.959	$LB^1_{VFL}, LB^2_{VFL}, LB^3_{VFL}$
	> 3000	> 3000	> 3000	SLB
17	27.580	97.107	80.717	$LBSS$
	419.347	1478.93	1894.091	$LB^1_{VFL}, LB^2_{VFL}, LB^3_{VFL}$
	> 3000	> 3000	> 3000	SLB
18	172.071	244.530	178.197	$LBSS$
	7984.230	7638.810	2694.640	LB^1_{VFL}, LB^3_{VFL}
	> 9000	> 9000	> 9000	SLB
20	774.506 (26)	508.405 (28)	1320.497 (26)	$LBSS$
	2850.320 (10)	346.510 (10)	1640.350 (10)	LB^2_{VFL}, LB^3_{VFL}
	> 9000	> 9000	> 9000	SLB

Table 2. The performance of branch-and-bound procedure (percentage of visited nodes) for different k values.

Jobs	PVN_{AA} ($k = 0.3$)	PVN_{VFL} ($k = 0.3$)	PVN_{AA} ($k = 0.5$)	PVN_{VFL} ($k = 0.5$)	PVN_{AA} ($k = 0.8$)	PVN_{VFL} ($k = 0.8$)
7	5.12×10^{-1}	7.15×10^{-1}	4.39×10^{-1}	7.00×10^{-1}	0.51×10^{0}	1.02×10^{0}
8	2.20×10^{-1}	2.66×10^{-1}	2.74×10^{-1}	2.70×10^{-1}	2.92×10^{-1}	3.18×10^{-1}
9	5.11×10^{-2}	7.74×10^{-2}	6.46×10^{-2}	7.02×10^{-2}	6.05×10^{-2}	9.81×10^{-2}
10	3.56×10^{-2}	1.76×10^{-2}	2.79×10^{-2}	2.04×10^{-2}	2.40×10^{-2}	2.64×10^{-2}
11	1.63×10^{-2}	0.48×10^{-2}	1.28×10^{-2}	0.61×10^{-2}	1.43×10^{-2}	0.76×10^{-2}
12	3.08×10^{-3}	0.78×10^{-3}	2.74×10^{-3}	1.35×10^{-3}	3.12×10^{-3}	1.51×10^{-3}
13	6.31×10^{-4}	1.91×10^{-4}	6.30×10^{-4}	4.15×10^{-4}	4.71×10^{-4}	4.14×10^{-4}
14	5.02×10^{-5}	5.93×10^{-5}	1.22×10^{-4}	0.67×10^{-4}	1.37×10^{-4}	1.61×10^{-4}
15	2.38×10^{-5}	2.24×10^{-5}	2.19×10^{-5}	1.75×10^{-5}	2.46×10^{-5}	2.69×10^{-5}
16	1.20×10^{-5}	0.20×10^{-5}	1.21×10^{-5}	0.19×10^{-5}	1.24×10^{-5}	0.86×10^{-5}
17	5.70×10^{-6}	0.62×10^{-6}	5.30×10^{-6}	1.85×10^{-6}	5.60×10^{-6}	1.48×10^{-6}
18	5.40×10^{-7}	2.06×10^{-7}	5.00×10^{-7}	1.99×10^{-7}	5.20×10^{-7}	2.17×10^{-7}
20	–	0.19×10^{-8}	–	0.17×10^{-8}	–	0.47×10^{-8}
		(26)		(28)		(26)

Table 3. The mean computation time for each class in problem $F_m|perm|\sum C_i$.

Jobs	$m = 5$	$m = 10$	Jobs	$m = 5$	$m = 10$
7	0.010	0.010	14	4.442	17.033
8	0.010	0.014	15	16.945	21.677
9	0.019	0.031	16	71.446	292.969
10	0.044	0.104	17	136.281	533.667 (27)
11	0.121	0.238	18	824.754	1329.541 (23)
12	0.242	0.538	20	2517.159 (23)	1991.824 (5)
13	1.554	2.781			

($n \in \{16, 17\}$ and $n \in \{18, 20\}$, respectively). We applied here the $LBSS$ and we set threshold α^* to thirty percent.

We have reported in Table 2 the mean percentage of visited nodes over the thirty instances of each class (n, k) of problems for our branch-and-bound (columns $PVN_{VFL}(k = 0.3)$, $PVN_{VFL}(k = 0.5)$ and $PVN_{VFL}(k = 0.8)$) and for Allahverdi and Al-Anzi's branch-and-bound (columns $PVN_{AA}(k = 0.3)$, $PVN_{AA}(k = 0.5)$, $PVN_{AA}(k = 0.8)$).

Table 2 shows that lower bounds LB^1_{VFL}, LB^2_{VFL} and LB^3_{VFL} are useful in reducing the number of visited nodes. They are really effective when compared to the performance in the study of Allahverdi and Al-Anzi [1] for instances with more than ten jobs.

Finally, in order to test furthermore the lower bounds as well as the branch-and-bound procedure, we decided to execute some tests concerning problem $F_m|perm|\sum C_i$ where $m \in \{5, 10\}$ and $n \in \{7,8,9,10,11,12,13,14,15,16,17,18,20\}$. As previously, processing times were generated between 1 and 100, time limits have been set to 1500, 3000 or 9000 s depending on the number of jobs ($\{7, 8, 9, 10, 11, 12, 13, 14, 15\}$,$\{16, 17\}$ or $\{18, 20\}$). Lower bound selection strategy $LBSS$ is applied to four lower bounds LB^1_{VFL}, LB^2_{VFL}, LB^{m-1}_{VFL} and LB^m_{VFL}. We have reported the mean computation time (in seconds) for each class of thirty instances in Table 3. The table shows that the branch-and-bound procedure with $LBSS$ can solve problem $F_m|perm|\sum C_i$ within acceptable time limit. The result confirms that computation times depend not only on the number of jobs but also on the number of machines.

7 Conclusions

We proposed a MaxPlus approach to tackle a m-machine flowshop problem with several additional constraints. The MaxPlus approach enables the transformation of a general flowshop problem into a matrix problem. Then some computations over these matrices allow us to highlight new lower bounds for the total completion time criterion, based on the resolution of a one-machine problem and an asymmetric traveling salesman problem. Despite the necessity of solving an $NP-hard$ problem, experimental results and comparison to a previously published research have shown the effectiveness of these lower bounds. Experimental results are expanded to general cases and the effectiveness is once more confirmed. A lower bound selection strategy has been proposed and has shown its usefulness in reducing computation times.

Our further research will aim at improving these lower bounds LB_{VFL}s as well as improving the branch-and-bound algorithm. The current strategy is not strong enough to shorten the computation time of the whole branch-and-bound algorithm when the number of jobs is increasing. In particular, the strategy has to be reformed. Moreover, tests can be executed to several constraints as $no-wait$, $min-max\ delay$, (S_{nsd}, R_{nsd}), limited stocks between machines or blocking constraints since these constraints only modify matrix T_i associated to job J_i and lower bounds LB_{VFL}s are still valid. The study can be also extended to the weighted total completion time criterion $\sum_{i=1}^{n} w_i C_i$.

References

1. Allahverdi, A., Al-Anzi, F.S.: A branch-and-bound algorithm for three-machine flowshop scheduling problem to minimize total completion time with separate setup times. Eur. J. Oper. Res. **169**(3), 767–780 (2006)
2. Augusto, V., Christophe, L., Bouquard, J.-L.: Résolution d'un flowshop avec delais minimaux et maximaux. In: MOSIM (2006)
3. Aydilek, H., Allahverdi, A.: Two-machine flowshop scheduling problem with bounded processing times to minimize total completion time. Comput. Math. Appl. **59**(2), 684–693 (2010)

4. Bouquard, J.-L., Christophe, L.: Two-machine flow shop scheduling problems with minimal and maximal delays. 4or **4**(1), 15–28 (2006)
5. Bouquard, J.-L., Lenté, C., Billaut, J.-C.: Application of an optimization problem in Max-Plus algebra to scheduling problems. Discrete Appl. Math. **154**(15), 2064–2079 (2006)
6. Brucker, P.: Scheduling Algorithms, 5th edn. Springer, Berlin (2006)
7. Carpaneto, G., Dell'amico, M., Toth, P.: Exact solution of large asymmetric traveling salesman problems. ACM Trans. Math. Softw. **21**(4), 394–409 (1995)
8. Cohen, G., Dubois, D., Quadrat, J.-P., Viot, M.: A linear system-theoretic view of discret-event processes and its use for performance evaluation in manufacturing. IEEE Trans. Autom. Control **30**, 210–220 (1985)
9. Croce, F.D., Narayan, V., Tadei, R.: The two-machine total completion time flow shop problem. Eur. J. Oper. Res. **90**, 227–237 (1996)
10. Emmons, H., Vairaktarakis, G.: Flow Shop Scheduling, 182nd edn. Springer, New York (2013)
11. Gaubert, S.: Théorie des systèmes linéaires dans les dioïdes. Ph.D. thesis (1992)
12. Gaubert, S., Maisresse, J.: Modeling and analysis of timed Petri nets using heaps of pieces. IEEE Trans. Autom. Control **44**(4), 683–698 (1999)
13. Giffler, B.: Schedule Algebras and their use in formulating general systems simulations. In: Industrial Schduling. Prentice Hall, New Jersey (1963)
14. Graham, R.L., Lawler, E.L., Lenstra, J.K., Rinnooy, A.H.G.: Kan. Optimization and approximation in deterministic sequencing and scheduling: a survey. Ann. Discrete Math. **5**(2), 287–326 (1979)
15. Gunawardena, J. (ed.): Idempotency. Publications of the Newton Institute, Cambridge (1998)
16. Hanen, C., Munier, A.: Cyclic Scheduling on Parallel Processors: An Overview. John Wiley, New York (1995)
17. Ignall, E., Schrage, L.: Application of branch-and-bound technique to some flow shop problems. Oper. Res. **13**(3), 400–412 (1965)
18. Johnson, S.M.: Optimal two- and three-stage production schedules with setup times included. Naval Res. Logistics **1**, 61–68 (1954)
19. Christophe, L.: Analyse Max-Plus de problèmes d'ordonnancement de type flowshop. Ph.D. thesis, Université François Rabelais de Tours (2001)
20. Lenté, C.: Mathématiques. Université François Rabelais de Tours, Ordonnancement et Santé. Habilitation à diriger des recherches (2011)
21. Pan, Q.-K., Ruiz, R.: A comprehensive review and evaluation of permutation flowshop heuristics to minimize flowtime. Comput. Oper. Res. **40**(1), 117–128 (2013)
22. Smith, W.E.: Various optimizers for single-stage production. Naval Res. Logistics Quart **3**(1–2), 59–66 (1956)
23. Ling-Huey, S., Lee, Y.-Y.: The two-machine flowshop no-wait scheduling problem with a single server to minimize the total completion time. Comput. Oper. Res. **35**(9), 2952–2963 (2008)
24. Trabelsi, W., Sauvey, C., Sauer, N.: Heuristics and metaheuristics for mixed blocking constraints flowshop scheduling problems. Comput. Oper. Res. **39**(11), 2520–2527 (2012)
25. Vo, N.-V., Christophe, L.: Equivalence between two flowshop problems - Max-Plus Approach. In: Proceedings of the 2nd International Conference on Operations Research and Enterprise Systems. SciTePress - Science and Technology Publications, Barcelona, pp. 174–177 (2013)

Author Index

Printed in the United States
By Bookmasters